Natural Bioactive Compounds Exerting Health Promoting Effects through Ameliorating Oxidative Stress

Natural Bioactive Compounds Exerting Health Promoting Effects through Ameliorating Oxidative Stress

Guest Editors

Baojun Xu
Bin Du

MDPI

Basel • Beijing • Wuhan • Barcelona • Belgrade • Novi Sad • Cluj • Manchester

Guest Editors

Baojun Xu
Guangdong Provincial Key Laboratory IRADS
BNU-HKBU United International College
Zhuhai
China

Bin Du
Hebei Key Laboratory of Natural Products Activity Components and Function
Hebei Normal University of Science and Technology
Qinhuangdao
China

Editorial Office

MDPI AG
Grosspeteranlage 5
4052 Basel, Switzerland

This is a reprint of the Special Issue, published open access by the journal *Antioxidants* (ISSN 2076-3921), freely accessible at: www.mdpi.com/journal/antioxidants/special_issues/L54NH0QEA5.

For citation purposes, cite each article independently as indicated on the article page online and using the guide below:

Lastname, A.A.; Lastname, B.B. Article Title. *Journal Name* **Year**, *Volume Number*, Page Range.

ISBN 978-3-7258-3170-8 (Hbk)
ISBN 978-3-7258-3169-2 (PDF)
https://doi.org/10.3390/books978-3-7258-3169-2

Contents

About the Editors

Baojun Xu

Dr. Xu is a chair professor at Beijing Normal University–Hong Kong Baptist University United International College (UIC, a full English-teaching college in China), a fellow of the Royal Society of Chemistry, a Zhuhai scholar distinguished professor, the department head of the Department of Life Sciences, the program director of the Food Science and Technology Program, and the author of over 370 peer-reviewed papers. Dr. Xu received a Ph.D. in Food Science from Chungnam National University, South Korea. He conducted postdoctoral research work at North Dakota State University (NDSU), Purdue University, and Gerald P. Murphy Cancer Foundation in the USA during 2005-2009. He did short-term visiting research at NDSU in 2012 and the University of Georgia in 2014, followed by visiting research during his sabbatical leave (7 months) at Pennsylvania State University in the USA in 2016. Dr. Xu is serving as associate editor-in-chief of *Food Science and Human Wellness*, associate editor of *Food Research International*, associate editor of *Food Frontiers*, and an editorial board member of around 10 international journals. He received the inaugural President's Award for Outstanding Research of UIC in 2016 and the President's Award for Outstanding Service of UIC in 2020. Dr. Xu has been listed in the world's top 2% of scientists by Stanford University for the past 5 consecutive years and has been listed as the best scientist in the world in the field of biology and biochemistry at Research.com in 2023 and 2024. Prof. Xu was named an inaugural highly ranked scholar (top 0.05%) by ScholarGPS™, ranked at #8 in food science and technology and at #14 in agricultural and natural sciences.

Bin Du

Dr. Du is a professor at the Hebei Normal University of Science and Technology, director of the Analysis and Testing Center, and vice dean of the Hebei Key Laboratory of Natural Products Activity Components and Function. Dr. Du served as principal investigator or co-investigator on several extramural research grants from funding agencies. His research interests primarily focus on the isolation and purification of phytochemicals (such as polysaccharides, phenolics, and flavones) in fruits, vegetables, and mushrooms. He also shows interest in the structure, physicochemical, and biological properties (such as anti-inflammatory and antioxidant activities) of these phytochemicals. He authored over 30 scientific peer-reviewed papers in different prestigious journals and also served as an invited scientific reviewer for many highly respected journals. Dr. Du was frequently invited to give oral presentations at different academic occasions and to present his studies at several national and international conferences. During his doctoral training, Dr. Du received travel grants from different agencies. Moreover, his work aims to understand how nutrition affects the functioning of the human body. Better understanding is key to developing strategies to improve human health and well-being, to lower disease risk, and to treat nutrition-related illnesses.

antioxidants

Editorial

Natural Bioactive Compounds Exerting Health-Promoting Effects by Ameliorating Oxidative Stress

Bin Du [1] and Baojun Xu [2,*]

[1] Hebei Key Laboratory of Natural Products Activity Components and Function, Hebei Normal University of Science and Technology, Qinhuangdao 066004, China; bindufood@aliyun.com

[2] Guangdong Provincial Key Laboratory IRADS, Department of Life Sciences, BNU-HKBU United International College, Zhuhai 519087, China

* Correspondence: baojunxu@uic.edu.cn

Oxidative stress is caused by various intrinsic and extrinsic factors. It is a widely accepted notion that dietary or dietary supplement interventions can decrease levels of oxidative stress [1]. Natural bioactive compounds (such as flavonoids, stilbenes, terpenes, alkaloids, saponins, polysaccharide, etc.) have been reported to exhibit a multitude of health-promoting effects including antioxidant activity. Previous studies have shown that the mechanisms involved in the amelioration of oxidative stress include the regulation of GPX4 and ferroptosis [2]. However, the underlying cellular signaling and molecular mechanisms of their effect in ameliorating oxidative stress are not fully understood.

This Special Issue aims at collate and summarize the observed antioxidant activity of natural bioactive compounds (such as saponins, ergothioneine, flavonoids, taurine, chlorogenic acid, and phenolic compounds) and extracts (including bean extracts, *Moringa oleifera* extracts, Massa Medicata Fermentata, goji-berry extracts and dietary nitrate) based on studies from in vitro models (cell culture models; the gastrointestinal digestion model) and in vivo models (the *Caenorhabditis elegans* model and zebrafish model). It contains 11 papers covering the above-mentioned topics, including 1 systematic review and meta-analysis, and 10 research papers.

Flavonoids and phenolic compounds are a large group of plant metabolites that have long garnered interest from researchers worldwide due to their functions in human health and well-being [3]. Four contributions in this Special Issue address these compounds. In one, the bioactivity and stress resistance mechanisms of flavonoids derived from Chinese sea buckthorn leaves (such as FCL) were investigated. The authors identified six novel flavonoids. The results showed that FCL possesses potent antioxidant properties, as evidenced by its ability to effectively scavenge DPPH free radicals and demonstrate ferric-reducing antioxidant power and oxygen radical absorbance capacity levels comparable with Trolox. Furthermore, FCL showed remarkable efficacy in reducing reactive oxygen species (ROS) levels and malondialdehyde (MDA) levels while enhancing the activities of key antioxidant enzymes, namely superoxide dismutase and catalase in *C. elegans* (Contribution 1). The second contribution evaluates the effect of digestion on the phenolic compounds and antioxidant capacity of *Artocarpus heterophyllus* Lam. (jackfruit) pulp using an in vitro gastrointestinal digestion model. The results showed that the total phenol content (TPC) was significantly higher after gastric digestion. Thirty phenolic compounds (hydroxybenzoic acids and derivatives, hydroxycinnamic acids and derivatives, and flavonoids) were identified. The antioxidant activities of the digested samples varied with the TPC, and there was a correlation between antioxidant activity and TPC (Contribution 2). The third contribution tests cost-effective staple foods by fortifying them with

check for updates

Received: 7 January 2025
Accepted: 9 January 2025
Published: 13 January 2025

Citation: Du, B.; Xu, B. Natural Bioactive Compounds Exerting Health-Promoting Effects by Ameliorating Oxidative Stress. *Antioxidants* **2025**, *14*, 85. https://doi.org/10.3390/antiox14010085

M. oleifera leaf powder (MOLP) and its phenolic-rich extract. Moreover, the antioxidant antibacterial, and antidiabetic properties of the phenolic extract from MOLP were observed The resulting extract had a TPC of 138.2 mg of gallic acid equivalents/g. ABTS and DPPH assays presented IC_{50} values of 115.2 mg/L and 544.0 mg/L, respectively. Furthermore the extract exhibited notable α-amylase inhibition and no cytotoxicity towards human fibroblasts (Contribution 3). Chlorogenic acid (CGA) is a widely distributed natural compound with many crucial pharmacological effects, including antioxidant, anti-inflammatory, antibacterial, antiviral, hypoglycemic, lipid lowering, anti-cardiovascular, antimutagenic, anticancer, and immunomodulatory effects [4]. Wang et al. assessed the effect and mechanisms of CGA in the alleviation of cytotoxicity induced by aflatoxin B1 (AFB1) in L-02 cells. These findings indicated that CGA (160 μM) significantly recovered cell viability and cell membrane integrity in AFB1-treated (8 μM) cells. Furthermore, it was found that CGA reduced AFB1-induced oxidative injury by neutralizing ROS and activating the nuclear factor erythroid 2-related factor 2 (Nrf2)/heme oxygenase-1 (HO-1) signaling pathway (Contribution 4).

A growing body of evidence indicates that legumes exhibit potent antioxidant properties that protect against various chronic diseases induced by oxidative stress [5]. In this Special Issue, a further two papers focus on bioactive bean extracts: Luo et al. further elucidated the anti-obesity effects of adzuki bean saponins (ABSs) in improving lipid metabolism and oxidative stress, exploring the effective ingredients and potential molecular mechanisms through UHPLC-QE-MS analysis, network pharmacology, bioinformatics, and in vitro experiments both in the 3T3-L1 cell line and HepG2 cell line. Their results showed that ABS can improve intracellular lipid accumulation, adipogenesis, oxidative stress, and mitochondrial damage caused by lipid accumulation including ROS generation, abnormal mitochondrial membrane potential, and ATP disorder (Contribution 5). In another contribution, eight bean extracts (BE1–BE8) were tested for their ability to inhibit lipid accumulation and oxidation in hepatocytes. The results indicated that the antioxidants BE2 and BE8 are potential candidates for the regulation of triacylglycerol (TAG) and TAG hydroperoxide (TGOOH) accumulation in fatty acid-induced lipid droplets (LDs) (Contribution 6).

In another contribution, the chondrocyte-protective functions of ergothioneine (EGT) were investigated using a model of oxidative stress-induced injury of chondrocytes by hydrogen peroxide (H_2O_2). Human chondrocytes were subjected to oxidative stress induced by H_2O_2 treatment; cell viability, the activity of lactate dehydrogenase (LDH) released into the medium, dead cell ratio, intracellular ROS production, and mitochondrial morphology were assessed. EGT improved chondrocyte viability and LDH activity in the medium and strongly suppressed the dead cell ratio. EGT also exerted protective effects on intracellular ROS production and mitochondrial morphology (Contribution 7).

Seol et al. explored the antioxidative and protective effects of taurine chloramine (Tau-Cl) in astrocytes. Tau-Cl increased the expression and nuclear translocation of nuclear factor E2-related factor (Nrf2) and the expression of Nrf2-regulated antioxidant genes, including heme oxygenase 1 (HO-1). They also found that Nrf2 activity is negatively regulated by Kelch-like ECH-associated protein 1 (Keap1). Tau-Cl decreased the level of the reduced thiol groups of Keap1, resulting in disruptions to the Keap1–Nrf2 complex (Contribution 8).

Furthermore, in an investigation of the antioxidant activity of Massa Medicata Fermentata (MMF), the ability of the hydrothermal extract of MMF to scavenge free radicals corresponding to biological oxidative stresses was demonstrated. In addition, the effects of MMF on the body mass index and cholesterol accumulation in adult zebrafish fed a high-cholesterol diet for three weeks were examined. MMF prevented DNA and lipid damage caused by oxidative stress, inhibited LDL oxidation, and reduced the expression of cytokines and related proteins (MAPK and NFκB), with prominent anti-oxidative pathway (NRF2-HO-1) activation properties (Contribution 9).

Li et al. employed a rapid and convenient method applicable in the green synthesis of gold nanoparticles (AuNPs) using goji berry (*Lycium barbarum*—LB) extracts. Moreover, they evaluated their structural analysis and evaluation of the induced antioxidant protection and anti-inflammatory effects of the synthesized gold nanoparticles in endothelial cells (HUVECs) exposed to hyperglycemia. The results of this study reveal that AuNPs with a predominantly spherical shape and an average size of 30 nm were obtained. The ultraviolet-visible spectrum showed a characteristic absorption band at $\lambda max = 536$ nm for the AuNPs. FTIR analysis revealed the presence of phenolic acids, flavonoids and carotenoids, acting as the capping and stabilizing agents of AuNPs. Both the *L. barbarum* extract and AuNPs were well tolerated by HUVECs, increased the antioxidant defense and decreased the production of inflammatory cytokines induced via hyperglycemia-mediated oxidative damage (Contribution 10).

Finally, this Special Issue includes one systematic review and meta-analysis of randomized controlled trials. Data from Liang et al. showed that the PRISMA guidelines were adhered to in the studies analyzed. MEDLINE and SPORTDiscus were searched from inception to January 2023. A paired analysis model for cross-over trials was incorporated to perform a random-effects meta-analysis for each performance outcome and to generate standardized mean differences (SMDs) between the inorganic nitrate (NO_3^-) and placebo supplementation conditions. This systematic review and meta-analysis included 27 and 23 studies, respectively. The time to reach peak power (SMD: 0.75, $p = 0.02$), mean power output (SMD: 0.20, $p = 0.02$), and total distance covered in the Yo-Yo intermittent recovery level 1 test (SMD: 0.17, $p < 0.0001$) were all improved after NO_3^- supplementation (Contribution 11).

In summary, the results of the above-mentioned studies will help to address the challenges remaining in this field in terms of the limited knowledge about natural bioactive compounds and their underlying cellular signaling and molecular mechanisms. These studies may be potential resources for the development of functional foods and drug discovery.

Author Contributions: B.D. wrote the introduction and the conclusion. B.X. wrote the central part, with comments on the cited papers and references. All authors have read and agreed to the published version of the manuscript.

Conflicts of Interest: The authors declare no conflicts of interest.

List of Contributions

1. Zhao, J.; Jiang, Y.; Bi, Y.; Wei, J. Antioxidant and stress resistance properties of flavonoids from Chinese sea buckthorn leaves from the Qinghai–Tibet Plateau. *Antioxidants* **2024**, *13*, 763. https://doi.org/10.3390/antiox13070763.
2. Cheng, M.; He, J.; Gu, Y.; Wu, G.; Tan, L.; Li, C.; Xu, F.; Zhu, K. Changes in phenolic compounds and antioxidant capacity of *Artocarpus heterophyllus* Lam. (jackfruit) pulp during in vitro gastrointestinal digestion. *Antioxidants* **2024**, *13*, 37. https://doi.org/10.3390/antiox13010037.
3. Ferreira, T.; Gomes, S.M.; Santos, L. Elevating cereal-based nutrition: *Moringa oleifera* supplemented bread and biscuits. *Antioxidants* **2023**, *12*, 2069. https://doi.org/10.3390/antiox12122069.
4. Wang, Q.; Liu, T.; Koci, M.; Wang, Y.; Fu, Y.; Ma, M.; Ma, Q.; Zhao, L. Chlorogenic acid alleviated AFB1-induced hepatotoxicity by regulating mitochondrial function, activating Nrf2/HO-1, and inhibiting noncanonical NF-κB signaling pathway. *Antioxidants* **2023**, *12*, 2027. https://doi.org/10.3390/antiox12122027.

5. Luo, J.; Luo, J.; Wu, Y.; Fu, Y.; Fang, Z.; Han, B.; Du, B.; Yang, Z.; Xu, B. Anti-obesity effects of adzuki bean saponins in improving lipid metabolism through reducing oxidative stress and alleviating mitochondrial abnormality by activating the PI3K/Akt/GSK3β/β-Catenin signaling pathway. *Antioxidants* **2024**, *13*, 1380. https://doi.org/10.3390/antiox13111380.
6. Dibwe, D.F.; Kitayama, E.; Oba, S.; Takeishi, N.; Chiba, H.; Hui, S.-P. Inhibition of lipid accumulation and oxidation in hepatocytes by bioactive bean extracts. *Antioxidants* **2024**, *13*, 513. https://doi.org/10.3390/antiox13050513.
7. Sakata, S.; Kunimatsu, R.; Tanimoto, K. Protective effect of ergothioneine against oxidative stress-induced chondrocyte death. *Antioxidants* **2024**, *13*, 800. https://doi.org/10.3390/antiox13070800.
8. Seol, S.-I.; Kang, I.S.; Lee, J.S.; Lee, J.-K.; Kim, C. Taurine chloramine-mediated Nrf2 activation and HO-1 induction confer protective effects in astrocytes. *Antioxidants* **2024**, *13*, 169. https://doi.org/10.3390/antiox13020169.
9. Jung, K.-M.; Yu, G.-R.; Kim, D.-H.; Lim, D.-W.; Park, W.-H. Massa Medicata Fermentata, a functional food for improving the metabolic profile via prominent anti-oxidative and anti-inflammatory effects. *Antioxidants* **2024**, *13*, 1271. https://doi.org/10.3390/antiox13101271.
10. David, L.; Morosan, V.; Moldovan, B.; Filip, G.A.; Baldea, I. Goji-berry-mediated green synthesis of gold nanoparticles and their promising effect on reducing oxidative stress and inflammation in experimental hyperglycemia. *Antioxidants* **2023**, *12*, 1489. https://doi.org/10.3390/antiox12081489
11. Alsharif, N.S.; Clifford, T.; Alhebshi, A.; Rowland, S.N.; Bailey, S.J. Effects of dietary nitrate supplementation on performance during single and repeated bouts of short-duration high-intensity exercise: A systematic review and meta-analysis of randomised controlled trials. *Antioxidants* **2023**, *12*, 1194. https://doi.org/10.3390/antiox12061194.

References

1. Gorrini, C.; Harris, I.S.; Mak, T.W. Modulation of oxidative stress as an anticancer strategy. *Nat. Rev. Drug Discov.* **2013**, *12*, 931–947. [CrossRef]
2. Kong, D.; Liu, Y.; Li, L.; Wang, H.; Li, K.; Zheng, G. Astaxanthin ameliorates oxidative stress in lens epithelial cells by regulating GPX4 and Ferroptosis. *Chem. Biol. Interact.* **2023**, *383*, 110684. [CrossRef] [PubMed]
3. Kuljarusnont, S.; Iwakami, S.; Iwashina, T.; Tungmunnithum, D. Flavonoids and other phenolic compounds for physiological roles, plant species delimitation, and medical benefits: A promising view. *Molecules* **2024**, *29*, 5351. [CrossRef] [PubMed]
4. Miao, M.; Xiang, L. Pharmacological action and potential targets of chlorogenic acid. *Adv. Pharmacol.* **2020**, *87*, 71–88. [CrossRef] [PubMed]
5. Ganesan, K.; Xu, B.J. Polyphenol-rich lentils and their health promoting effects. *Int. J. Mol. Sci.* **2017**, *18*, 2390. [CrossRef] [PubMed]

antioxidants

Article

Anti-Obesity Effects of Adzuki Bean Saponins in Improving Lipid Metabolism Through Reducing Oxidative Stress and Alleviating Mitochondrial Abnormality by Activating the PI3K/Akt/GSK3β/β-Catenin Signaling Pathway

Jinhai Luo [1,†], Jincan Luo [2,3,†], Yingzi Wu [1], Yu Fu [3], Zhonghao Fang [2,3], Bincheng Han [1], Bin Du [4], Zifeng Yang [2,3,5,6,*] and Baojun Xu [1,*]

[1] Guangdong Provincial Key Laboratory IRADS and Department of Life Sciences, BNU-HKBU United International College, Zhuhai 519087, China; luojinhai@uic.edu.cn (J.L.); wuyingzi@uic.edu.cn (Y.W.); q030013010@mail.uic.edu.cn (B.H.)
[2] National Clinical Research Center for Respiratory Disease, State Key Laboratory of Respiratory Disease, Guangzhou Institute of Respiratory Health, the First Affiliated Hospital of Guangzhou Medical University, Guangzhou 510120, China; jcluo9208@163.com (J.L.); fzhh0769@stu.gzhmu.edu.cn (Z.F.)
[3] Guangzhou National Laboratory, International Bio-Island, Guangzhou 510005, China; fu_yu@gzlab.ac.cn
[4] Hebei Key Laboratory of Natural Products Activity Components and Function, Hebei Normal University of Science and Technology, Qinhuangdao 066004, China; bindufood@aliyun.com
[5] Guangzhou Key Laboratory for Clinical Rapid Diagnosis and Early Warning of Infectious Diseases, Guangzhou 511436, China
[6] State Key Laboratory of Quality Research in Chinese Medicine, Macau University of Science and Technology, Taipa, Macau SAR 999078, China
* Correspondence: yang_zifeng@gzlab.ac.cn (Z.Y.); baojunxu@uic.edu.cn (B.X.); Tel.: +86-020-37103057 (Z.Y.); +86-756-2620636 (B.X.)
† These authors contributed equally to this work.

Citation: Luo, J.; Luo, J.; Wu, Y.; Fu, Y.; Fang, Z.; Han, B.; Du, B.; Yang, Z.; Xu, B. Anti-Obesity Effects of Adzuki Bean Saponins in Improving Lipid Metabolism Through Reducing Oxidative Stress and Alleviating Mitochondrial Abnormality by Activating the PI3K/Akt/GSK3β/β-Catenin Signaling Pathway. *Antioxidants* **2024**, *13*, 1380. https://doi.org/10.3390/antiox13111380

Academic Editors: Alessandra Napolitano and Sunmin Park

Received: 10 October 2024
Revised: 1 November 2024
Accepted: 7 November 2024
Published: 11 November 2024

Abstract: Obesity is a chronic and complex disease defined by the excessive deposition of fat and is highly associated with oxidative stress. Adzuki bean saponins (ABS) showed anti-obesity activity in our previous in vivo study; however, the active saponins of adzuki beans and potential mechanisms are still unclear. This research aims to elucidate the anti-obesity effects of ABS in improving lipid metabolism and oxidative stress, exploring the effective ingredients and potential molecular mechanisms through UHPLC-QE-MS analysis, network pharmacology, bioinformatics, and in vitro experiments both in the 3T3-L1 cell line and HepG2 cell line. The results indicate that ABS can improve intracellular lipid accumulation, adipogenesis, oxidative stress, and mitochondrial damage caused by lipid accumulation including ROS generation, abnormal mitochondrial membrane potential, and ATP disorder. Fifteen saponin components were identified with the UHPLC-QE-MS analysis. The network pharmacology and bioinformatics analyses indicated that the PI3K/Akt signaling pathway is associated with the bioactive effect of ABS. Through Western blotting and immunofluorescence analysis, the anti-obesity effect of ABS is achieved through regulation of the PI3K/Akt/GSK3β/β-catenin signaling pathway and activation of downstream transcription factor c-Myc in the lipid accumulation cell model, and regulation of β-catenin signaling and inhibition of downstream transcription factor C/EBPα in the adipocyte cell model. These results illustrate the biological activity of ABS in improving fat metabolism and oxidative stress by restoring mitochondrial function through β-catenin signaling, the PI3K/Akt/GSK3β/β-catenin signaling pathway, laying the foundation for its further development.

Keywords: adzuki bean; obesity; lipid metabolism; network pharmacology; mitochondrial function

1. Introduction

Obesity is a chronic and complex disease characterized by excessive fat deposition, which can harm health [1]. Obesity tends to lead to other metabolic diseases, including type 2 diabetes, hypertension, and cardiovascular diseases [2]. With changing eating habits, obesity and related metabolic diseases caused by high-fat diets are increasingly prevalent globally [3]. The most apparent characteristic of individuals with obesity is the dysregulation of lipid and energy metabolism, and mitochondrial dysfunction further exacerbates the imbalance of lipid and energy metabolism [4,5]. Lipid metabolism disorders can disrupt the balance of lipid peroxidation products, which may impair mitochondrial function and lead to a decrease in mitochondrial membrane potential. Mitochondrial membrane potential is critical for maintaining cellular energy homeostasis, and its reduction due to oxidative stress can impair the electron transport chain and ATP production. This interplay between lipid metabolism, oxidative stress, and mitochondrial function can exacerbate cellular damage and contribute to the pathogenesis of various diseases. Therefore, lipid metabolism imbalance, energy metabolism imbalance, and mitochondrial dysfunction are vital factors inducing obesity, which are potential targets for alleviating diet-induced obesity [6].

In the last decade, the most commonly used obesity drugs have been sibutramine and orlistat [7]. Sibutramine was banned by the United States Food and Drug Administration in 2010 due to its significant side effects. Orlistat has also been reported to cause adverse reactions such as gastrointestinal disorders, liver damage, allergic reactions, and abnormal endocrine system reactions. In addition, researchers have found out that glucagon-like peptide-1 receptor agonists such as semaglutide and dual glucose-dependent insulinotropic polypeptide and glucagon-like peptide-1 receptor agonists such as tirzepatide are currently the best obesity drugs available [8,9]. It is still essential that scientists find more potential compounds from natural products to provide more choices to address obesity. Adzuki bean (*Vigna angularis* L.) is mainly produced in China and some other East Asian countries. Adzuki bean has been used as food for thousands of years, and it has various biological activities, such as anti-cancer [10,11], anti-diabetes [12,13], anti-oxidation [14], and liver protection [15,16].

Our previous research extracted saponins from adzuki beans and identified them using HPLC-DAD-ESI-MS^n [17]. Subsequently, our previous research validated adzuki bean saponin extract's anti-obesity and lipid-lowering biological activities in a high-fat diet obesity mice model especially in the liver and adipose tissue [18]. A high-fat diet can lead to lipid metabolism and energy metabolism disturbances, resulting in lipid accumulation [19]. Moreover, a high-fat diet is a common way to induce obesity and can increase free fatty acid (FFA) levels in the blood and liver instead of triglycerides [20]. However, the potential molecular mechanisms by which adzuki bean saponin's extract achieves its anti-obesity effects and the energy metabolism of intracellular mitochondria and intracellular lipid metabolism are still unclear.

The WHO has pointed out that obesity is a chronic complex disease defined by excessive fat deposits that can impair health [1]. The excessive fat deposits include adipocytes differentiated from stem cells and lipid accumulation in organs, especially in the liver [21]. In addition, the liver is the crucial place of lipid metabolism, which is the major issue in this study. An in vitro FFA-induced lipid accumulation cell model in HepG2 was widely used for studying the anti-obesity effect of the bioactive compound from food or herbs [22–31]. Recently, this FFA-induced HepG2 cell model was also used to validate the network pharmacology results of the anti-obesity effect of Chenpi [25] and mulberry [24]. Moreover, for studying adipogenesis, the classical adipocyte cell model (3T3-L1) was also employed in this research to study the anti-adipogenesis effect [32]. Based on the above, this study also used two different cell models focusing on the potential role and molecular mechanism of adzuki bean saponins in achieving anti-obesity biological activity by improving adipogenesis, lipid metabolism, and mitochondrial energy in metabolism disorders.

Currently, the combination of LC-MS analysis, network pharmacology, and bioinformatics has received increasing attention in studying the molecular mechanism of phytochemicals [33,34]. In this study, we used the extraction methods of previous studies to extract adzuki bean saponins (ABS) from adzuki beans and identified the soybean saponin components using UHPLC-QE-MS analysis [17]. Subsequently, we conducted network pharmacology prediction and bioinformatics to explore their molecular mechanisms of action. Finally, we combined the in vitro intracellular lipid accumulation cell model and adipocyte cell model to focus on mitochondrial function, energy metabolism, and adipogenesis to elucidate the molecular mechanism of adzuki bean saponins' anti-obesity effect.

2. Materials and Methods

2.1. Materials

The 8-well glass chamber slide for oil red staining and slide scanning was purchased from Millipore (St. Louis, MO, USA). The 96-well clear-glass-bottom and black-edge plate for cellular ATP measurement and specific fluorescent-signal determination was purchased from Cellvis (Sunnyvale, CA, USA) for fluorescence and luminescence measurement by multi-function microplate reader. The 35-millimeter glass and clear-bottom (15 mm) black-edge culture dish was purchased from Biosharp (Beijing, China) for imaging under super-resolution microscope. The minimum essential medium (MEM), high-glucose Dulbecco's Modified Eagle Medium (DMEM), 100× penicillin–streptomycin solution, 3T3-L1 cell line, and HepG2 cell line were purchased from Procell Life Science & Technology Co., Ltd. (Wuhan, China). The Adipogenesis Assay was purchased from Sigma-Aldrich (St. Louis, MO, USA). Phosphate-buffered solution (PBS), dimethyl sulfoxide (DMSO) solution, and Oil Red O staining storage solution were purchased from Beijing Solarbio Science & Technology Co., Ltd. (Beijing, China). Fetal bovine serum (FBS) was bought from Shanghai XP Biomed Ltd. (Shanghai, China). The cell counting kit-8 (CCK-8) reagent kit, CellTiter-LumiTM Plus II Cellular ATP level determination Kit, DPAI staining solution, Triton-X 100 stock solution, BeyoColor™ Pre-stained color protein marker (10–170 kD), the mitochondrial membrane potential assay kit with JC-1 (C2006) kit, the secondary anti-rabbit IgG (H+L) antibodies of Western blotting assay, the blocking reagent, as well as the SignalUp™ 1st and QuickBlock™ 2nd antibody dilution reagent for Immunofluorescence Assay were purchased from Beyond Biotech Inc. (Shanghai, China). The RIPA lysis buffer, phenylmethanesulfonyl fluoride, 10X TBST solution, and pre-stained protein marker IV were purchased from Wuhan Servicebio Technology Co., Ltd. (Wuhan, China). The triglyceride test kit, non-fat powdered milk, Epigallocatechin gallate (EGCG), and the AB-8 resin column were purchased from Shanghai Yuanye Bio-Technology Co., Ltd. (Shanghai, China). The sodium oleate/palmitic acid reagent kit was purchased from Xi'an Kunchuang Science and Technology Develop Co., Ltd. (Xian, China). The BCA Protein Quantification Kit was purchased from Yeasen Biotechnology Co., Ltd. (Shanghai, China). The Omni-Easy™ protein sample loading buffer (Denaturing, Reducing, 5×) was purchased from Epizyme Biotech Co., Ltd. (Shanghai, China). The primary antibodies against PI3K (4257), p-PI3K (17366), Akt (4691), p-Akt (4060), GSK-3β (12456), p-GSK-3β (5558), GAPDH (5174S), and c-Myc (5605T) were obtained from Cell Signaling Technology (Danvers, MA, USA). The primary antibody against β-catenin (AF6266) and C/EBPα (AF6333) was purchased from Affinity Biosciences (Cincinnati, OH, USA). The SuperKine™ West Femto Maximum Sensitivity Substrate was purchased from Abbkine Scientific Co., Ltd. (Wuhan, China). The 2nd antibody, immunofluorescent antibody, Fluorescein (FITC)-conjugated Affinipure Goat Anti-Rabbit IgG(H+L) was purchased from Proteintech (Wuhan, China). The 4% paraformaldehyde for cell fixation was purchased from Biosharp (Beijing, China). The $DiBaC_4$ fluorescent membrane potential indicator was purchased from GLPBio (Montclair, CA, USA). Hochest 33342, used for fluorescent nucleic acid staining, Carboxy-H_2DCFDA, used as a general oxidative stress fluorescent indicator, and ProLong™ Gold Antifade Mountant, a mounting medium with photobleaching resistance capability, were purchased from Invitrogen (Carlsbad, CA, USA). Adzuki beans (*Vigna angularis* L.) were bought from agricultural market in Guangxi Province.

2.2. Preparation of Adzuki Bean Saponins

Based on the methods of our previous research [17,18], the adzuki bean saponins were prepared. The adzuki beans were weighed at 1 kg and ground to powder. Subsequently, the powdered substance underwent a triple extraction process using 10 L of 70% ethanol solution. The resultant mixture was combined, filtered, and then subjected to a concentration process to eliminate the ethanol content. The leftover liquid was subsequently treated with three rounds of extraction using 3 L of petroleum ether at room temperature, with the aqueous fraction being retained for further investigation. The aqueous fraction was further extracted three times with 3 L of *n*-butanol at room temperature, and the organic layer was collected for subsequent processing. Then, the distilled water was added to the organic phase and n-butanol was removed through vacuum evaporation. The obtained aqueous solution was added to the AB-8 resin column. Next, it was washed with water and 45% ethanol, and then, the eluent was discarded. Finally, 80% ethanol was added to the AB-8 resin column and the eluent of the enriched saponins was collected. The eluents were ABS after they were freeze-dried.

2.3. UHPLC-QE-MS Analysis

The ABS constituent was analyzed and identified through the application of UHPLC-QE-MS. The ABS was mixed with an 80% methanol solution in water and then subjected to centrifugal force at a rate of 10,614× *g* for 15 min at a temperature of 4 °C. The supernatant was taken and filtered through a 0.22 μm microporous membrane. The liquid chromatography–tandem mass spectrometry (LC-MS/MS) analysis was conducted using an ultra-high-performance liquid chromatography (UHPLC) system (Vanquish, Thermo Fisher Scientific, Waltham, MA, USA) in conjunction with a Waters UPLC BEH C_{18} column characterized by a particle size of 1.7 μm and dimensions of 2.1 mm by 100 mm. The volume of sample injected into the system was 5 μL, with a flow rate set at 0.5 mL per minute. The mobile phase was a mixture of 0.1% formic acid in water (solvent A) and 0.1% formic acid in acetonitrile (solvent B), and the gradient elution program followed a multi-step linear progression: starting from 85% A and decreasing to 25% A over the first 11 min; then reducing to 2% A by the 12th min; maintaining at 2% A until the 14th min; followed by a rapid increase to 85% A within 0.1 min; and finally holding at 85% A for 2 min for the remainder of the analysis. For the mass spectrometry part, a Q Exactive Focus mass spectrometer interfaced with Xcalibur (Version 4.1) software was utilized to capture both MS and MS/MS data in the ion-driven acquisition (IDA) mode. The mass range scanned spanned from 100 to 1500 m/z, with the top three precursor ions from each cycle being selected for further fragmentation and analysis. The sheath gas flow rate was set at 45 arbitrary units (Arb), the auxiliary gas flow rate at 15 Arb, and the capillary temperature was maintained at 400 degrees Celsius. The resolution for full MS was 70,000 and for MS/MS it was 17,500. The collision energy was modulated in neutral collision energy (NCE) mode at three different levels: 15, 30, and 45. The spray voltage was adjusted to 4.0 kV in positive mode or −3.6 kV in negative mode. Mass spectrometry-based identification and structural elucidation of ABS constituents, including their primary and secondary spectral data, were conducted by the BIOTREE TCM database of Shanghai BIOTREE Biological Technology Co., Ltd. (Shanghai, China) [35,36].

2.4. Network Pharmacology Analysis

The compound structures obtained from UHPLC-QE-MS analyses were converted to Smiles format and were input into the Swiss Target Prediction (http://www.swisstargetprediction.ch/? (accessed on 24 Aprial 2024)) and SEA (https://sea.bkslab.org/ (accessed on 24 Aprial 2024)) database to predict the compound target based on a probability greater than zero [37,38]. The obesity-related targets were downloaded from the GEO. The GEO database (https://www.ncbi.nlm.nih.gov/geo/ (accessed on 4 June 2024)) included GSE25401, which includes transcriptomic data from white adipose tissue of 30 adults with obesity and 26 adults without obesity. To choose DEGs between normal group and obesity group, the "Limma" package of R language was utilized, and the filtering criteria were set to p value < 0.05. Finally, the result was visualized using volcano plot [39]. Moreover, the lipid metabolism-related target was downloaded from the GeneCards (https://www.genecards.org/ (accessed on 1 June 2024)) using the keyword "Lipid metabolism" and screened by the median [40]. The overlapping targets of compound targets, obesity targets, and lipid metabolism targets obtained using the Venny website (https://jvenn.toulouse.inra.fr/app/example.html (accessed on 2 June 2024)) were considered potential targets for ABS to cure obesity by improving lipid metabolism [41].

The standardization of all collected targets was performed using the UniProt database (https://www.uniprot.org/ (accessed on 1 June 2024)), employing the selection criteria of "human" and "reviewed". The protein–protein interaction (PPI) analysis was performed by the String Database with the constraints of "Homo sapiens" and "confidence > 0.4". The network was constructed by Cytoscape 3.9.1 software, the contribution of the node was assessed by evaluating the topology parameters degree, betweenness, and closeness. The KEGG and GO enrichment analyses were conducted by Metascape (https://metascape.org/ (accessed on 4 June 2024)) [42]. Dot plot was generated by the bioinformatics online platform SRplot (https://www.bioinformatics.com.cn (accessed on 4 June 2024)).

2.5. *Bioinformatics Analysis*

To identify gene signatures that were uniquely associated with obesity, we utilized three distinct machine learning algorithms: support vector machine with recursive feature elimination (SVM-RFE), random forest, and least absolute shrinkage and selection operator (LASSO) [33]. For LASSO analysis, the cv.glmnet function from the R software package glmnet (version 4.1-8) was employed, with a turn/penalty option and 10-fold cross-validation to ensure reliable results. The random forest algorithm was applied using the "Random Forest" tool, focusing on stable image values and minimal error for optimal performance. A supervised approach named random forest recursive feature elimination (RF-RFE) was then used to prioritize immune- and adipocyte-related genes following weight loss. Features were ranked based on their prediction performance and relative importance, with genes exhibiting a value greater than 0.25 considered as gene signatures. Notably, the SVM-RFE-based approach exceeded the performance of the linear algorithm discriminant analysis. The pair mean square error method was employed to discern significant from redundant features in the SVM-RFE framework. For all algorithms, 10-fold cross-validation was conducted to ensure the reliability of feature selection. The overlapping gene sets identified by these three algorithms were further narrowed down to those exhibiting strong obesity-associated characteristics. Our findings were visually represented using the jvenn online tool (https://jvenn.toulouse.inra.fr/ (accessed on 4 June 2024)), providing a comprehensive and intuitive overview of the identified gene signatures [43]. Through these rigorous and comprehensive analyses, we aimed to gain deeper insights into the genetic markers of obesity and their associated biological processes.

Utilizing the CIBERSORT algorithm, we aimed to decipher the intricate link between the infiltration levels of immune cells and obesity-associated hub genes. This algorithm, fed with the normalized gene expression matrix, offered insights into the immune cell composition infiltrating the tissue. By submitting the data to the CIBERSORT web portal4 (http://CIBERSORT.stanford.edu/ (accessed on 4 June 2024)), we leveraged the LM22 reference gene expression signature—which permitted 1000 permutations—to identify and categorize 22 distinct immune cell types. These included macrophages (M0, M1, and M2), multiple subsets of T cells (CD8+, naïve CD4+, memory resting CD4+, memory activated CD4+, Tfh cells, regulatory, and gamma delta), natural killer (NK) cells (resting and activated), mast cells (resting and activated), B cells (naïve and memory), dendritic cells (resting and activated), monocytes, plasma cells, neutrophils, and eosinophils. To ensure the reliability of our findings, we employed strict filtering criteria, accepting only those with a CIBERSORT p value < 0.05. This filtering step culled out spurious results, leaving us with a focused set of data for further exploration. The resulting matrix, directly integrating the immune cell subsets, offered a comprehensive overview of the immune landscape. To aid in interpretation, we leveraged R packages such as "corplot", "vioplot", "ggplot2", and "glmnet" to visually represent the CIBERSORT findings. This visual representation not only enhanced the comprehensibility of the data but also provided valuable insights into the intricate relationships between immune cells and obesity-related genes [33].

Using the R package "pROC", we generated receiver operating characteristic (ROC) curves and computed the area under the curve (AUC) values to evaluate the diagnostic potential of obesity samples against controls in the GSE25401 dataset [39].

2.6. Cell Culture and Treatment

HepG2 cells were cultured in MEM medium supplemented with 10% fetal bovine serum and 1% 100 × penicillin–streptomycin solution at 37 °C and 5% CO_2. This model involves incubating HepG2 cells with 0.75 mM free fatty acid (FFA) mixture containing 2:1 oleate–palmitate. More specifically, the HepG2 cells were treated with MEM essential medium for 24 h. Next, the HepG2 cells were treated with MEM essential medium containing 750 μM of FFA (free fatty acid; sodium oleate–sodium palmitate, 2:1) for 24 h before harvesting the cells for further experiments. For the treatment group, the different dosages of ABS were added with the FFA together.

The 3T3-L1 cells were cultured in high-glucose DMEM medium supplemented with 10% fetal bovine serum and 1% 100 × penicillin–streptomycin solution at 37 °C and 5% CO_2. This model involves incubating 3T3-L1 cells with adipogenesis differentiation medium which includes methylisobutylxanthine, dexamethasone, and insulin (MDI). There are two kinds of adipogenesis differentiation media, A and B. The adipogenesis differentiation medium A (MDI A) has 0.5 mM methylisobutylxanthine, 1 μM dexamethasone, and 10 μg/mL insulin in DMEM medium with 10% FBS; the adipogenesis differentiation medium B (MDI B) has 10 μg/mL insulin in DMEM medium with 10% FBS. When the 3T3-L1 cells were almost fully grown, the adipogenesis differentiation medium was used to process the cells for modeling. The MDI A was used to incubate the 3T3-L1 cell for 3 days with or without the different dosages of ABS; the MDI B was added to the 3T3-L1 cell for 1 day with or without the different dosages of ABS. Four days is a modeling cycle and normally, the whole adipocyte differentiation needs 3–4 cycles (12–16 days).

2.7. Cell Viability Analysis

Using the CCK-8 method to detect the effect of ABS on cell viability, HepG2 cells were cultured into 96-well plates and exposed to different concentrations of ABS (0, 0.05, 0.1, 0.15, 0.2, and 0.4 mg/mL) for 24 h. After removing the previous medium and adding 100 μL of MEM essential culture medium, 10 μL of CCK-8 reagent was added to each well. The 96-well plate was then incubated at 37 °C in the dark for 30 min, and the absorbance at 450 nm was measured in the FLUOstar Omega Microplate Reader (BMG Labtech, Ortenberg, Germany).

The 3T3-L1 cells were cultured into 96-well plates and exposed to different concentrations of ABS (0, 0.05, 0.1, 0.2, 0.3, and 0.4 mg/mL) and EGCG (0, 4, 8, 16, and 32 μM) for 24 h. The further processing was the same as above.

2.8. Intracellular TG Assay

After 24 h of cell modeling, the supernatant was discarded and the cells were washed twice with PBS. Then, the 80 μL frozen RIPA solution with a final concentration of 2 μg/mL of aprotinin, 5 μg/mL of leupeptin, 1 μg/mL of pepstatin A, 5 mM of NaF, 1mM of Na_3VO_4, and 1 mM of PMSF was added to each well and placed on ice for 30 min. The cell lysates were later used for the determination of the intracellular triglyceride (TG) levels. The TG levels were measured using a triglyceride assay kit according to the manufacturer's instructions. The BCA protein kit was used to standardize the intracellular TG content according to the manufacturer's instructions.

2.9. Oil Red O Staining

The working solution of Oil Red O (ORO) preparation followed the user guidelines provided by the manufacturer as below: before staining, 6 parts of saturated ORO dye stock solution and 4 parts of distilled water were well-mixed and left at 4 °C overnight statically, filtered once with qualitative filter paper the next day, and filtered for the second time after 24 h at 4 °C to obtain ORO working solution. The HepG2 cells (~225,000 cells per well) were cultured in an 8-well glass chamber slide and treated the same as above. When the time point was reached, the medium with or without ABS was removed, and cells were washed thrice with 1× PBS. Following, live cells of either the control, experimental, or drug treatment group were treated with 4% paraformaldehyde lasting for 30 min fixation at room temperature (RT, 25 °C). Then, cells were washed three times by 1× PBS and permeation by 60% isopropyl alcohol. Afterward, cells were treated with 30-minute staining through ORO working solution at RT. Afterward, cells in the chamber were washed with the prepared 60% isopropyl alcohol lasting for 5 min, which was then discarded for six repeats. Immediately next, cells were washed with 1 × PBS six times to remove the rest of the ORO. After removal of separation capsule for the 8-well glass chamber slide, 1 drop of antifade reagent per region was applied for the fitness of the slide and coverslip before imaging and scanning under SLIDEVIEW VS200 research Slice Scanner (Olypums, Tokyo, Japan) with 40× objective.

The 3T3-L1 cells were cultured in an 8-well glass chamber slide and treated the same as mentioned above before. When the time point was reached, the further processing was the same as above.

2.10. Nile Red Staining

The HepG2 cells (~900,000 cells per well) were cultured in a black glass-bottom cell culture dish and treated the same as above. When the time point was reached, the medium with or without ABS was removed, and cells were washed thrice with 1× PBS. Following, live cells were co-stained by Nile Red at 3 μM and Hoechst 33,342 at 1 μM at 37 °C in a 5% CO_2 humidified incubator for 30 min. Next, cells were washed with 1× PBS six times and maintained in 1× PBS at the same volume before imaging. After washing in 1× PBS 6 times, images were visualized by applying ZEISS Elyra 7 with Lattice SIM^2 super-resolution fluorescent microscope (ZEISS, Germany) applying a 63× oil-immersion objective.

The 3T3-L1 cells were cultured in a black glass-bottom cell culture dish and treated the same as mentioned above. When the time point was reached, the further processing was the same as above.

2.11. Ptychographic Quantitative Phase Imaging

The HepG2 cells (~600,000 cells per well) were cultured in a 6-well plate. Modeling and treatment were the same as above. Once FFA mixture or treatment was added, observation under bright field and quantitative phase imaging (QPI) through a 650 nm laser was initiated for a period of 24 h, and images were taken every 0.5 h while cells were kept at 37 °C in 5% CO_2.

2.12. ROS Staining

The HepG2 cells (~900,000 cells per well) were cultured in black glass-bottom cell culture dishes and treated the same as above. When the time point was reached, the medium with or without ABS was removed; subsequently, cells were washed thrice with the commercial 1× PBS. Following, live cells were co-stained by Carboxy-H_2DCFDA at 5 μM and Hoechst 33342 at 5 μM at 37 °C in a 5% CO_2 humidified incubator for 30 min. After, cells were washed with 1× PBS six times and maintained in 1× PBS at the same volume before imaging. After washing in 1× PBS 6 times, images were visualized by applying ZEISS Elyra 7 with Lattice SIM^2 super-resolution fluorescent microscope (ZEISS, Oberkochen, Germany) with a 10× objective.

For ROS level changes in bulk culture, cells (~9000 cells per well) were cultured in a 96-well, clear-glass-bottom and black-edge plate, and treated as outlined above. When the time point was reached, medium with or without ABS was removed, and cells were washed thrice by 1× PBS before resuspension by the same volume of 1× PBS. The cells were digested with trypsin and suspended in the 96-well plate. Carboxy-H_2DCFDA (Invitrogen, Carlsbad, CA, USA) was introduced into the suspended HepG2 cell culture for staining at 5 μM at 37 °C in a 5% CO_2 humidified incubator for 30 min. Afterward, dye-containing medium was removed by low-speed centrifugation at 400× *g*, 5 min, keeping at 4 °C, and cells were washed thrice by 1× PBS and resuspended by same volume of 1× PBS. Then, cells in a 96-well black bottom plate were applied for relative fluorescent unit (RFU) (Carboxy-H_2DCFDA, 488 ex/515 em) determination using a microplate reader with monochromator (Agilent, BioTek Synergy HTX, Santa Clara, CA, USA). RFU of Carboxy-H_2DCFDA was normalized by diving to cell number.

2.13. Membrane Potential Measurement

The HepG2 cells (~900,000 cells per well) were cultured in a black glass-bottom cell culture dish and treated same as above. When the time point was reached, medium with or without ABS was removed, and cells were washed thrice by 1× PBS. Following, live cells were co-stained by DiBAC$_4$ at 5 μM and Hoechst 33,342 at 5 μM at 37 °C in a 5% CO_2 humidified incubator for 30 min. Next, cells were washed by 1× PBS six times and maintained at 1× PBS at same volume before imaging. After washing in 1× PBS 6 times, images were visualized by applying ZEISS Elyra 7 with Lattice SIM^2 super-resolution fluorescent microscope (ZEISS, Oberkochen, Germany) with a 63× oil-immersion objective.

For membrane potential changes in bulk culture, cells (~9000 cells per well) were cultured in a 96-well, clear-glass-bottom and black-edge plate, and treated as outlined above. When the time point was reached, medium with or without ABS was removed, and cells were washed thrice by 1× PBS before resuspension by same volume of 1× PBS. The cells were digested with trypsin and suspended in the 96-well plate. DiBaC$_4$ was introduced into the suspended HepG2 cell culture for staining at 5 μM at 37 °C in a 5% CO_2 humidified incubator for 30 min. Afterward, dye-containing medium was removed by centrifugation at 400× g, 5 min, 4 °C, and cells were washed thrice by 1× PBS and resuspended by same volume of 1× PBS. Following, cells in a 96-well black-bottom plate were applied for RFU (DiBAC4, 488 ex/515 em) determination by a microplate reader with a monochromator (Agilent, BioTek Synergy HTX, Santa Clara, CA, USA). RFU of DiBAC$_4$ was normalized by diving to cell number.

2.14. Mitochondrial Membrane Potential Measurement

The HepG2 cells (~900,000 cells per well) were cultured in a black glass-bottom cell culture dish and treated the same as outlined above. When the time point was reached, the medium with or without ABS was removed, and cells were washed thrice with by 1× PBS. Following, live cells were co-stained by JC-1 at 5 μg/mL and Hoechst 33,342 at 1 μM for 30 min at 37 °C in a 5% CO_2 humidified incubator. Next, cells were washed with 1× PBS six times and maintained in 1× PBS at the same volume before imaging. After washing in 1× PBS 6 times, images were visualized by applying ZEISS Elyra 7 with Lattice SIM^2 super-resolution fluorescent microscope (ZEISS, Oberkochen, Germany) with 63× oil-immersion objective.

2.15. Intracellular ATP Determination

Intracellular ATP level was measured by commercial Cellular ATP level determination Kit following the manufacturer's instructions. Cells (~9000 cells per well) were cultured in a 96-well, clear-glass-bottom and black-edge plate, and treated as outlined above. When the time point was reached, medium with or without ABS was removed, and cells were washed thrice with 1× PBS before resuspension by same volume of 1× PBS. Following, cells were incubated at room temperature (RT) lasting for 15 min after the introduction of cellular ATP indicator before measurement by a microplate reader with a monochromator (Thermo Fisher, Waltham, Varioskan). The relative luminescent unit (RLU) was normalized by cell number.

2.16. Western Blotting Analysis

HepG2 cells were cultured in a 6-well plate (8×10^5 cells/well) and treated with FFA and ABS based on the description in Section 2.11. The culture medium was discarded and washed with 1 mL 1× PBS solution, and 80 μL frozen RIPA solution containing a final concentration of 2 μg/mL of aprotinin, 5 μg/mL of leupeptin, 1 μg/mL of pepstatin A, 5 mM of NaF, 1mM of Na_3VO_4, and 1 mM of PMSF was added to each well and placed on ice for 30 min. In addition, the nuclear protein sample for the Western blot was prepared based on the instructions of the Nuclear and Cytoplasmic Protein Extraction Kit. The supernatants were collected and the BCA protein assay kit was used to determine the total protein concentration of each sample. The sample was dissolved in 5x protein sample loading buffer, vortexed, and boiled at 100 °C for 10 min. Protein samples were separated by 10% SDS-PAGE gel. All imprints were transferred onto the PVDF membrane by electrophoresis and the membrane was sealed at room temperature for 2 h using TBST buffer containing 5% non-fat milk. The PVDF membrane was incubated with the corresponding primary antibody (PI3K, p-PI3K, Akt, p-Akt, GSK-3β, p-GSK-3β, β-catenin, c-Myc, Lamin B1, and GAPDH; 1:1000 dilution) at 4 °C for twelve hours. The PVDF membrane was washed thrice by TBST buffer for 15 min each time. Next, the PVDF membrane was incubated with the secondary anti-rabbit IgG (H+L) antibodies (1:5000 dilution) at room temperature for 1 h. The PVhDF membrane was washed by TBST buffer thrice for 15 min each time. The protein strips on the PVDF membrane were scanned by the Image Quant LAS 500 imaging system (GE Healthcare Bio-Sciences AB, Sweden) and visualized by Image J software v1.8.0 (National Institutes of Health, Rockville, MD, USA). The protein strips were standardized by GAPDH or Lamin B1, and all results were independently repeated three times.

2.17. Immunofluorescence Analysis

The HepG2 cells (~900,000 cells per well) were cultured in black glass-bottom cell culture dish (Cat: BS-15-GJM-B) (Biosharp, Hefei, China) and treated the same as outlined above. When the time point was reached, medium with or without ABS was removed, and afterward, cells were washed six times with the commercially sourced 1× PBS. Cells were then fixed with 4% paraformaldehyde for at least 10 min. Subsequently, cells in the culture dish were permeabilized by 0.25% Triton X-100 diluted in 1× PBS for 15 min, and then blocked with commercial QuickBlock mixture for 1.5 h at room temperature. Afterward, the blocking solution was removed, and cells were incubated with the primary antibody, β-catenin, which was diluted into a commercial primary antibody diluent at 1:100 for 24 h at 4 °C statically. When the time point was reached, cells were washed six times before incubation with secondary antibody—Fluorescein (FITC)-conjugated Affinipure Goat Anti-Rabbit IgG (H+L)—which was diluted into commercial secondary antibody diluent at 1:80 for 1 h at 37 °C statically with protection from light. Afterward, the incubation solution was removed, and the cells in the culture dish were washed six times with the commercially sourced 1× PBS before counter-staining with commercial DAPI staining solution, dilution at 1:1000, for 15 min at room temperature (RT) with protection from light. This was followed by washing in PBS 6 times, and cells were visualized by applying ZEISS Elyra 7 with Lattice SIM^2 super-resolution fluorescent microscope (ZEISS, Oberkochen, Germany) applying a 63× oil-immersion objective.

The 3T3-L1 cells were cultured in a black glass-bottom cell culture dish and treated the same as mentioned above. When the time point was reached, the further processing was the same as outlined above; β-catenin and C/EBPα were chosen as the primary antibodies.

2.18. Microscope and Image Analysis

The observations of intracellular ROS (Carboxy-H_2DCFDA) or Membrane potential (DiBaC$_4$)/Chromosome status (Hochest 33342) determination were performed under ZEISS Elyra 7 with Lattice SIM^2 super-resolution fluorescent microscope (ZEISS, Oberkochen, Germany) equipped with 10× air or 63× oil-immersion objective. During image acquisition, cells were in a humidified chamber maintained at 37 °C in the presence of 5% CO_2. Two-dimensional original images with co-staining were captured by ZEISS Black edition. Images were reconstructed, processed, and exported, applying the ZEISS Blue edition.

For slide scanning figure processing (ORO staining), images were processed through OlyVIA (Olympus, Version Number: 4.1.1).

For cellular fluorescence measurement, image batch processing was conducted through Fiji (ImageJ, version 2.14.0); detailed code articles are included in the Supplementary Materials.

2.19. Statistical Analysis

All tests for this study were conducted at least three times. One-way analysis of variance (ANOVA) was used for statistical analysis, and the data are represented as mean ± SD. Tukey's test was used for analysis on Graphpad Prism 9.5.0 (Boston, MA, USA) and SPSS 26.0 software (IBM, Endicott, NY, USA), with statistical significance set at $p < 0.05$.

3. Results

3.1. Identification of the Active Compounds of ABS

The UHPLC-QE-MS analysis was used to analyze the components of ABS. The identification was carried out based on the mass spectrum of the sample, including the retention time, acquisition mode, MS spectra, adduct ions of $[M+H]^+$, $[M-H]^-$, $[M+FA-H]^-$, isotope information, and secondary fragment information (Supplementary Materials, Figures S1–S15). Fifteen saponin compounds were identified (Table 1 and Supplementary Materials, Figures S16 and S17) including azukisaponin I, azukisaponin II, azukisaponin III, azukisaponin IV, azukisaponin V, AZ I, AZ II, AZ III, AZ IV, soyasaponin I, dehydrosoyasaponin I, soyasaponin V, soyasaponin VI, soyasaponin Bd, and soyasaponin γg. In addition, our previous research also identified azukisaponin I, azukisaponin II, azukisaponin III, azukisaponin IV, and azukisaponin V through HPLC–DAD–ESI–MS^n [17]. This is consistent with the latest identification results by UHPLC-QE-MS.

Table 1. Compounds identified in adzuki bean saponin extraction by UHPLC-QE-MS.

No.	Compound Name	Adduct Ions	Formula	Measured Value	Molecular Weight Theoretical Value	Molecular Weight Error Value	Retention Time (s)	MS2 (M/Z)	Peak Value	ppm
1	Azukisaponin I	$[M+H]^+$	C42H68O13	780.46977	780.466	0.00377	400.814	441.37077, 85.02816, 423.35864, 141.01762, 83.04881	163,436,780.79455	2.85482
2	Azukisaponin II	$[M+H]^+$	C42H68O14	796.46826	796.4609	0.00736	400.319	441.37107, 423.3586, 141.01764, 95.08560, 109.10067	437,806,024.86214	2.82806
3	Azukisaponin III	$[M+H]^+$	C42H66O15	810.44692	810.4402	0.00672	475.707	455.35282, 95.08558, 437.34381, 141.01765, 203.17869	55,259,318.69737	1.13432
4	Azukisaponin IV	$[M-H]^-$	C48H76O20	972.48704	972.493	0.00596	379.409	971.48099, 971.50971, 513.10683, 72.99311, 113.02456	760,441,902.26418	2.10397

Table 1. *Cont.*

No.	Compound Name	Adduct Ions	Formula	Measured Value	Molecular Weight Theoretical Value	Molecular Weight Error Value	Retention Time (s)	MS2 (M/Z)	Peak Value	ppm
5	Azukisaponin V	[M-H]$^-$	C48H78O18	942.50914	942.5188	0.00966	400.667	941.49680, 941.52420, 85.02957, 71.01351, 101.02394	748,285,192.51553	0.91286
6	AZ I	[M-H]$^-$	C48H74O18	938.47971	938.4875	0.00779	459.139	937.48260, 937.50982, 497.11238, 113.02460, 72.99308	14,844,651.78951	0.31438
7	AZ II	[M-H]$^-$	C54H82O23	1098.51519	1098.5247	0.00951	349.7	1097.53284, 1097.49836, 85.02943, 113.02472, 221.06508	20,708,578.64993	4.38042
8	AZ III	[M-H]$^-$	C54H82O22	1082.51984	1082.5298	0.00996	444.2375	1081.53791, 1081.47045, 497.11181, 113.02467, 321.08450	67,337,840.04083	0.15001
9	AZ IV	[M-H]$^-$	C54H84O22	1084.53716	1084.5454	0.00824	437.536	1083.53099, 1083.49717, 85.02949, 113.02458, 221.06527	93,817,390.43866	2.62045
10	Soyasaponin I	[M+H]$^+$	C48H78O18	942.52503	942.5188	0.00623	398.4	441.37103, 423.35907, 85.02816, 141.01755, 599.38996	507,128,589.07886	1.08788
11	Dehydro-soyasaponin I	[M-H]$^-$	C48H76O18	940.49239	940.5032	0.01081	386.514	939.49993, 939.52724, 113.02476, 71.01352, 99.00856	273,042,021.70692	2.78004
12	Soyasaponin V	[M+FA-H]$^-$	C48H78O19	958.47793	958.5137	0.03577	394.75	957.50802, 957.47992, 85.02954, 92.68074, 221.06830	25,799,920.53848	2.06398
13	Soyasaponin VI	[M-H]$^-$	C54H84O21	1068.54070	1068.5505	0.00980	443.487	1067.54184, 1067.47569, 85.02939, 113.02483, 593.26701	128,371,685.59563	0.65605
14	Soyasaponin Bd	[M-H]$^-$	C48H76O19	956.49142	956.4981	0.00668	405.108	955.48588, 955.51389, 113.02480, 72.99307, 513.10690	405,142,018.01550	1.49019
15	Soyasaponin γg	[M-H]$^-$	C48H74O17	922.48554	922.4926	0.00706	465.995	921.49279, 921.51932, 71.01358, 113.02483, 101.02401	78,978,193.68895	0.58421

3.2. Network Pharmacology

The obesity-related datasets (GSE25401) were collected from the NCBI GEO public database, which included 56 patients in total (30 in the control group and 26 in the obesity group). We used the "limma" package to identify differentially expressed genes (DEGs) across the two groups based on a genetic screening requirement of a p value < 0.05. We identified 2841 DEGs, including 2487 upregulated genes and 354 downregulated genes (Figure 1A). The structures of the 15 bioactive compounds from ABS were used to collect targets from the SwissTargetPrediction and SEA databases. Moreover, the targets of lipid metabolism were collected from the DisGeNet database. The 80 interacting targets are included in the Venn diagram between the drug target, obesity, and lipid metabolism (Figure 1B).

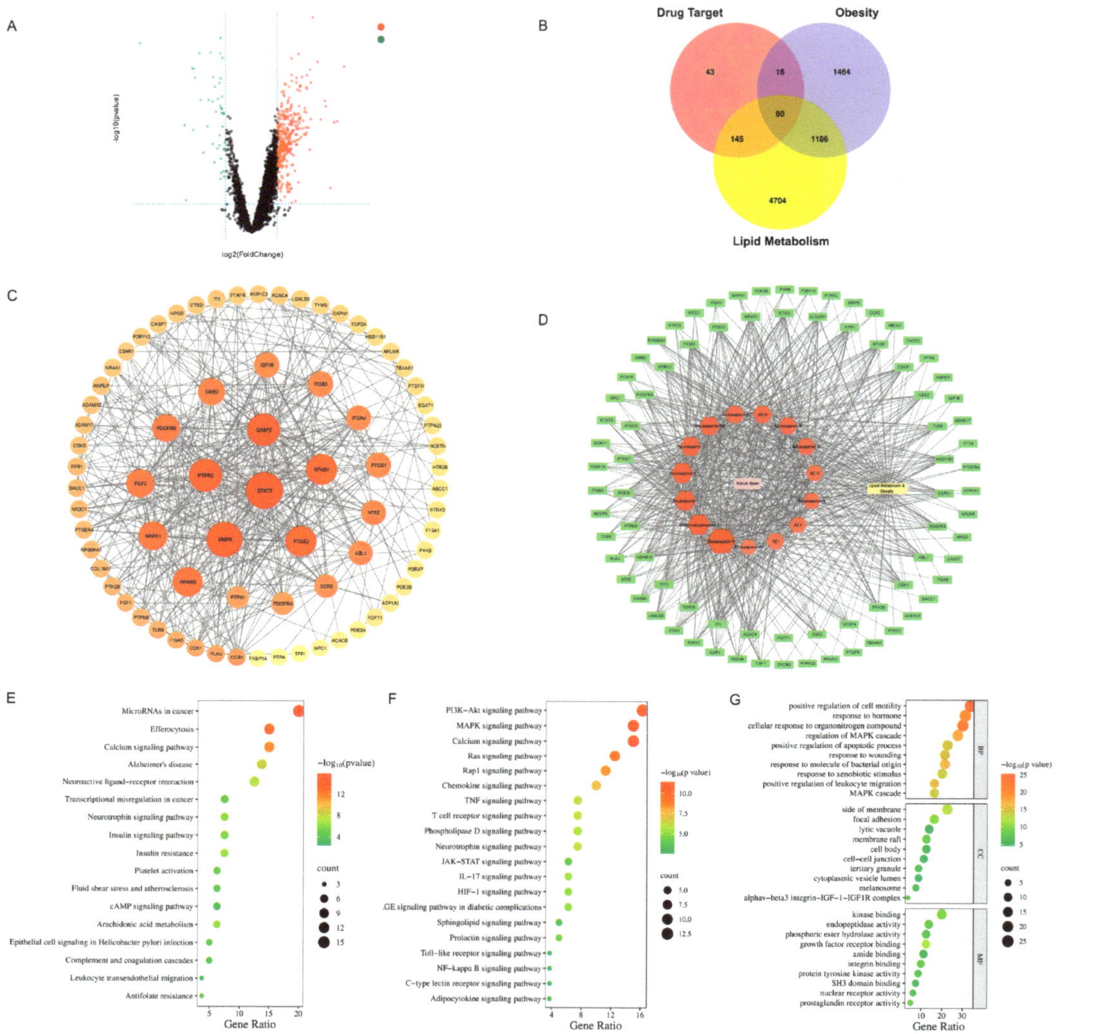

Figure 1. The network pharmacology analysis of ABS. (**A**) DEGs screening and presentation as Volcano plots. (**B**) The Venn diagram between the drug targets, obesity targets, and lipid metabolism. (**C**) The PPI network of the interacting targets. (**D**) The drug–compounds–targets–disease network. (**E**) The dot plot of KEGG pathway enrichment analysis. (**F**) The dot plot of KEGG signaling pathway enrichment analysis. (**G**) The dot plot of GO enrichment analysis. BP, biological process; CC, cellular component; MF, molecular function.

The interacting targets are the anti-obesity targets of ABS that improve lipid metabolism. The anti-obesity target of each compound is shown in Table 2. The PPI analysis was conducted in the string database based on the 80 interacting targets and displayed as a network figure (Figure 1C). The results showed that the core targets with top 10 degree values are STAT3, MMP9, CASP3, PTPRC, NFKB1, PTGS2, PPARG, MAPK1, FGF2, and PDGFRB. The drug–compounds–targets–disease network is constructed in Figure 1D. Soyasaponin V, dehydrosoyasaponin I, azukisaponin V, soyasaponin I, and azukisaponin II may play an important role in the anti-obesity effect of ABS by improving lipid metabolism based on the network pharmacology.

Table 2. The targets and structures of adzuki bean saponins.

Pubchem CID/SID	Name	Structure	Targets
14103656	Azukisaponin I	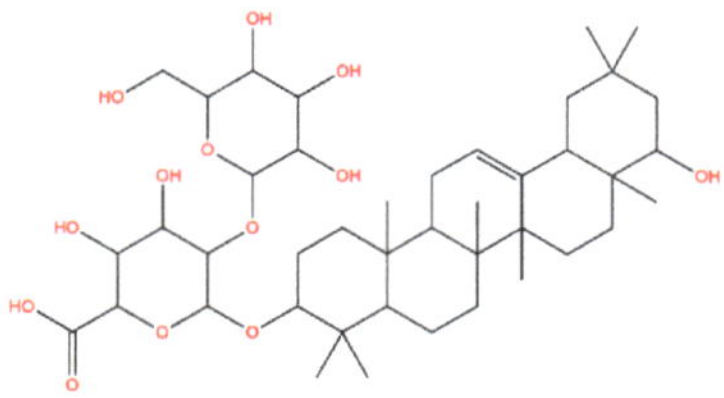	37
475872498	Azukisaponin II	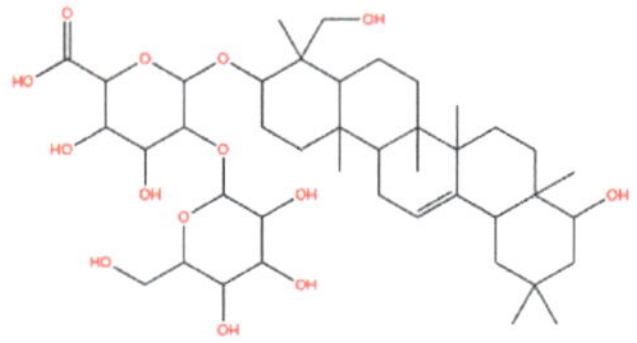	42
441909	Azukisaponin III	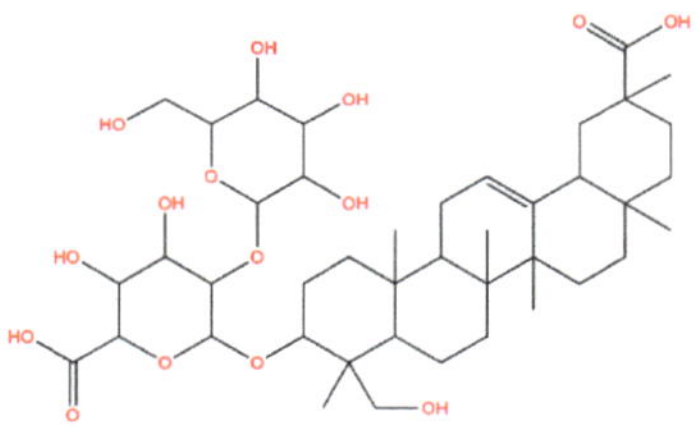	36
163194451	Azukisaponin IV	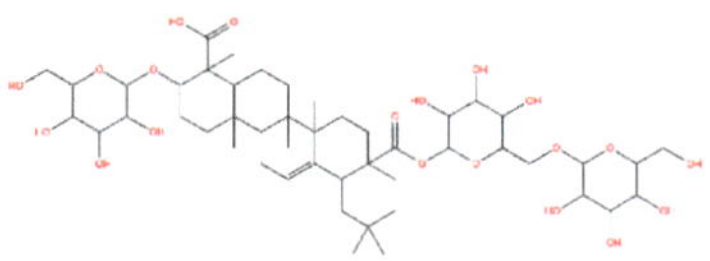	30
5087640	Azukisaponin V	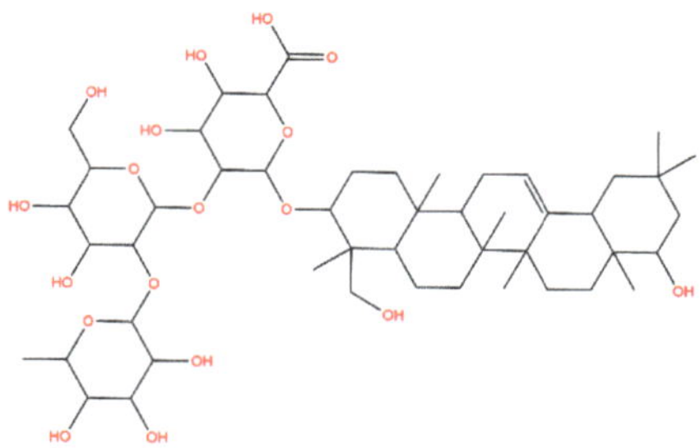	45
131751192	AZ I	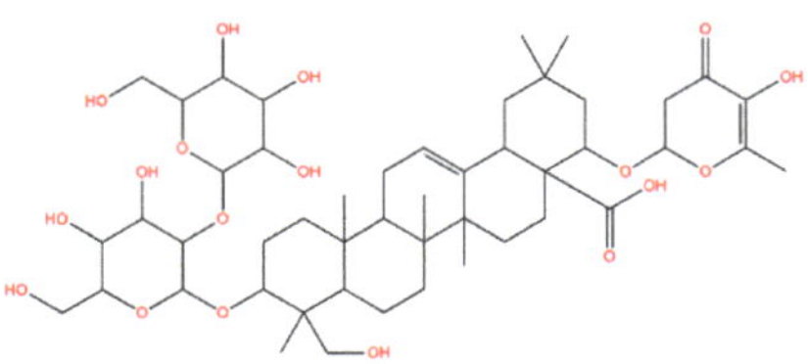	33

Table 2. *Cont.*

Pubchem CID/SID	Name	Structure	Targets
131751576	AZ II	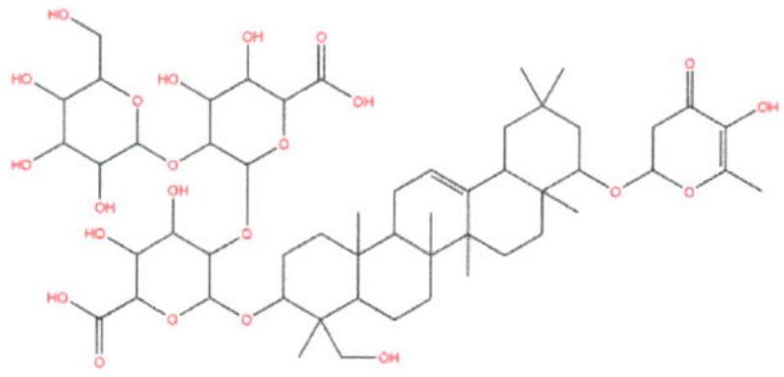	33
131751577	AZ III	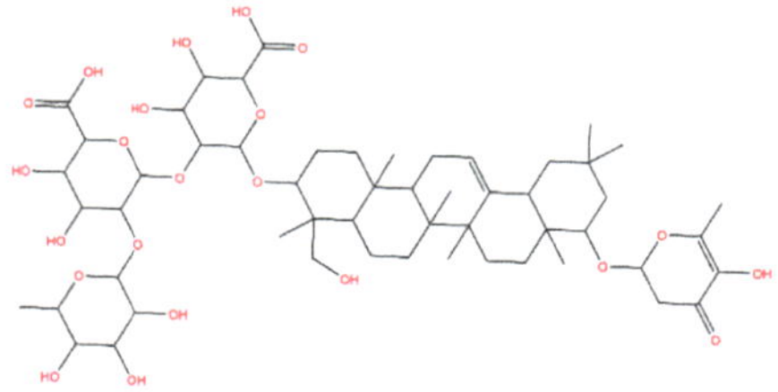	37
85058456	AZ IV	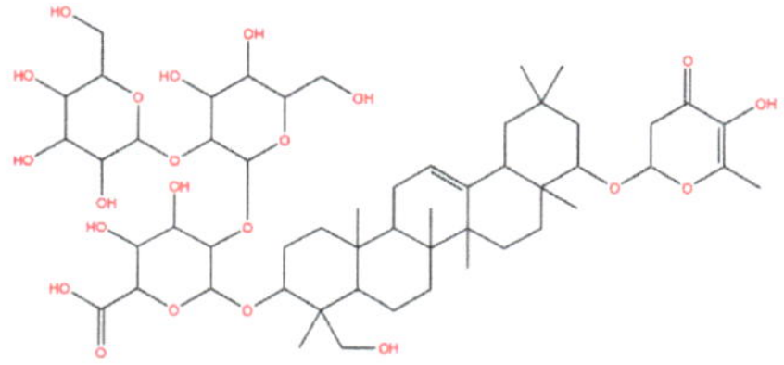	33
122097	Soyasaponin I	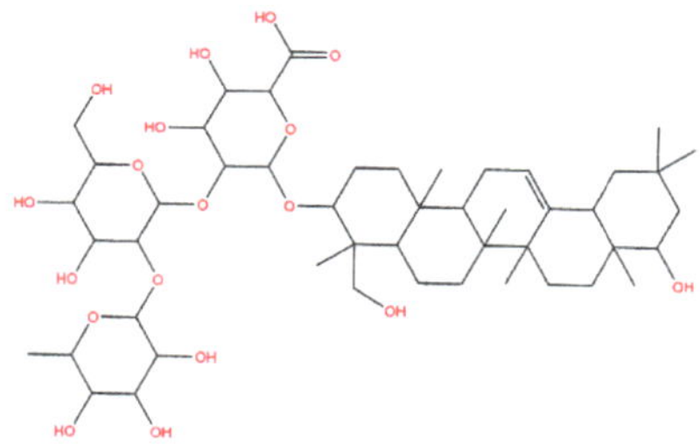	45
656760	Dehydrosoyasaponin I	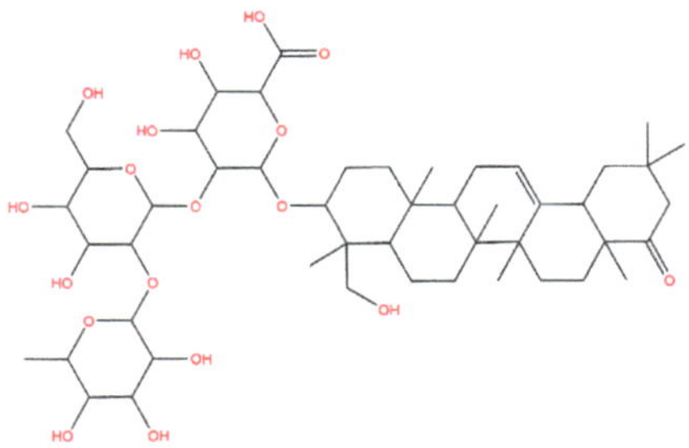	47

Table 2. *Cont.*

Pubchem CID/SID	Name	Structure	Targets
15608234	Soyasaponin V	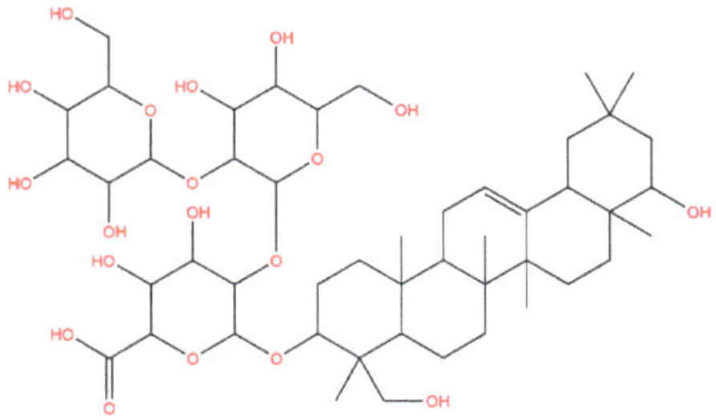	62
164453	Soyasaponin VI	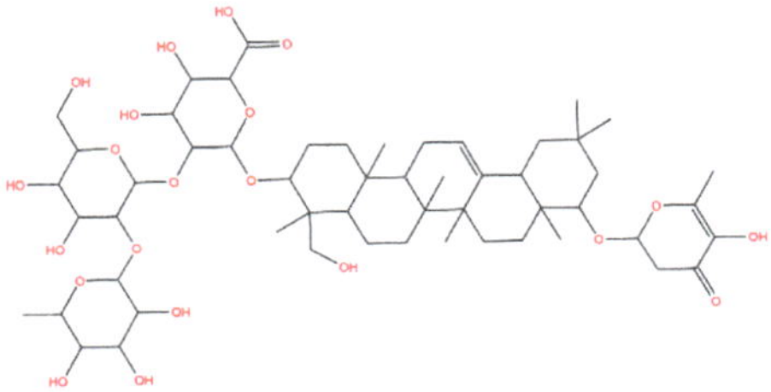	36
178247	Soyasaponin Bd	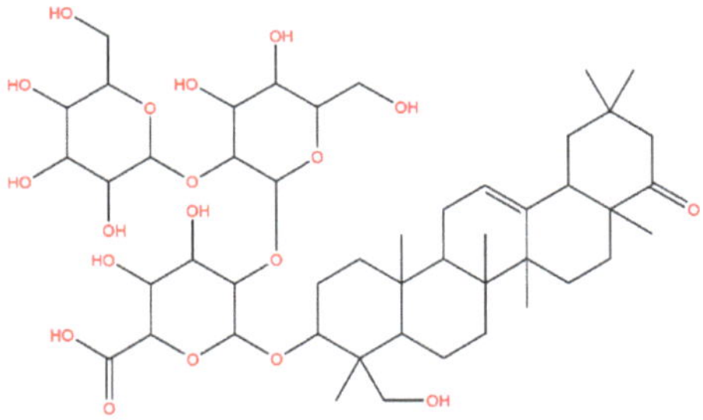	40
447274937	Soyasaponin γg	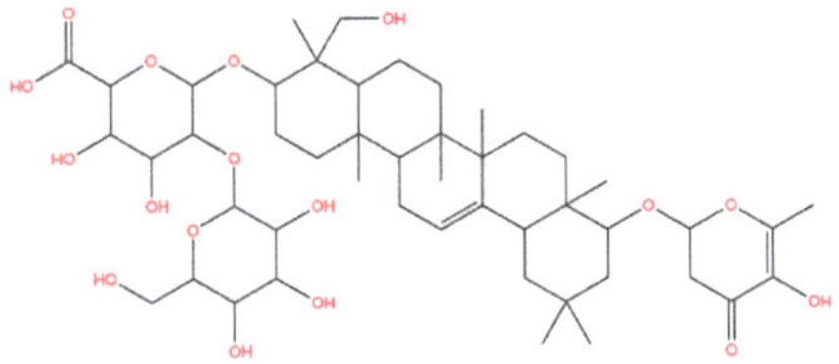	33

To explore the molecular mechanism of the anti-obesity effect of ABS in improving lipid metabolism, the 80 interacting targets were used to perform the KEGG and GO enrichment analyses. The KEGG enrichment analysis indicated that the PI3K-Akt signaling pathway and MAPK signaling pathway are associated with the anti-obesity effect of ABS (Figure 1E,F). This implied that ABS might exert its anti-obesity through the involvement of multiple pathways, especially the PI3K-Akt signaling pathway. Moreover, the GO analysis was conducted and indicated that ABS primarily participates in positive regulation of cell motility, responses to hormones, and cellular responses to organonitrogen compounds, as shown in Figure 1G. The cellular components mainly focused on the side of the membrane, focal adhesion, and lytic vacuole. Moreover, the effects of ABS on molecular functions mainly included kinase binding, endopeptidase activity, and phosphoric ester hydrolase activity.

3.3. Bioinformatics

Three algorithms were used to pick signature genes from the intersection genes of adzuki bean saponins and obesity. The LASSO algorithm is a regression analysis technique that uses regularization to increase prediction accuracy. Using the LASSO technique, 15 signature genes were found after 10 cross-validations, including *HSD11B1*, *FGF1*, *FGF2*, *F13A1*, *ACACB*, *PDE3A*, *ABCC1*, *STAT3*, *APLNR*, *PDE3B*, *MMP9*, *FDFT1*, *PTPN22*, *TACR3*, and *ITGA5* (Figure 2A,B). Using the random forest algorithm, the 15 signature genes with the highest scores of importance identified include *FGF2*, *ABCC1*, *F13A1*, *PTGFR*, *P2RY12*, *MMP9*, *PDE3A*, *VEGFA*, *NR3C1*, *PDE3B*, *FDFT1*, *ACACB*, *PTPN22*, *FGF1*, and *HSD11B1* (Figure 2D,E). By employing the SVM-RFE algorithm (Figure 2C), the classifier showed the minimum error with 19 features, including *F13A1*, *FGF2*, *FGF1*, *MMP9*, *PTGFR*, *PDE3A*, *P2RY12*, *HSD11B1*, *ABCC1*, *ACACB*, *APLNR*, *ITGAV*, *FDFT1*, *PTPN22*, *PDE3B*, *ACACA*, *VEGFA*, *NTRK3*, and *NR3C1*. Following the identified gene intersection using the three algorithms, 11 characteristic genes were finally identified, including *F13A1*, *FGF2*, *FGF1*, *MMP9*, *PDE3A*, *HSD11B1*, *ABCC1*, *ACACB*, *FDFT1*, *PTPN22*, and *PDE3B* (Figure 2F).

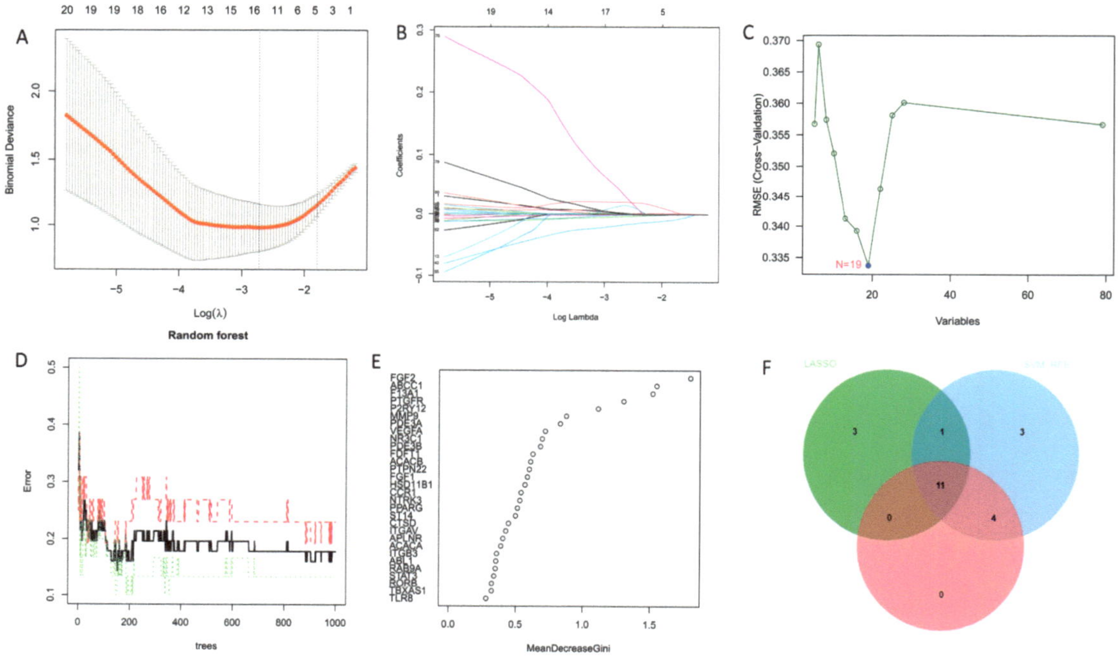

Figure 2. Selection of characteristic genes of intersection genes. (**A**) Ten-time cross-verification of tuning parameters in the LASSO model. Each curve refers to one gene. Solid vertical lines indicate the standard error (SE) of the partial likelihood deviation and dotted vertical line marks the optimal lambda value. (**B**) In the LASSO coefficient profile, adjustment of feature selection in the minimum absolute shrinkage and selection operator model. (**C**) The SVM-RFE algorithm was used for selecting features. (**D**) Examining the correlation between the count of trees in a random forest and the error rate. The red dotted line indicates the Training Error, which is how the model behaves on the training set. The black implementation indicates the Validation Error, which is how the model behaves on the validation set. The green dotted line indicates the Test Error, which is how the model behaves on the test set (**E**) Ranking of genes based on their relative importance. (**F**) A Venn diagram depicting genes common to the LASSO, random forest, and SVM-RFE methods.

To further elucidate the relationship between the identified genes and immune cell dynamics in obesity, we performed a correlation analysis (Figure 3). Our results indicate that certain genes, specifically *HSD11B1*, *PTPN22*, *ABCC1*, *MMP9*, *FGF1*, and *F13A1*, exhibit a positive correlation with M0 and M1 macrophages, which are known to play a role in the early stages of adipose tissue inflammation and contribute to the resolution of inflammation. Conversely, these genes show a negative correlation with other immune cells, such as naive B cells, plasma cells, CD8 T cells, CD4 T cells, regulatory T cells, and activated NK cells, which might suggest a complex interplay between these genes and the immune response in obesity.

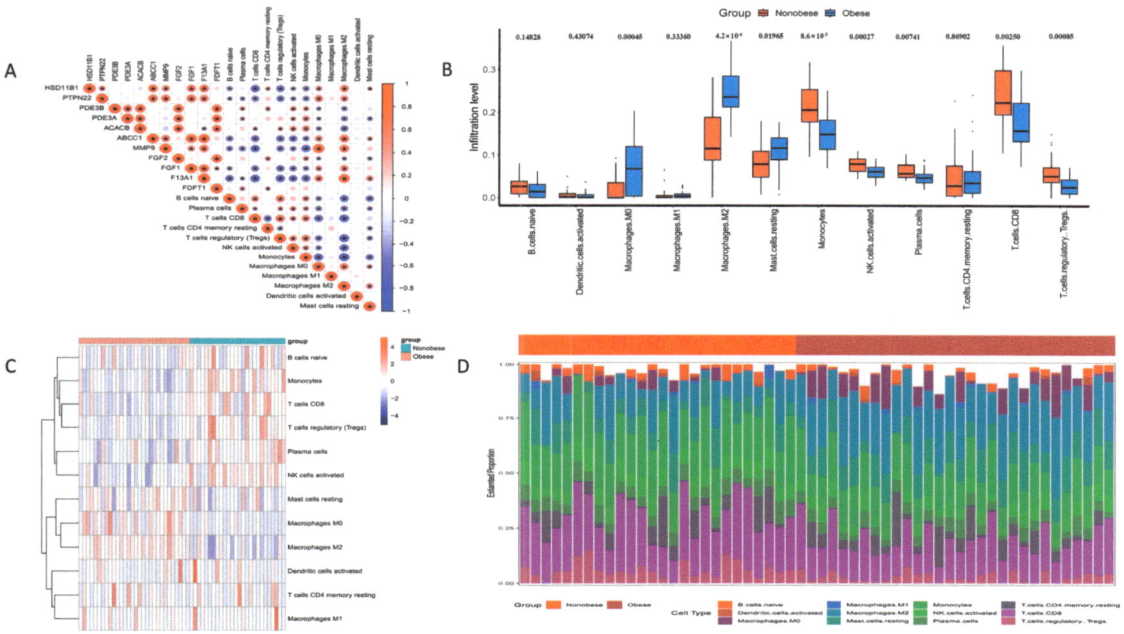

Figure 3. Immune infiltration study of character genes. (**A**) demonstrates a robust correlation among 11 genes, where red denotes a positive correlation, blue indicates a negative correlation, and "*" signifies a statistically significant correlation. Red color represents positive correlation, blue color represents negative correlation, and "*" indicates statistically correlation (**B**) illustrates variations in immune cell infiltration across different samples. (**C**) shows the disparity in immune cell infiltration between groups with and without obesity, which was statistically assessed, with "ns" indicating that the difference was statistically insignificant. (**D**) indicates the difference in immune cell infiltration between the groups with and without obesity.

The observed correlations suggest that these genes may have a regulatory role in modulating the immune environment within adipose tissue during obesity. The positive correlation with M0 and M1 macrophages could imply a potential role in promoting a more anti-inflammatory state, while the negative correlation with other immune cells might indicate an involvement in dampening excessive immune responses that could contribute to obesity-related inflammation. These findings open avenues for further research into the immunomodulatory effects of these genes and their potential as therapeutic targets.

To further investigate the diagnostic potential of the characteristic genes, an ROC analysis was performed. An AUC value of more than 0.75 for a gene indicated a good diagnostic value for weight remission. The results were as follows: *ABCC1* (AUC = 0.846, Figure 4A); *ACACB* (AUC = 0.783, Figure 4B); *F13A1* (AUC = 0.877, Figure 4C); *FDFT1* (AUC = 0.708, Figure 4D); *FGF1* (AUC = 0.829, Figure 4E); *FGF2* (AUC = 0.777, Figure 4F); *HSD11B1*

(AUC = 0.797, Figure 4G); *PDE3A* (AUC = 0.774, Figure 4H); *PTPN22* (AUC = 0.776, Figure 4I); *MMP9* (AUC = 0.847, Figure 4J); and *PDE3B* (AUC = 0.723, Figure 4H). Thus, *ABCC1*, *ACACB*, *F13A1*, *FGF1*, *FGF2*, *HSD11B1*, *PDE3A*, *PTPN22*, and *MMP9* have better obesity diagnostic values.

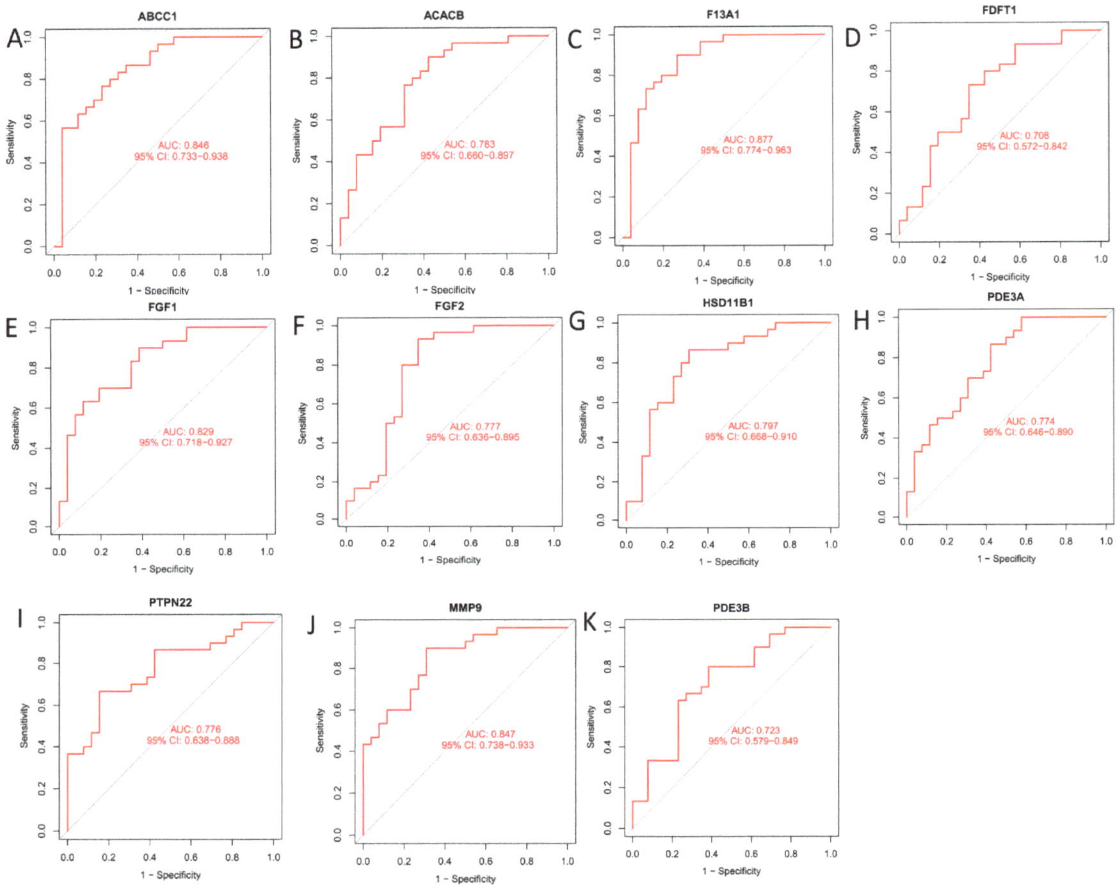

Figure 4. Diagnostic potential of characteristic genes in obesity. (**A**–**K**) ROC curves estimating the diagnostic performance of characteristic genes. (**A**) ABCC1, (**B**) ACACB, (**C**) F13A1, (**E**) FGF1, (**F**) FGF2, (**G**) HSD11B1, (**H**) PDE3A, (**I**) PTPN22, (**J**) MMP9 have better diagnostic potential for obesity.

3.4. *Effect of ABS on HepG2 Cell Viability*

The CCK8 method was used to evaluate the effect of ABS on cell viability (Figure 5A). HepG2 cells did not show significant inhibitory effects when treated with ABS at concentrations from 0 to 0.2 mg/mL for 24 h compared with the control group. However, when the concentration of ABS reached 0.4 mg/mL, the viability of HepG2 cells significantly decreased ($p < 0.0005$). Therefore, in this study's subsequent experiments, the ABS concentration was chosen to be below 0.2 mg/mL.

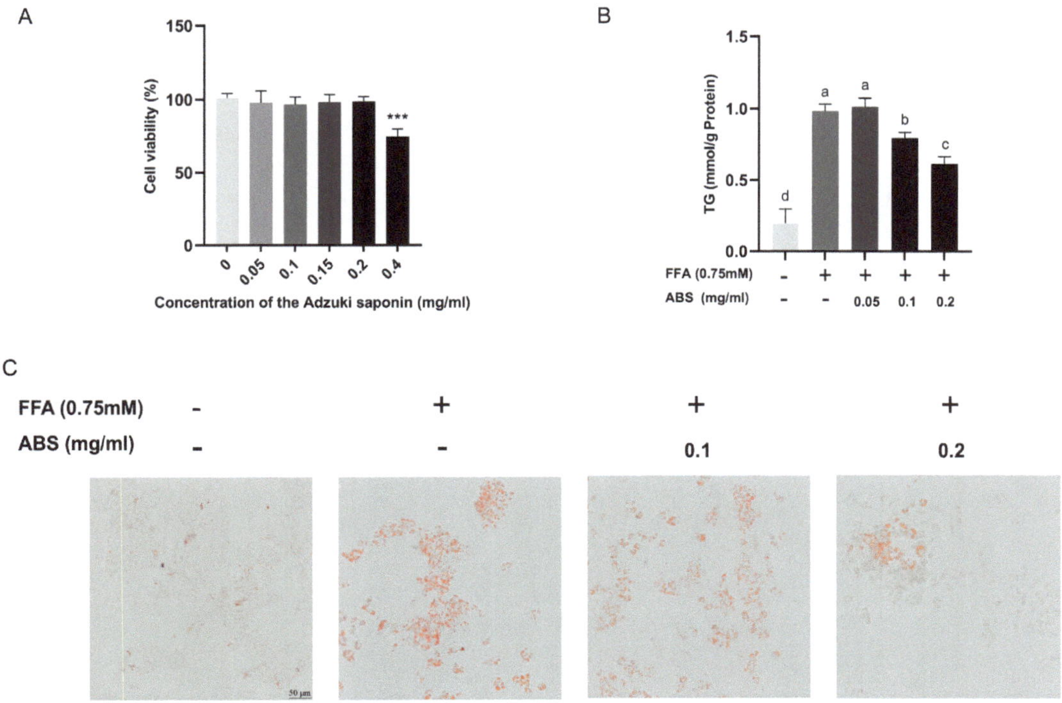

Figure 5. The anti-obesity effect of ABS on HepG2 cells. (**A**) The effect of ABS on the cell viability of HepG2 cells. *** $p < 0.001$. (**B**) Effect of ABS on the triglyceride (TG) content in HepG2 cells. (**C**) Oil Red O (ORO) staining of free fatty acid (FFA)-induced HepG2 cellular model with or without treatment of adzuki bean saponins (ABS) at different doses. Figures are captured through scanning system of VS200 slide scanner system (Olypums, Tokyo, Japan) with a 40× objective. Scale bar, 50 µm. All results are presented as mean ± S.D and the experiments were repeated as triplicates. Bars with different letters (a–d) significantly ($p < 0.05$) differ according to ANOVA and Tukey's multiple range test. FFA, free fatty acid; ABS, adzuki bean saponins.

3.5. ABS Inhibit Intracellular Lipid Accumulation

To explore the inhibition effect of lipid accumulation of ABS, we used different dosages to treat the modeled HepG2 cell. We determined the TG content (Figure 5B). The results showed that the TG content in the model group increased significantly compared with the control group and ABS can dose-dependently decrease the intracellular TG content in the dosages between 0.1 mg/mL and 0.2 mg/mL. In addition, the 0.05 mg/mL ABS treatment did not show any effect. The dosages in our further study are set as up to 0.1 mg/mL and 0.2 mg/mL. We also used ORO staining to evaluate lipid droplet generation (Figure 5C). The results also showed that both 0.1 mg/mL and 0.2 mg/mL ABS can decrease lipid droplet generation, which is similar in the intracellular TG content testing. Moreover, we also conducted the Nile Red and Hochest 33342 double staining (Figure 6). Nile Red (NR), a lipophilic staining, is of an intensive red fluorescent signal (~552 ex/635 em) after interaction with lipids. Since NR binding with the lipid bilayer of a cell membrane also induces a strong red signal, and considering the fact that NR binding with an intracellular lipid droplet induces a specific green fluorescent signal [44] (~488 ex/528 em), we used two super-high-speed cameras and a long-pass (LP) filter (LP560) to capture both the green (488 ex/495~550 em) and red (561 ex/570~620 em) signals simultaneously after excitation with both a 488 and 561 nm laser. Therefore, we can clearly observe the total lipid distribution and FFA-induced lipid droplet status at the same time. Meanwhile, Hochest staining for the chromosomes was applied for location under microscope (405 ex/420~480 em) (Figure 6A).

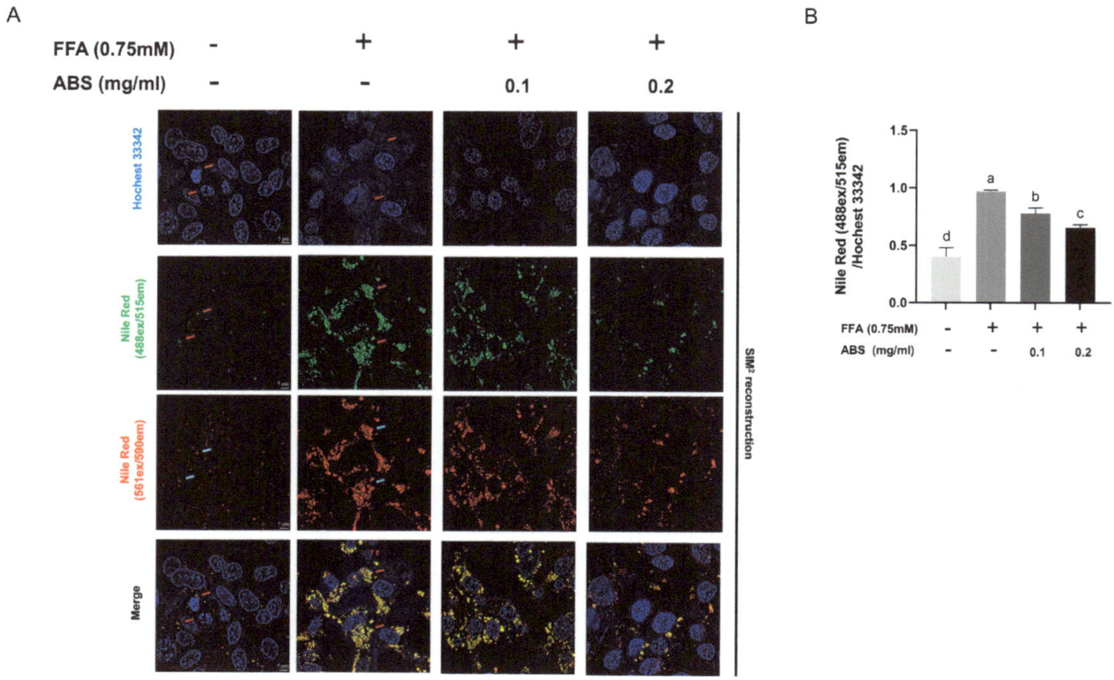

Figure 6. ABS alleviates FFA-mediated lipid aggregation. (**A**) Fluorescent images of HepG2 cells co-stained with Hochest 33342 (blue) and Nile Red (green and red) taken by ZEISS Elyra 7 with Lattice structured illumination microscopy $(SIM)^2$ super-resolution fluorescent microscope equipped with a 63× oil-immersion objective Red Arrow: Fluorescent aggregation with strong green and red signal. Scale bars, 5 μm. (**B**) Semi-quantitative analysis based on original images taken by SIM^2 of fluorescence signal and lipid droplet by Image J. All results are presented as mean ± S.D and the experiments were repeated as triplicates. Bars with different letters (a–d) significantly ($p < 0.05$) differ according to ANOVA and Tukey's multiple range test. FFA, free fatty acid; ABS, adzuki bean saponins.

After SIM^2 reconstruction with the removal of the background signal, it is clearly displayed that lipid distribution is distributed uniformly (red signal distribution evenly), and only a few small lipid droplets are found (round, green, ball-like signal distributions in small amounts). However, after FFA introduction, intracellular lipid droplets (green) are significantly enhanced—with a bigger size and stronger fluorescent intensity—alongside a disappearance of the even distribution of the "red signal", both of which indicate lipid accumulation. Due to FFA introduction, in a dose-dependent manner, the ball-like aggregation (green) of lipid droplets significantly decreased alongside a recovery of the red fluorescent signal's even distribution, indicating the phenotypic rescue of lipid metabolic disorder. Meanwhile, the semi-quantitative analysis of lipid droplets (green signal/blue signal) (Figure 6B) supported our observation at the single-cell level (Figure 6A).

Besides fluorescence and fluorescent imaging systems with high-magnification objectives (63 or 40×), to solidly verify our finding, we applied the latest quantitative phase imaging (QPI) system to conduct a label-free, long-duration, time-lapse, live-cell destiny tracking method to discuss the cellular morphological change due to FFA-inducement and treatment with/without ABS, to avoid any false-positive results or phototoxicity due to laser excitation and strong background of the bright field. Accordingly, we clearly observed unique bright foci formation or fluorescence uneven distribution at the single-cell level (10× objective) due to the time-lapse imaging, while treatment with ABS alleviates this phenotypic change. Since lipid droplet formation will change the refractive index (RI) of cytoplasm, and imaging by QPI relies on the RI, it implies that FFA-inducement facilitates lipid droplet formation, but this could be alleviated by ABS introduction (Figure 7). All these findings were consistent with our results of the ORO staining and intracellular TG content testing in our previous research.

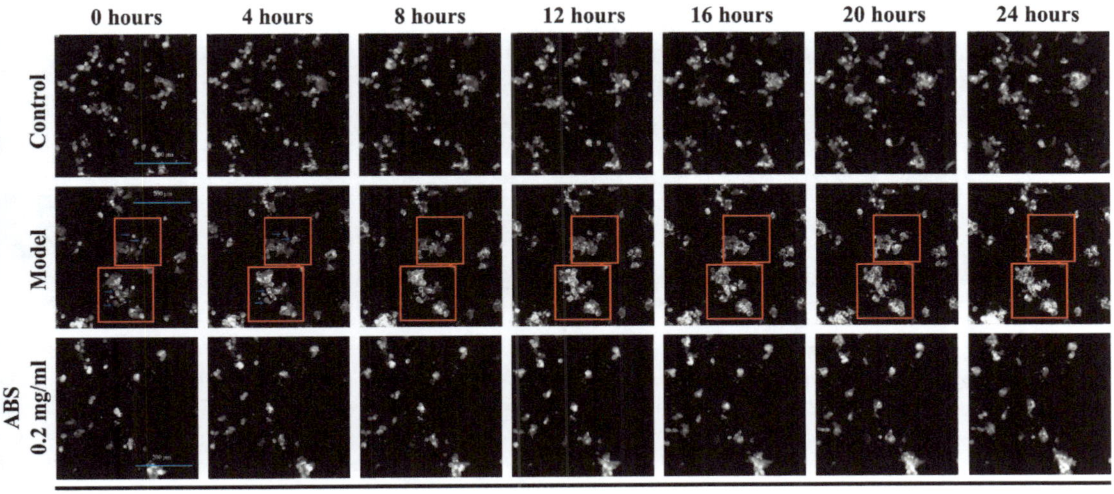

Figure 7. Quantitative phase imaging (QPI) of 24-hour observation (captured every 0.5 hours) of FFA-induced model with or without treatment of ABS performed using a Livecyte2 microscope with a 10× objective. Control is the HepG2 cell treated with the MEM medium; model is the HepG2 cell treated with the MEM medium containing the free fatty acid; ABS 0.2 mg/mL is the HepG2 cell treated with the MEM medium containing the free fatty acid and 0.2 mg/mL dosage of the adzuki bean saponins. Blue Arrow: Time-lapse change of single cell with fluorescence re-distribution.

3.6. ABS Inhibit ROS Generation

To study the ROS generation in our ABS treatment group, we used the Carboxy-H_2DCFDA and Hochest 33,342 co-staining analysis (Figure 8A). In the normal cells, few green foci (ROS signal) are found, while FFA introduction facilitates bright green foci formation intracellularly—some of which are co-located with blue signal aggregation (chromosome) (red arrow), while others are distributed in the cytoplasm—both indicating that ROS stress the cellular survival status like DNA condensation. ABS introduction alleviates the amount of green fluorescent signal and bright foci in a dose-dependent manner at the single-cell level. Meanwhile, the semi-quantitative analysis of the staining was conducted, and the results showed that ROS increase significantly in comparison to the model group, and the treatment of 0.1 mg/mL and 0.2 mg/mL ABS can decrease the ROS level significantly. Interestingly, the ABS dosage at either 0.1 mg/mL or 0.2 mg/mL did not show as statistically significant (Figure 8B). Furthermore, we also conducted research on the ROS level at the cellular level (bulk culture) through the microreader using the same dye (Carboxy-H_2DCFDA) (Figure 8C). This result is also similar to the semi-quantitative analysis. ROS can damage the mitochondrial membrane, leading to a decrease in its potential, which is essential for ATP synthesis and cellular energy production. Thus, the change in ROS level indicated that the anti-obesity effect of ABS may be associated with mitochondrial energy metabolism since this is an important source of ROS driven by the TCA cycle at either the single-cell level or bulk culture grade.

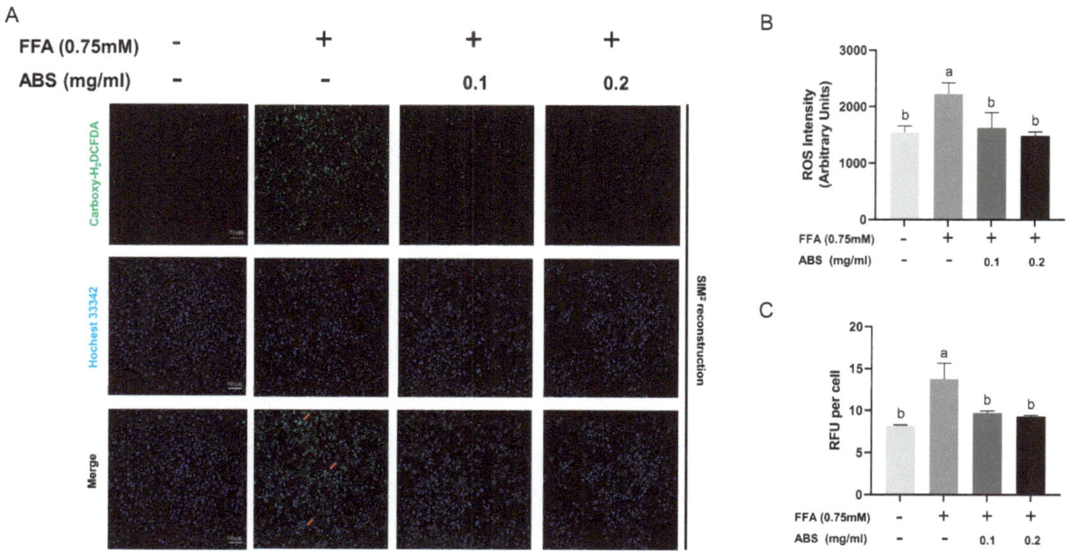

Figure 8. ABS alleviates FFA-facilitated intracellular ROS accumulation. (**A**) Fluorescent images of HepG2 cells co-stained with Hochest 33342 (blue) and Carboxy-H2DCFDA (green) taken by ZEISS Elyra 7 equipped with Lattice structured illumination microscopy (SIM)^2 super-resolution fluorescent microscope applying a 10× objective. Scale bars, 50 μm. Red Arrow: Fluorescent aggregation with strong green signal (**B**) Semi-quantitative analysis based on original images taken by SIM^2 of fluorescence signal of ROS intensity by Image J. (**C**) Normalized relative fluorescent unit (RFU) representing ROS accumulation (Carboxy-H_2DCFDA) in bulk culture in 96-well plates, which is divided by the cell number. All results are presented as mean ± S.D and the experiments were repeated as triplicates. Bars with different letters (a–b) significantly ($p < 0.05$) differ according to ANOVA and Tukey's multiple range test. FFA, free fatty acid; ABS, adzuki bean saponins.

3.7. ABS Improve Mitochondrial Energy Metabolism

In line with previous study on the ROS level, the results showed that the curing effect of ABS may be associated with mitochondrial energy metabolism. We used DiBaC$_4$ and Hochest 33342 co-staining to indirectly study changes in the mitochondrial membrane potential (Figure 9A). When mitochondrial function is impaired, a decrease in mitochondrial membrane potential will indirectly facilitate depolarization of the cell membrane potential (Δψ). Accordingly, DiBAC$_4$ can indirectly reflect changes in mitochondrial membrane potential as it enters depolarized cells alongside fluorescence enhancement. Our results show that the DiBAC$_4$ fluorescence in the model group increases (green) co-located mito-DNA (blue, small area) alongside both the green and blue signal aggregation and condensation, indicating that FFA drives mitochondrial DNA stress and potential abnormality. However, treatment with 0.2 mg/mL ABS can decrease both the green and blue fluorescence aggregation at the single-cell level, implying an alleviation in the depolarization of the mitochondrial membrane potential and mito-DNA stress at the single-cell level. In contrast, the 0.1 mg/mL ABS did not show a bioactive effect. Furthermore, the signal cellular RFU of the DiBAC$_4$ dye (bulk culture) was conducted and also confirmed this finding (Figure 9C).

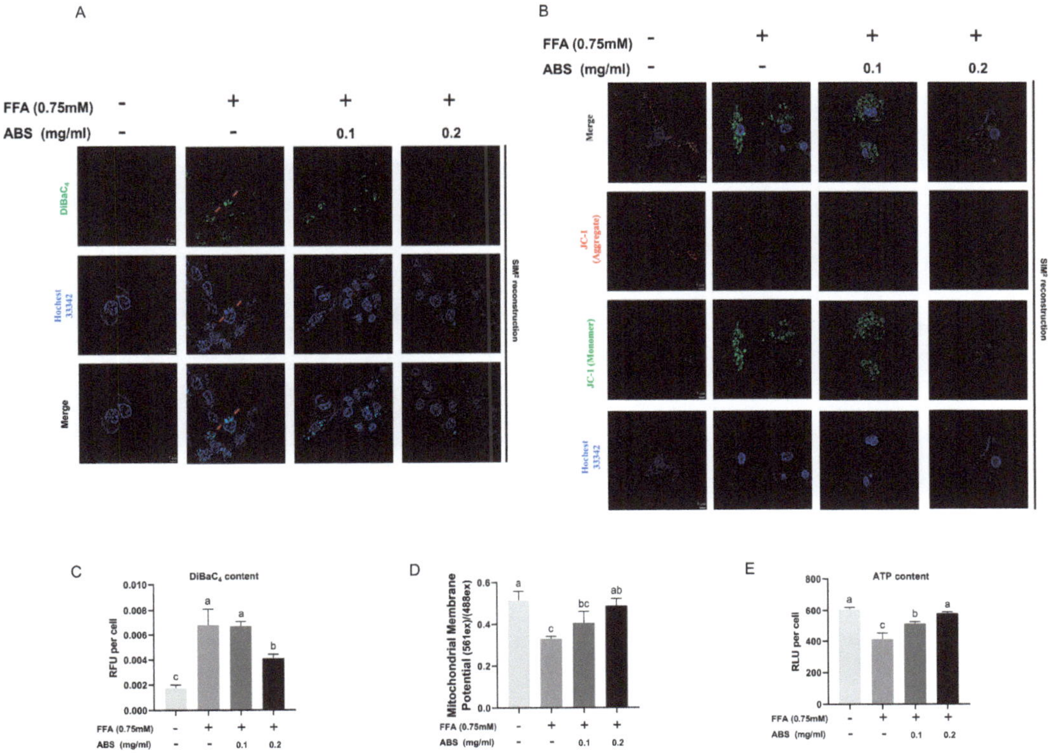

Figure 9. ABS alleviates FFA-facilitated mitochondrial membrane potential loss and ATP level abnormality. (**A**) Fluorescent images of HepG2 cells co-stained with Hochest 33,342 (blue) and DiBaC$_4$ (green) taken by ZEISS Elyra 7 equipped with Lattice structured illumination microscopy (SIM)^2 super-resolution fluorescent microscope with a 63× oil-immersion objective. Scale bars, 5 μm. Red Arrow: Fluorescent aggregation with strong green and blue signal. (**B**) Fluorescent images of HepG2 cells co-stained with Hochest 33,342 (blue) and JC-1 (green or red) taken by ZEISS Elyra 7 equipped with Lattice structured illumination microscopy (SIM)^2 super-resolution fluorescent microscope with a 63× oil-immersion objective. Scale bars, 5 μm. (**C**) Normalized relative fluorescent unit (RFU) representing membrane potential change (DiBaC$_4$) in bulk culture in 96-well plates, which is divided by the cell number. (**D**) Semi-quantitative analysis based on original images taken by SIM^2 of fluorescence signal of JC-1 intensity by Image J. (**E**) Normalized relative luminescent unit (RLU) representing ATP level in bulk culture in 96-well plates, which is divided by the cell number. All results are presented as mean ± S.D, and the experiments were repeated as triplicates. Bars with different letters (a–c) significantly ($p < 0.05$) differ according to ANOVA and Tukey's multiple range test. FFA, free fatty acid; ABS, adzuki bean saponins.

Since DiBAC$_4$ is a general membrane potential determination probe, to further verify our hypothesis, we also applied JC-1 (5,5′,6,6′-tetrachloro-1,1′,3,3′-tetraethylbenzimidazolylcarbocyanine iodide) to directly study changes in mitochondrial membrane potential (Figure 9B). JC-1 is the most commonly used mitochondrial membrane potential measurement-specific dye [45,46]. After SIM^2 reconstruction with the removal of the background signal, the results clearly show that green and red fluorescence are co-located, indicating the normality of the mitochondrial membrane potential. However, due to the FFA introduction, we clearly see a green fluorescence dispersion, which is without a strong red signal, indicating a $\Delta\psi$ decrease and following leakage of JC-1 monomer into the cytoplasm. This phenotype can be rescued with ABS treatment (Figure 9D). Meanwhile, the semi-quantification analysis of the JC-1 aggregate (red) to JC-1 monomer (green) also supported the phenotypic rescue. Moreover, ATP is a significant product of mitochondrial energy metabolism. Therefore, the ATP content was determined and shown as the signal cellular RLU (Figure 9E). This result showed that the FFA can decrease the ATP content in the bulk cell and disrobe the mitochondrial energy metabolism, which can be rescued by ABS treatment in a dose-dependent manner. Taking together, we applied two different dyes and methods to confirm that the treatment of ABS did facilitate the rescue of the mitochondrial membrane potential's decrease, and it did improve mitochondrial energy metabolism.

3.8. Effect of ABS on PI3K/Akt/GSK3β/β-Catenin Signaling Pathway in HepG2 Cells

To validate the prediction of network pharmacology, we conducted Western blotting to study the protein level change in the PI3K/Akt/GSK3β signaling pathway and the result was standardized by GAPDH (Figure 10A). The lipid accumulation induced by FFA inhibited the PI3K, Akt, and GSK3β phosphorylation in the model group compared with the control group; however, treating with 0.2 mg/mL ABS upregulated the PI3K, Akt, and GSK3β phosphorylation significantly (Figure 10B–D). This result is consistent with our previous predictions of network pharmacology.

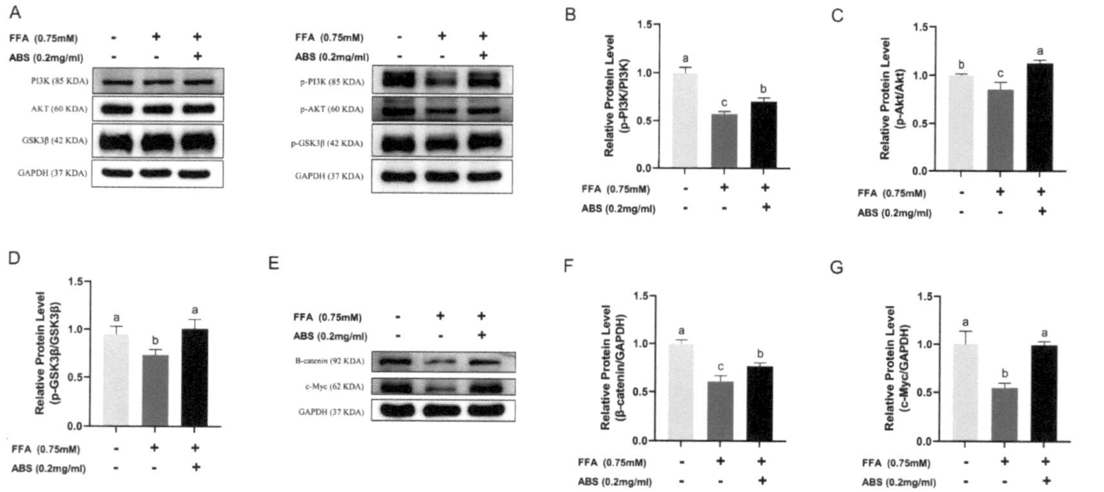

Figure 10. The regulation effect of ABS on the PI3K/Akt/GSK3β/β-catenin signaling pathway. (**A**) The PI3K/Akt/GSK3β signaling pathway-related proteins and phosphorylated proteins. (**B**) The protein level of p-PI3K/PI3K. (**C**) The protein level of p-Akt/Akt. (**D**) The protein level of p-GSK3β/GSK3β. (**E**) The proteins of β-catenin, its downstream transcriptional factor c-Myc, and GAPDH. (**F**) The protein level of β-catenin/GAPDH. (**G**) The protein level of c-Myc/GAPDH. All results are presented as mean ± S.D and the experiments were repeated as triplicates. Bars with different letters (a–c) significantly ($p < 0.05$) differ according to ANOVA and Tukey's multiple range test. FFA, free fatty acid; ABS, adzuki bean saponins.

Furthermore, we want to explore the deeper molecular mechanisms of how this signaling pathway boosts mitochondrial function and we performed the Western blot to determine the protein level of β-catenin and c-Myc, which is downstream of the PI3K/Akt/GSK3β signaling pathway and associated with mitochondrial function (Figure 10E). The semi-quantitative analysis of the Western blot showed that 0.2 mg/mL ABS treatment can effectively improve the inhibition effect of lipid accumulation in the protein levels of β-catenin and c-Myc. The results of the Western blot showed that 0.2 mg/mL

ABS treatment could effectively improve the protein levels of β-catenin and c-Myc, which were downregulated in the model group compared with the control group (Figure 10F,G).

3.9. Effect of ABS in the Nuclear Translocation of β-Catenin

Based on our previous results, the total protein level of β-catenin increased with the treatment of 0.2 mg/mL ABS and its downstream transcription factor c-Myc was also upregulated. Accordingly, we conducted the immunofluorescence analysis at the single-cell level (Figure 11A) as well as the Western blot of a bulk culture (Figure 11C) to further study—at both the micro- and macro-perspective—the distribution and expression of the nuclear protein level of β-catenin. The results of the immunofluorescence analysis showed that the green fluorescent (β-catenin) signaling is downregulated in the model group compared with the control group and can be upregulated by the 0.1 mg/mL and 0.2 mg/mL ABS. Moreover, based on the co-staining of DAPI and β-catenin, in the model group, the green fluorescent signal (β-catenin) overlaps with the blue fluorescent signal (nucleus), representing protein located in the nucleus; this is less bright compared to the model group indicating that the lipid accumulation inhibited the nuclear translocation of β-catenin. According to our semi-quantitative analysis (Figure 11B), the ABS treatment reduced the decrease in the normalized nuclear protein level of β-catenin (β-catenin/DAPI) in the model group of our study, which recovered to the level of the normal control group.

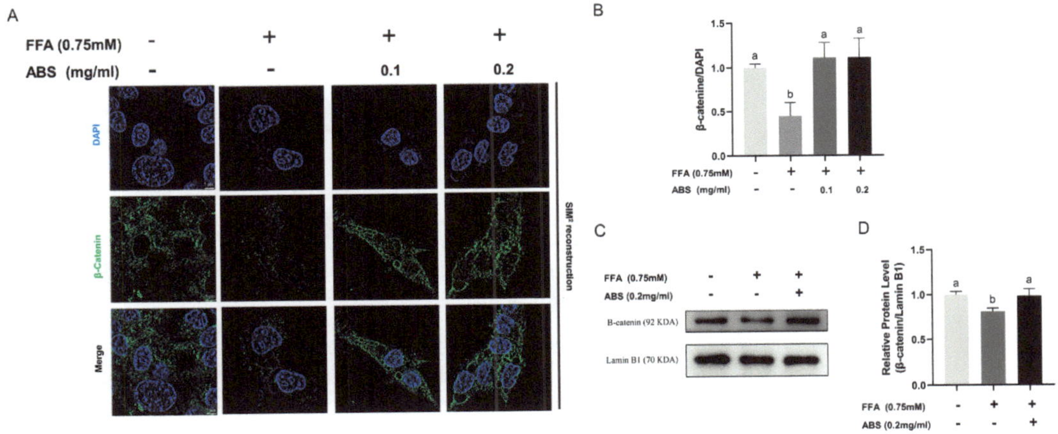

Figure 11. The regulation effect of ABS on nuclear translocation of β-catenin. (**A**) Fluorescent images of HepG2 cells containing immuno-fluorescent signal of β-catenin (green) and counter-stained with DAPI (blue), taken by ZEISS Elyra 7 equipped with Lattice structured illumination microscopy SIM^2 super-resolution fluorescent microscope with a 63× oil-immersion objective. Scale bars, 5 μm. (**B**) Semi-quantitative analysis based on original images taken by SIM^2 of relative green fluorescence signal (β-catenin), normalized by dividing to blue fluorescence signal (DAPI) by Image J. (**C**) The nuclear proteins of β-catenin and Lamin B1. (**D**) The relative protein levels of β-catenin/Lamin B1. All results are presented as mean ± S.D and the experiments were repeated as triplicates. Bars with different letters (a,b) significantly ($p < 0.05$) differ according to ANOVA and Tukey's multiple range test. FFA, free fatty acid; ABS, adzuki bean saponins.

Moreover, we also conducted Western blotting of the nuclear protein of β-catenin that was normalized by the Lamin B1 (Figure 11C). The results imply that the ABS treatment can both reduce the downregulation of the nuclear protein level of β-catenin in the model group and can restore the protein level of β-catenin to the same level as the control group (Figure 11D).

3.10. Effect of ABS and EGCG on 3T3-L1 Cell Viability

The classical adipocyte cell 3T3-L1 was also used to study the anti-obesity effect of ABS. The CCK8 method was used to evaluate the effect of ABS and EGCG on cell viability (Figure 12A,B). Moreover, EGCG is treated as a positive drug for studying the anti-adipogenesis effect of ABS [47]. The 3T3-L1 cells did not show significant inhibitory effects when treated with ABS at concentrations from 0 to 0.2 mg/mL and EGCG at concentrations from 0 to 16 µM for 24 h compared with the control group. However, when the concentration of ABS reached 0.3 mg/mL, the viability of 3T3-L1 cells significantly decreased ($p < 0.0005$). Therefore, in this study's subsequent experiments in this cell line, the ABS and EGCG concentrations were chosen to be below 0.2 mg/mL and 16 µM.

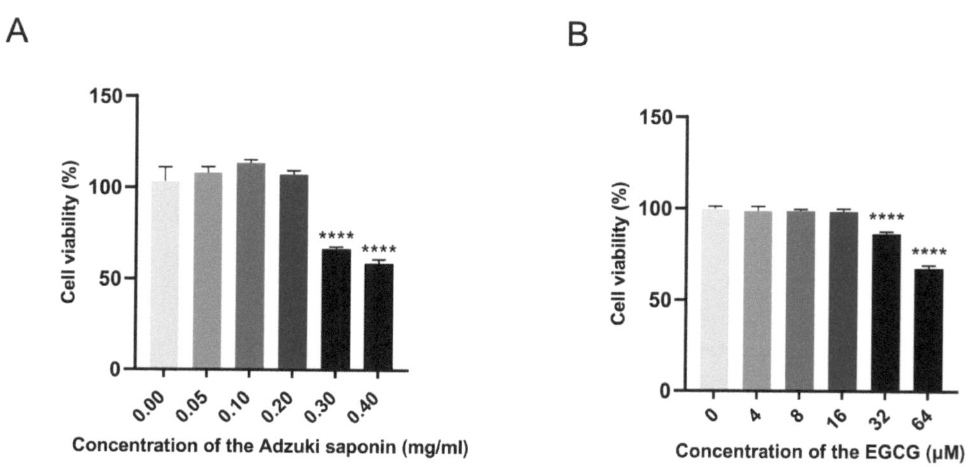

Figure 12. (**A**) The effect of ABS on the cell viability of 3T3-L1 cells. (**B**) The effect of EGCG on the cell viability of 3T3-L1 cells. **** $p < 0.0001$. All results are presented as mean ± S.D and the experiments were repeated as triplicates. EGCG, epigallocatechin gallate.

3.11. Effect of ABS on Adipogenesis

To solidly verify our finding of the concentration-dependent anti-obesity effect of ABS, we changed our cell model from the lipid accumulation cell model (HepG2) to the adipogenesis cell model (3L3-L1). We conducted Nile Red staining (Figure 13A,B), observed via a super-solution microscope, and Oil Red staining (Figure 13C), viewed by a slide scanner. Accordingly, in a concentration-dependent manner, either via the semi-quantification results of the normalized green fluorescent signal representing lipid accumulation or the single-cell level oversedation through fluorescence, the results clearly show that ABS treatment alleviated MDI-induced adipogenesis compared to the positive control, EGCG, and negative control. A similar result was also observed on the bright field imaging of the ORO-stained intracellular lipid distribution. Accordingly, a change in cell line or imaging system with different principals did not affect the MDI-induced adipogenesis, which was alleviated by introducing concentration-dependent ABS. Taken together, our results can very solidly support the anti-obesity effect of ABS.

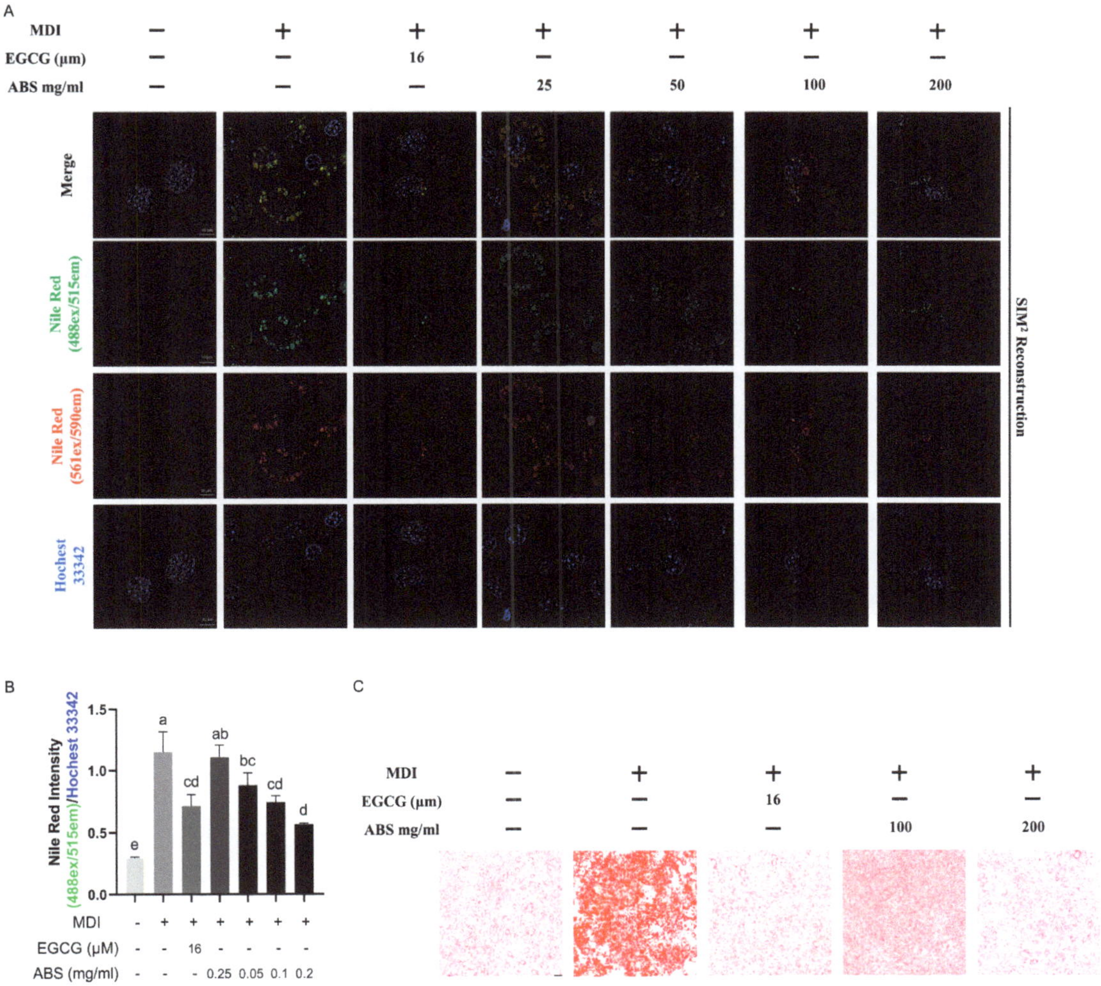

Figure 13. ABS alleviates MDI-induced adipogenesis. (**A**) Fluorescent images of 3T3-L1 cells co-stained with Hochest 33,342 (blue) and Nile Red (green and red) taken by ZEISS Elyra 7 with Lattice structured illumination microscopy (SIM)^2 super-resolution fluorescent microscope equipped with a 63× oil-immersion objective. Scale bars, 5 μm. (**B**) Semi-quantitative analysis based on original images taken by SIM^2 of fluorescence signal and lipid droplet by Image J. (**C**) ORO staining of MDI-induced 3T3-L1 cellular model with or without treatment of ABS or EGCG. All results are presented as mean ± S.D and the experiments were repeated as triplicates. Bars with different letters (a–e) significantly ($p < 0.05$) differ according to ANOVA and Tukey's multiple range test. Scale bar, 50 μm. MDI, adipogenesis differentiation medium; EGCG, epigallocatechin gallate; ABS, adzuki bean saponins.

3.12. Effect of ABS on β-Catenin Signaling and Its Downstream Transcriptional Factor CEBP/α and PPARγ in Adipocytes

To study the underlying molecular mechanism of the anti-obesity effect of ABS in adipocytes, we also conducted the immunofluorescence analysis of β-catenin, C/EBPα, and PPARγ in the 3L3-L1 cell line (Figures 14A,B and 15A–D). Similar to our findings in HepG2, MDI-introduction induced β-catenin downregulation, while this phenotype can be reversed by a concentration-dependent addition of ABS (Figure 14A). Moreover, the

ABS treatment can also improve the nuclear translocation of the β-catenin in adipocytes. For the *CEBP/α* and PPARγ intracellular expression level, MDI modeling facilitates its expression level in the nucleus region, which is reflected through the overlapping region of blue (DAPI) and green (CEBP/α and PPARγ) fluorescence, while treatment with ABS could alleviate the upregulation facilitated by MDI modeling (Figure 15A,C). Also, both semi-quantifications of IF also supported this result (Figures 14B and 15B,D).

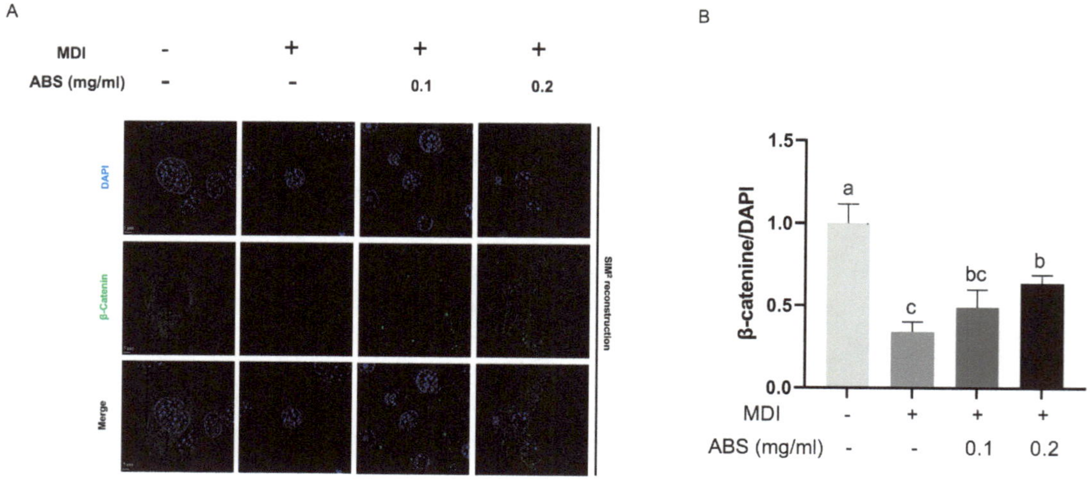

Figure 14. The regulation effect of ABS on protein level of β-catenin. (**A**) Fluorescent images of 3T3-L1 cells containing immuno-fluorescent signal of β-catenin (green) and counter-stained with DAPI (blue) taken by ZEISS Elyra 7 equipped with Lattice structured illumination microscopy (SIM)^2 super-resolution fluorescent microscope with a 63× oil-immersion objective. Scale bars, 5 μm. MDI, adipogenesis differentiation medium; ABS, adzuki bean saponins. (**B**) Semi-quantitative analysis based on original images taken by SIM^2 of relative green fluorescence signal (β-catenin), normalized by dividing to blue fluorescence signal (DAPI) by Image J. Bars with different letters (a–c) significantly ($p < 0.05$) differ according to ANOVA and Tukey's multiple range test.

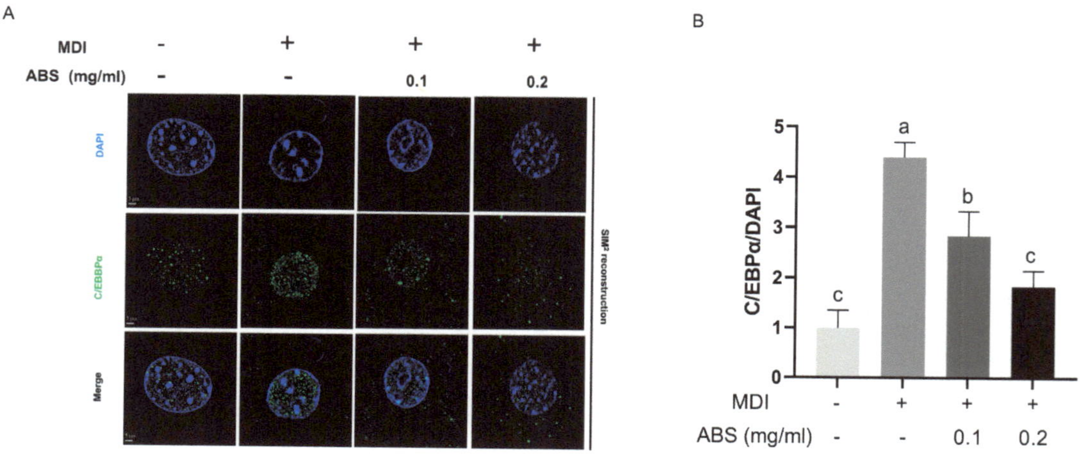

Figure 15. *Cont.*

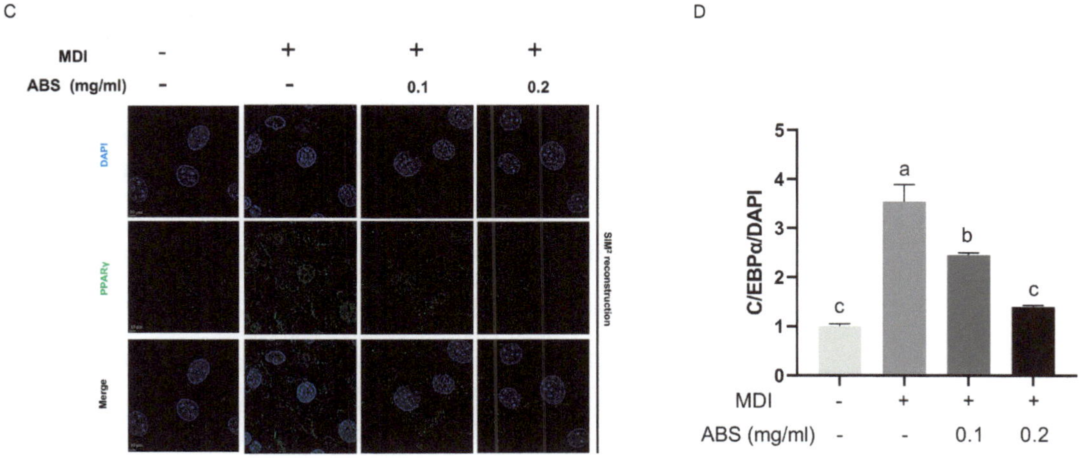

Figure 15. The regulation effect of ABS on protein level of transcriptional factor C/EBPα and PPARγ. (**A**) Fluorescent images of 3T3-L1 cells containing immuno-fluorescent signal of CEBP/α (green) and counter-stained with DAPI (blue) taken by ZEISS Elyra 7 equipped with Lattice structured illumination microscopy (SIM)^2 super-resolution fluorescent microscope with a 63× oil-immersion objective. Scale bars, 5 μm. (**B**) Semi-quantitative analysis based on original images taken by SIM^2 of relative green fluorescence signal (CEBP/α), normalized by dividing to blue fluorescence signal (DAPI) by Image J. (**C**) Fluorescent images of 3T3-L1 cells containing immuno-fluorescent signal of PPARγ (green) and counter-stained with DAPI (blue) taken by ZEISS Elyra 7 equipped with Lattice structured illumination microscopy (SIM)^2 super-resolution fluorescent microscope with a 63× oil-immersion objective. Scale bars, 10 μm. (**D**) Semi-quantitative analysis based on original images taken by SIM^2 of relative green fluorescence signal (PPARγ), normalized by dividing to blue fluorescence signal (DAPI) by Image J. All results are presented as mean ± S.D and the experiments were repeated as triplicates. Bars with different letters (a–c) significantly ($p < 0.05$) differ according to ANOVA and Tukey's multiple range test. MDI, adipogenesis differentiation medium; ABS, adzuki bean saponins.

4. Discussion

Obesity is a chronic disease caused by the excessive accumulation of fat that affects health. Mitochondrial dysfunction can also disrupt energy metabolism and lead to lipid accumulation. Mitochondrial biosynthesis is a biological process that increases mitochondrial mass to meet cellular energy needs. The impaired mitochondrial biosynthesis caused by mitochondrial dysfunction is related to the pathogenesis of obesity [48]. The c-Myc is closely related to mitochondria and can upregulate mitochondrial biosynthesis and mitochondrial function [49]. The upregulation of c-Myc expression can counteract obesity and insulin resistance caused by a high-fat diet [50]. Therefore, if drugs can achieve anti-obesity by targeting c-Myc to regulate mitochondrial function and improve lipid metabolism, it is very promising.

Network pharmacology is a prospective approach that can reveal the potential therapeutic mechanisms of traditional Chinese medicine [33,34]. Our previous study extracted saponin components from adzuki beans, identified six saponin components, and preliminarily validated their anti-obesity biological activity in vivo through a high-fat diet-induced obesity mice model [18]. This preliminarily confirmed that ABS can achieve anti-obesity biological activity by improving lipid metabolism but in-depth molecular mechanisms still needed to be elucidated. In this study, a total of fifteen saponin components were identified in the ABS through UHPLC-QE-MS analysis for further study. This study aims to integrate the UHPLC-QE-MS analysis, network pharmacology methods, and bioinformatics to preliminarily elucidate the potential molecular mechanisms of ABS in achieving anti-obesity effects by improving lipid metabolism. The result of network pharmacology showed that the PI3K/Akt signaling pathway may be associated with the anti-obesity effect of ABS by improving lipid metabolism, which will require further validation of its specific regulation by the ABS.

When it comes to the network pharmacology and bioinformatics analyses, we collected DEGS from obesity-related datasets (GSE25401) that included transcriptomic data from white adipose tissue of 30 adults with obesity and 26 adults without obesity using the "Limma" package with a p value < 0.05; 2841 DEGs, 2487 upregulated genes, and 354 downregulated genes were identified. Then, we used three machine learning algorithms including SVM-RFE, random forest algorithm, and least absolute shrinkage selection operator (LASSO) to identify 11 characteristic genes, including *F13A1*, *FGF2*, *FGF1*, *MMP9*, *PDE3A*, *HSD11B1*, *ABCC1*, *ACACB FDFT1*, *PTPN22*, and *PDE3B*.

The immunological aspects of obesity are complex and multi-faceted and the CIBERSORT algorithm was employed for the evaluation of the relationship between the infiltration level of immune cells and obesity hub genes. Our study's bioinformatics analysis and subsequent correlation studies indicate that certain genes may have immunomodulatory functions influencing the infiltration and activity of various immune cells in adipose tissue. The positive correlation of genes such as *HSD11B1*, *PTPN22*, *ABCC1*, *MMP9*, *FGF1*, and *F13A1* with M0 and M1 macrophages suggests a possible role in shaping the immune response toward a more beneficial phenotype in obesity. This could have implications for the development of therapies that aim to modulate immune cell activity to mitigate obesity and its associated inflammation. Furthermore, the negative correlation with other immune cells underscores the need for a nuanced understanding of the immune system's role in metabolic disorders. It is possible that these genes could be involved in a feedback mechanism to balance immune responses and maintain homeostasis within the adipose tissue microenvironment. Although our study was not focused on immune cells, these insights contribute to the growing body of research that recognizes the importance of the immune system in metabolic regulation and provides a foundation for future studies to explore novel immunotherapeutic approaches for obesity.

It has been found that tissue corticosterone concentrations in human adipose tissue are persistently low but *ABCC1* mRNA is upregulated in obesity. *ABCC1*, but not *ABCB1*, is expressed in human adipose tissue, and inhibiting *ABCC1* raises intracellular corticosterone but not cortisol and stimulates glucocorticoid-responsive gene transcription in human adipocytes [51]. Acetyl-CoA carboxylase beta, produced by the *ACACB* gene, is crucial in fatty acid oxidation. Research has shown that prevalent variants of the *ACACB* gene correlate with obesity and, separately, with type 2 diabetes in postmenopausal women, suggesting a critical function of acetyl-CoA carboxylase beta in these metabolic energy disorders [52]. The differential expression of *F13A1* (ΔHeavy–Lean) was found to relate to 47 genes related to the immunological response, leucocyte and neutrophil activation, cytokine response, and signaling (by gene enrichment analysis). Research also indicates that deficiency of *F13A1* in mice modifies adipose tissue cellularity, increasing both small and large adipocytes, reducing macrophage infiltration, and enhancing insulin sensitivity in obese adipose tissue. This suggests that *FXIII-A* responds to weight gain and could negatively impact the health of adipose tissue [53]. Angiogenic factors have been linked to obesity. A study obtained subcutaneous white adipose tissue samples from 45 children, aged between 0 and 9 years, who were undergoing elective surgeries, to explore the links between angiogenic factors and individual as well as tissue characteristics. They found that age positively correlates with *FGF1* and *FGF2* but negatively with *ANGPT2*, especially marking significant variations within the first two years of life. Additionally, *FGF1*, *FGF2*, and *ANGPT1* were positively associated with adipocyte size [54]. The enzyme 11-beta hydroxysteroid dehydrogenase type 1 (*HSD11B1*) converts inactive cortisone into active cortisol, a process facilitated by hexose-6-phosphate dehydrogenase (H6PD). The production of cortisol through this reaction could elevate intra-abdominal cortisol levels, potentially playing a role in the development of obesity and metabolic syndrome (MetS). Research indicates that individuals with obesity may exhibit reduced intra-abdominal expression of the *VAT HSD11B1* gene, potentially as a compensatory mechanism to mitigate central and overall adiposity by lowering intra-abdominal cortisol levels [55]. *PDE3A* plays a critical role in obesity by regulating cAMP levels in adipocytes, affecting their function and metabolism. In obesity, inflammation increases MCP-1 production in adipocytes, attracting monocytes/macrophages and exacerbating inflammation. Inhibiting *PDE3A* elevates cAMP levels, activates PKA, upregulates MKP-1, and reduces ERK and p38 phosphorylation, thereby lowering MCP-1 production. This mechanism helps mitigate adipocyte inflammation, potentially improving insulin resistance and metabolic complications [56]. *PTPN22* is linked to obesity through its role in immune and inflammatory responses. Polymorphisms in the *PTPN22* gene, specifically the +1858C/T variant, are associated with various autoimmune diseases, suggesting a potential connection to obesity [57]. MMP9 contributes considerably to obesity by regulating extracellular matrix remodeling. In people with obesity, elevated MMP9 levels are associated with an increase in body mass index (BMI) and waist circumference. In addition, the MMP9/TIMP1 ratio has been associated with endothelial dysfunction, which is frequent in obesity-related cardiovascular disorders. Although studies on MMP9 levels in obesity have yielded varied results, it is well-acknowledged that MMP9 activity relates to inflammation and vascular abnormalities. Thus, MMP9 could be used as a biomarker to identify obesity-related metabolic and cardiovascular problems [58]. Among these nine candidate genes, the protein expression of ACACB is higher in the adipose tissue and liver, while HSD11B1 protein expression is high in the liver, based on the Human Protein Atlas (HPA) database. This suggests that *ACACB* and *HSD11B1* could be the key genes of interest in future research. Although the primary aim of this study was not to investigate the tissue-specific expression of the identified genes, we provide in the Supplementary Materials (Figures S26–S34) an overview of the tissue distribution of the candidate genes based on the Human Protein Atlas database. This information may serve as a resource for future studies aiming to delve deeper into the roles of these genes in different tissues.

The in vitro HepG2 cell experiments and our results of TG content testing, ORO staining, Nile Red staining, and quantitative phase images found that the ABS can improve lipid metabolism to reduce the lipid accumulation in the cell significantly at the single-cell level. This result is similar to those of natural products that can decrease lipid accumulation in the HepG2 cell [22]. Lipid metabolism disorders can increase ROS production, which in turn can impair mitochondrial function by reducing membrane potential, disrupting energy production. Therefore, we not only focused on the imbalance in lipid metabolism but also on mitochondrial energy metabolic chaos and membrane potential abnormality. Our results showed that ABS can decrease ROS accumulation, alleviate the membrane potential loss of mitochondria, facilitate improvement of the mitochondrial damage, and rescue the ATP homeostasis imbalance. Taken together, we can confirm that ABS achieve an anti-obesity effect of improving lipid metabolism through alleviation of the mitochondrial abnormality in energetic metabolism and the related consequences like potential loss.

The PI3K/Akt signaling pathway is highly associated with obesity [59]; therefore, this is a promising pathway for studying obesity and energy metabolism and natural compounds have been found as promising PI3K/Akt modulators [60]. The network pharmacology highlighted this signaling pathway and our results indicate that ABS can upregulate the phosphorylation of PI3K, Akt, and GSK3β in the ABS treatment group compared to the model group, and this change is similar to other natural products improving obesity [60]. Moreover, our study elucidated that ABS achieves an anti-obesity effect that is also associated with mitochondrial function and energy metabolism. We combined those results together and further analyzed the downstream protein change including β-catenin and c-Myc. Interestingly, ABS treatment can obviously improve the expression both of β-catenin and c-Myc as well as the nuclear translocation of β-catenin. The nuclear translocation of β-catenin can upregulate the protein level of downstream transcription factor c-Myc [61]. Moreover, the up-regulation of c-Myc can upregulate mitochondrial biosynthesis and mitochondrial function, counteracting obesity and insulin resistance caused by a high-fat diet [49,50]. Taken together, ABS achieves an anti-obesity effect by improving lipid metabolism and mitochondrial function through the PI3K/Akt/GSK3β/β-catenin signaling pathway, and ABS are targeting the downstream of this signaling pathway c-Myc.

Moreover, through in vitro adipocyte experiments, our results indicated that ABS could improve adipogenesis to reduce lipid droplets generated in the cell significantly, as displayed in the ORO staining and Nile Red staining. The MDI-induced 3T3-L1 cell is the most classical adipocyte cell model used for studying the anti-obesity effect [62]. For the underlying molecular mechanism, ABS has a similar regulation effect of β-catenin, which can alleviate the changes induced by MDI of the downregulation of the protein level in β-catenin and inhibition of its nuclear translocation. Normally, a transcription factor in adipogenesis, CCAAT/Enhancer-Binding Protein α (C/EBPα), can be upregulated by the MDI induction for the development of obesity [63]; moreover, this transcription factor can be downregulated by the nuclear translocation of β-catenin [64]. Based on our results, ABS can improve obesity by influencing adipogenesis through the regulation of β-catenin signaling and its downstream transcription factor in adipogenesis, C/EBPα.

In addition, our research verified the anti-obesity activity of ABS and identified nine potential genes that are related to obesity using bioinformatics analysis. After studying the distribution of these genes in various tissues using the HPA database, we consider that *ACACB* and *HSD11B1*, two genes that are significantly expressed in liver or adipose tissue, deserve more attention. In future trials, researchers can use protein stability methods like CETSA or DARTS to access protein–ABS binding to strengthen the research foundation of ABS against obesity. Finally, in further in vivo studies, the effect of ABS on the intestinal system should be studied for its underlying mechanism.

5. Conclusions

In this study, we first identified 15 saponin components of ABS using UHPLC-QE-MS and analyzed the potential mechanism of ABS achieving anti-obesity bioactivity, by improving fat metabolism, using network pharmacology ensemble bioinformatics. We validated in the HepG2 cell model that ABS can improve fat metabolism and enhance mitochondrial function to reduce lipid accumulation. Based on network pharmacology predictions, the PI3K/Akt signaling pathway was identified as having a significant correlation with the anti-obesity efficacy of ABS. Combined with the boosting effect of ABS on mitochondrial function, subsequent experimental verification confirmed that ABS regulates the PI3K/Akt/GSK3β/β-catenin signaling pathway-related proteins and downstream transcription factor c-Myc in the lipid accumulation cell model (HepG2 cell line). Moreover, ABS can also achieve an anti-obesity effect in adipocytes (3T3-L1 cell line) through the regulation of β-catenin signaling and its downstream transcription factor in adipogenesis, C/EBPα. Taken together, ABS can achieve anti-obesity effects by improving lipid accumulation, lipid metabolism, mitochondrial abnormality in liver cells, and adipogenesis in adipocytes. A further in vivo study of ABS should be conducted to explore its mechanism both in the liver and white adipose tissue. More specifically, the liver is a target organ of ABS for improving lipid metabolism and mitochondrial abnormality; equally, white adipose tissue is a target organ of ABS in inhibiting adipogenesis. These findings provide substantial support for ABS to achieve anti-obesity effects by improving fat metabolism and boosting mitochondrial function normalization, laying the foundation for its further development.

Supplementary Materials: The following supporting information can be downloaded at https://www.mdpi.com/article/10.3390/antiox13111380/s1: Figure S1. The MS/MS spectrum of azukisaponin I; Figure S2. The MS/MS spectrum of azukisaponin II; Figure S3. The MS/MS spectrum of azukisaponin III; Figure S4. The MS/MS spectrum of azukisaponin IV; Figure S5. The MS/MS spectrum of azukisaponin V; Figure S6. The MS/MS spectrum of AZ I; Figure S7. The MS/MS spectrum of AZ II; Figure S8. The MS/MS spectrum of AZ III; Figure S9. The MS/MS spectrum of AZ IV; Figure S10. The MS/MS spectrum of soyasaponin I; Figure S11. The MS/MS spectrum of dehydrosoyasaponin I; Figure S12. The MS/MS spectrum of soyasaponin V; Figure S13. The MS/MS spectrum of soyasaponin VI; Figure S14. The MS/MS spectrum of soyasaponin Bd; Figure S15. The MS/MS spectrum of soyasaponin γg; Figure S16. The negative ion diagram of UHPLC-QE-MS of ABS; Figure S17. The positive ion diagram of UHPLC-QE-MS of ABS; Figure S18. The original figures used for reconstruction through ZEISS Black edition referring to Figure 6A; Figure S19. The original figures used for reconstruction through ZEISS Black edition referring to Figure 8A; Figure S20. The original figures used for reconstruction through ZEISS Black edition referring to Figure 9A; Figure S21. The original figures used for reconstruction through ZEISS Black edition referring to Figure 9B; Figure S22. The original figures used for reconstruction through ZEISS Black edition referring to Figure 11A; Figure S23. The original figures used for reconstruction through ZEISS Black edition referring to Figure 13A; Figure S24. The original figures used for reconstruction through ZEISS Black edition referring to Figure 14A; Figure S25. The original figures used for reconstruction through ZEISS Black edition referring to Figure 15A; Figure S26. The original figures used for reconstruction through ZEISS Black edition referring to Figure 15C; Figure S27. Protein expression of HSD11B1 in the Human Protein Altas (HPA) database; Figure S28. Protein expression of ACACB in the HPA database; Figure S29. Protein expression of F13A1 in the HPA database; Figure S30. Protein expression of FGF2 in the HPA database; Figure S31. Protein expression of FGF1 in the HPA database; Figure S32. Protein expression of MMP9 in the HPA database; Figure S33. Protein expression of PDE3A in the HPA database; Figure S34. Protein expression of ABCC1 in the HPA database; Figure S35. Protein expression of PTPN22 in the HPA database.

Author Contributions: Conceptualization, J.L. (Jinhai Luo), J.L. (Jincan Luo) and B.X.; methodology, J.L. (Jinhai Luo), J.L. (Jincan Luo) and Y.F.; software, B.X.; validation, J.L. (Jinhai Luo), J.L. (Jincan Luo), B.D. and Z.F.; formal analysis, J.L. (Jinhai Luo), J.L. (Jincan Luo) and Y.W.; investigation, J.L. (Jinhai Luo), J.L. (Jincan Luo) and Y.W.; resources, B.X.; data curation, J.L. (Jinhai Luo), J.L. (Jincan Luo), Y.W. and B.H.; writing—original draft preparation, J.L. (Jinhai Luo), J.L. (Jincan Luo), B.D. and Y.W.; writing—review and editing, Y.F., Z.F., B.D., Z.Y. and B.X.; visualization, Y.F. and B.H.; supervision, Z.Y. and B.X.; project administration, B.X.; funding acquisition, Z.Y. and B.X. All authors have read and agreed to the published version of the manuscript.

Funding: This project was jointly supported by the National Natural Science Foundation of China (Project code: 32172198), National Multidisciplinary Innovation Team Project of Traditional Chinese Medicine (ZYYCXTU-D-202201 and ZYYCXTU-D-202206), Emergency Key Program of Guangzhou National Laboratory (Grant Number EKPG21-06), and the Guangdong Provincial Key Laboratory IRADS (2022B1212010006, R0400001-22).

Institutional Review Board Statement: Not applicable.

Informed Consent Statement: Not applicable.

Data Availability Statement: The datasets used and analyzed in the current study are available from the corresponding author on reasonable request.

Acknowledgments: We would like to thank the Advanced Bioimaging Core Facility of Guangzhou National Laboratory for our ZEISS Elyra 7 with Lattice SIM^2 super-resolution fluorescent microscope (ZEISS, Germany) (Structured Illumination Microscopy) and SLIDEVIEWTM VS200 Research Slide Scanner (Olympus, Japan) and Livecyte2 microscope (Phasefocus Limited, Sheffield, UK). We would like to thank Junwen Li, an engineer from Guangzhou Zhijing Digital Technologies Co., Ltd., for kind help with using the slide scanner (SLIDEVIEWTM VS200, Olympus, Japan) and the following image processing (OlyVIA, Version Number: 4.1.1). We would like to thank Qiaolian CHEN, a research assistant, and Chunguang YANG, a Ph.D. candidate, both of which are under the supervision of Zifeng YANG, for their kind help in sample preparation for cellular immunofluorescence imaging.

Conflicts of Interest: The authors declare no conflicts of interest.

Abbreviations

ABS, adzuki bean saponins; AUC, area under the ROC curve; BP, biological processes; CC, closeness centrality; CCK8, cell counting kit-8; DMSO, dimethyl sulfoxide; DEGs, differentially expressed genes; EGCG, epigallocatechin gallate; FBS, fetal bovine serum: GO, Gene Ontology; KEGG, Kyoto Encyclopedia of Genes and Genomes; LASSO, least absolute shrinkage selection operator; MDI, adipogenesis differentiation medium with methylisobutylxanthine, dexamethasone, and insulin; MEM, minimum essential medium; MF, molecular functions; ORO, Oil Red O; PBS, phosphate-buffered solution; PPI, protein–protein interaction; RF-RFE, random forest–recursive feature elimination; RFU, relative fluorescent unit; RLU, relative luminescent unit; ROC, receiver operating characteristic; TG, triglyceride.

References

1. WHO. Obesity and Overweight. World Health Organisation Newsroom Fact Sheets. 2024. Available online: https://www.who.int/news-room/fact-sheets/detail/obesity-and-overweight (accessed on 1 May 2024).
2. Tang, C.; Kong, L.; Shan, M.; Lu, Z.; Lu, Y. Protective and ameliorating effects of probiotics against diet-induced obesity: A review. *Food Res. Int.* **2021**, *147*, 110490. [CrossRef] [PubMed]
3. Jayarathne, S.; Stull, A.J.; Park, O.H.; Kim, J.H.; Thompson, L.; Moustaid-Moussa, N. Protective effects of anthocyanins in obesity-associated inflammation and changes in gut microbiome. *Mol. Nutr. Food Res.* **2019**, *63*, 1900149. [CrossRef] [PubMed]
4. Bozi, L.H.; Campos, J.C.; Zambelli, V.O.; Ferreira, N.D.; Ferreira, J.C. Mitochondrially-targeted treatment strategies. *Mol. Asp. Med.* **2020**, *71*, 100836. [CrossRef] [PubMed]
5. Yoo, S.-M.; Park, J.; Kim, S.-H.; Jung, Y.-K. Emerging perspectives on mitochondrial dysfunction and inflammation in Alzheimer's disease. *BMB Rep.* **2020**, *53*, 35. [CrossRef] [PubMed]
6. Tang, C.; Wang, Y.; Chen, D.; Zhang, M.; Xu, J.; Xu, C.; Liu, J.; Kan, J.; Jin, C. Natural polysaccharides protect against diet-induced obesity by improving lipid metabolism and regulating the immune system. *Food Res. Int.* **2023**, *172*, 113192. [CrossRef]
7. Luo, J.; Yu, Z.; Tovar, J.; Nilsson, A.; Xu, B. Critical review on anti-obesity effects of phytochemicals through Wnt/β-catenin signaling pathway. *Pharmacol. Res.* **2022**, *184*, 106461. [CrossRef]
8. Huang, C. The role of semaglutide in weight management: An analysis of clinical studies for overweight and obese adults. *J. Educ. Humanit. Soc. Sci.* **2024**, *27*, 551–558. [CrossRef]
9. Sidik, S. New Obesity Drugs Pormise To Be Cheaper And More Effective. *Nature* **2023**, *619*, 19. [CrossRef]
10. Itoh, T.; Itoh, Y.; Hibasami, H.; Katsuzaki, H.; Imai, K.; Furuichi, Y.; Komiya, T. Isolation of a substance inducing apoptosis of cultured human gastric cancer cells from a hot-water extract of adzuki (*Vigna angularis*). *J. Jpn. Soc. Nutr. Food Sci.* **2005**, *58*, 281–287. [CrossRef]
11. Itoh, T.; Itoh, Y.; Mizutani, M.; Fujishiro, K.; Furuichi, Y.; Komiya, T.; Hibasami, H. A hot water extract of adzuki (*Vigna angularis*) induces apoptosis in cultured human stomach cancer cells. *J. Jpn. Soc. Food Sci. Technol.* **2002**, *49*, 339–344. [CrossRef]
12. Itoh, T.; Kita, N.; Kurokawa, Y.; Kobayashi, M.; Horio, F.; Furuichi, Y. Suppressive effect of a hot water extract of adzuki beans (*Vigna angularis*) on hyperglycemia after sucrose loading in mice and diabetic rats. *Biosci. Biotechnol. Biochem.* **2004**, *68*, 2421–2426. [CrossRef] [PubMed]
13. Itoh, T.; Kobayashi, M.; Horio, F.; Furuichi, Y. Hypoglycemic effect of hot-water extract of adzuki (*Vigna angularis*) in spontaneously diabetic KK-Ay mice. *Nutrition* **2009**, *25*, 134–141. [CrossRef] [PubMed]
14. Yao, Y.; Cheng, X.; Wang, L.; Wang, S.; Ren, G. Biological potential of sixteen legumes in China. *Int. J. Mol. Sci.* **2011**, *12*, 7048–7058. [CrossRef] [PubMed]
15. Ohba, K.; Nirei, M.; Watanabe, S.; Han, K.-H.; Hashimoto, N.; Shimada, K.-I.; Sekikawa, M.; Chiji, H.; Fukushima, M. Effect of an adzuki bean extract on hepatic anti-oxidant enzyme mRNAs in D-galactosamine-treated rats. *Biosci. Biotechnol. Biochem.* **2005**, *69*, 1988–1991. [CrossRef]
16. Zhao, Q.; Hou, D.; Laraib, Y.; Xue, Y.; Shen, Q. Comparison of the effects of raw and cooked adzuki bean on glucose/lipid metabolism and liver function in diabetic mice. *Cereal Chem.* **2021**, *98*, 1081–1090. [CrossRef]
17. Liu, R.; Cai, Z.; Xu, B. Characterization and quantification of flavonoids and saponins in adzuki bean (*Vigna angularis* L.) by HPLC–DAD–ESI–MS n analysis. *Chem. Cent. J.* **2017**, *11*, 93. [CrossRef]
18. Liu, R.; Zheng, Y.; Cai, Z.; Xu, B. Saponins and flavonoids from adzuki bean (*Vigna angularis* L.) ameliorate high-fat diet-induced obesity in ICR mice. *Front. Pharmacol.* **2017**, *8*, 687. [CrossRef]
19. Cho, A.-S.; Jeon, S.-M.; Kim, M.-J.; Yeo, J.; Seo, K.-I.; Choi, M.-S.; Lee, M.-K. Chlorogenic acid exhibits anti-obesity property and improves lipid metabolism in high-fat diet-induced-obese mice. *Food Chem. Toxicol.* **2010**, *48*, 937–943. [CrossRef]
20. Liu, J.; Han, L.; Zhu, L.; Yu, Y. Free fatty acids, not triglycerides, are associated with non-alcoholic liver injury progression in high fat diet induced obese rats. *Lipids Health Dis.* **2016**, *15*, 27. [CrossRef]
21. Ebbert, J.O.; Jensen, M.D. Fat depots, free fatty acids, and dyslipidemia. *Nutrients* **2013**, *5*, 498–508. [CrossRef]
22. Yao, Y.; Li, Z.; Qin, B.; Ju, X.; Wang, L. Evaluation of the intracellular lipid-lowering effect of polyphenols extract from highland barley in HepG2 cells. *Food Sci. Hum. Wellness* **2024**, *13*, 454–461. [CrossRef]
23. Wang, Z.; Ge, S.; Li, S.; Lin, H.; Lin, S. Anti-obesity effect of trans-cinnamic acid on HepG2 cells and HFD-fed mice. *Food Chem. Toxicol.* **2020**, *137*, 111148. [CrossRef] [PubMed]
24. Li, R.; Wang, C.; Chen, Y.; Li, N.; Wang, Q.; Zhang, M.; He, C.; Chen, H. A combined network pharmacology and molecular biology approach to investigate the active ingredients and potential mechanisms of mulberry (*Morus alba* L.) leaf on obesity. *Phytomedicine* **2021**, *92*, 153714. [CrossRef] [PubMed]

25. Luo, J.; Yan, W.; Chen, Z.; Xu, B. Elucidating the anti-obesity phytochemicals in Chenpi and their molecular mechanisms. *Food Sci. Hum. Wellness* **2024**, accepted. [CrossRef]
26. Saeki, T.; Yamamoto, S.; Akaki, J.; Tanaka, T.; Nakasone, M.; Ikeda, H.; Wang, W.; Inoue, M.; Manse, Y.; Ninomiya, K. Ameliorative effect of bofutsushosan (fangfengtongshengsan) extract on the progression of aging-induced obesity. *J. Nat. Med.* **2024**, *78*, 576–589. [CrossRef]
27. Xiang, P.; Du, Y.; Chen, G.; Mao, Y.; Li, S.; Li, Q.; Yang, Y.; Li, X.; Wang, D. Dietary achievable dose of protocatechuic acid, a metabolite of flavonoids, inhibits high-fat diet-induced obesity in mice. *Mol. Nutr. Food Res.* **2024**, *68*, 2300451. [CrossRef]
28. Wang, W.; Chen, S.; Xu, S.; Liao, G.; Li, W.; Yang, X.; Li, T.; Zhang, H.; Huang, H.; Zhou, Y. Jianpi Shengqing Huazhuo Formula improves abnormal glucose and lipid metabolism in obesity by regulating mitochondrial biogenesis. *J. Ethnopharmacol.* **2024**, *319*, 117102. [CrossRef]
29. Gulzar, F.; Ahmad, S.; Singh, S.; Kumar, P.; Sharma, A.; Tamrakar, A.K. NOD1 activation in 3T3-L1 adipocytes confers lipid accumulation in HepG2 cells. *Life Sci.* **2023**, *316*, 121400. [CrossRef]
30. Zhao, J.; Sun, Y.; Yuan, C.; Li, T.; Liang, Y.; Zou, H.; Zhang, J.; Ren, L. Quercetin ameliorates hepatic fat accumulation in high-fat diet-induced obese mice via PPARs. *Food Func.* **2023**, *14*, 1674–1684. [CrossRef]
31. Sun, Y.; Hou, X.; Li, L.; Tang, Y.; Zheng, M.; Zeng, W.; Lei, X. Improving obesity and lipid metabolism using conjugated linoleic acid. *Vet. Med. Sci.* **2022**, *8*, 2538–2544. [CrossRef]
32. Eor, J.Y.; Kwak, M.-J.; Park, M.Y.; Choi, S.-w.; Whang, K.-Y.; Kim, Y. Anti-obesity effect of seaweed algae (*Ascophyllum nodosum*) extracts in the 3T3-L1 cell line and a high-fat diet-induced obesity rat model. *Food Biosci.* **2024**, *59*, 104176. [CrossRef]
33. Wu, Y.; Luo, J.; Xu, B. Network Pharmacology and bioinformatics study of geniposide regulating oxidative stress in colorectal cancer. *Int. J. Mol. Sci.* **2023**, *24*, 15222. [CrossRef] [PubMed]
34. Banerjee, S.; Bhattacharjee, P.; Kar, A.; Mukherjee, P.K. LC–MS/MS analysis and network pharmacology of Trigonella foenum-graecum–A plant from Ayurveda against hyperlipidemia and hyperglycemia with combination synergy. *Phytomedicine* **2019**, *60*, 152944. [CrossRef] [PubMed]
35. Guo, F.; Yao, L.; Zhang, W.; Chen, P.; Hao, R.; Huang, X.; Jiang, J.; Wu, S. The therapeutic mechanism of Yuye decoction on type 2 diabetes mellitus based on network pharmacology and experimental verification. *J. Ethnopharmacol.* **2023**, *308*, 116222. [CrossRef]
36. Luo, H.; Wang, M.; Xu, K.; Peng, Q.; Zou, B.; Yin, S.; Yu, C.; Ren, L.; Li, P.; Tang, L. Effect of fushengong decoction on pten/pi3k/akt/nf-κb pathway in rats with chronic renal failure via dual-dimension network pharmacology strategy. *Front. Pharmacol.* **2022**, *13*, 807651. [CrossRef]
37. Daina, A.; Zoete, V. Testing the predictive power of reverse screening to infer drug targets, with the help of machine learning. *Commun. Chem.* **2024**, *7*, 105. [CrossRef]
38. Keiser, M.J.; Roth, B.L.; Armbruster, B.N.; Ernsberger, P.; Irwin, J.J.; Shoichet, B.K. Relating protein pharmacology by ligand chemistry. *Nat. Biotechnol.* **2007**, *25*, 197–206. [CrossRef]
39. Hua, Y.; Xie, D.; Zhang, Y.; Wang, M.; Wen, W.; Sun, J. Identification and analysis of key genes in adipose tissue for human obesity based on bioinformatics. *Gene* **2023**, *888*, 147755. [CrossRef]
40. Safran, M.; Rosen, N.; Twik, M.; BarShir, R.; Stein, T.I.; Dahary, D.; Fishilevich, S.; Lancet, D. The genecards suite. In *Practical Guide to Life Science Databases*; Springer: Singapore, 2021; pp. 27–56.
41. Bardou, P.; Mariette, J.; Escudié, F.; Djemiel, C.; Klopp, C. jvenn: An interactive Venn diagram viewer. *BMC Bioinform.* **2014**, *15*, 1–7. [CrossRef]
42. Zhou, Y.; Zhou, B.; Pache, L.; Chang, M.; Khodabakhshi, A.H.; Tanaseichuk, O.; Benner, C.; Chanda, S.K. Metascape provides a biologist-oriented resource for the analysis of systems-level datasets. *Nat. Commun.* **2019**, *10*, 1523. [CrossRef]
43. Wang, Y.; Huang, Z.; Xiao, Y.; Wan, W.; Yang, X. The shared biomarkers and pathways of systemic lupus erythematosus and metabolic syndrome analyzed by bioinformatics combining machine learning algorithm and single-cell sequencing analysis. *Front. Immunol.* **2022**, *13*, 1015882. [CrossRef] [PubMed]
44. Stuhr, N.L.; Nhan, J.D.; Hammerquist, A.M.; Van Camp, B.; Reoyo, D.; Curran, S.P. Rapid lipid quantification in Caenorhabditis elegans by oil red O and Nile red staining. *Bio-protocol* **2022**, *12*, e4340. [CrossRef] [PubMed]
45. Perelman, A.; Wachtel, C.; Cohen, M.; Haupt, S.; Shapiro, H.; Tzur, A. JC-1: Alternative excitation wavelengths facilitate mitochondrial membrane potential cytometry. *Cell Death Dis.* **2012**, 3, e430. [CrossRef] [PubMed]
46. Connolly, N.M.; Theurey, P.; Adam-Vizi, V.; Bazan, N.G.; Bernardi, P.; Bolaños, J.P.; Culmsee, C.; Dawson, V.L.; Deshmukh, M.; Duchen, M.R. Guidelines on experimental methods to assess mitochondrial dysfunction in cellular models of neurodegenerative diseases. *Cell Death Differ.* **2018**, *25*, 542–572. [CrossRef]
47. Lee, H.-W.; Karki, R.; Han, J.-H. Inhibition of the RPS6KA1/FoxO1 signaling axis by hydroxycitric acid attenuates HFD-induced obesity through MCE suppression. *Phytomedicine* **2024**, *128*, 155551. [CrossRef]
48. Heinonen, S.; Buzkova, J.; Muniandy, M.; Kaksonen, R.; Ollikainen, M.; Ismail, K.; Hakkarainen, A.; Lundbom, J.; Lundbom, N.; Vuolteenaho, K. Impaired mitochondrial biogenesis in adipose tissue in acquired obesity. *Diabetes* **2015**, *64*, 3135–3145. [CrossRef]
49. Graves, J.A.; Wang, Y.; Sims-Lucas, S.; Cherok, E.; Rothermund, K.; Branca, M.F.; Elster, J.; Beer-Stolz, D.; Van Houten, B.; Vockley, J. Mitochondrial structure, function and dynamics are temporally controlled by c-Myc. *PLoS ONE* **2012**, *7*, e37699. [CrossRef]
50. Riu, E.; Ferre, T.; Hidalgo, A.; Mas, A.; Franckhauser, S.; Otaegui, P.; Bosch, F. Overexpression of c-myc in the liver prevents obesity and insulin resistance. *FASEB J.* **2003**, *17*, 1715–1717. [CrossRef]

51. Nixon, M.; Mackenzie, S.D.; Taylor, A.I.; Homer, N.Z.; Livingstone, D.E.; Mouras, R.; Morgan, R.A.; Mole, D.J.; Stimson, R.H.; Reynolds, R.M. ABCC1 confers tissue-specific sensitivity to cortisol versus corticosterone: A rationale for safer glucocorticoid replacement therapy. *Sci. Transl. Med.* **2016**, *8*, ra109–ra352. [CrossRef]
52. Riancho, J.; Vázquez, L.; Garcia-Perez, M.; Sainz, J.; Olmos, J.; Hernandez, J.; Pérez-López, J.; Amado, J.; Zarrabeitia, M.; Cano, A. Association of ACACB polymorphisms with obesity and diabetes. *Mol. Genet. Metab.* **2011**, *104*, 670–676. [CrossRef]
53. Kaartinen, M.T.; Arora, M.; Heinonen, S.; Rissanen, A.; Kaprio, J.; Pietiläinen, K.H. Transglutaminases and obesity in humans: Association of F13A1 to adipocyte hypertrophy and adipose tissue immune response. *Int. J. Mol. Sci.* **2020**, *21*, 8289. [CrossRef] [PubMed]
54. Gaebler, N.; Haggenmüller, B.; Kapapa, M.; Serra, A.; Tews, D.; Funcke, J.-B.; Brandt, S.; Ioannidis, V.; Schön, M.; Möller, P. Age-and BMI-associated expression of angiogenic factors in white adipose tissue of children. *Int. J. Mol. Sci.* **2019**, *20*, 5204. [CrossRef] [PubMed]
55. Chedid, M.F.; do Nascimento, F.V.; de Oliveira, F.S.; de Souza, B.M.; Kruel, C.R.; Gurski, R.R.; Canani, L.H.; Crispim, D.; Gerchman, F. Interaction of HSD11B1 and H6PD polymorphisms in subjects with type 2 diabetes are protective factors against obesity: A cross-sectional study. *Diabetol. Metab. Syndr.* **2019**, *11*, 78. [CrossRef] [PubMed]
56. Zhang, X.; Gao, Y.; Liu, Z.; Li, W.; Kang, Y.; Li, X.; Xu, Z.; Peng, C.; Qi, Y. Salicylate sodium suppresses monocyte chemoattractant protein-1 production by directly inhibiting phosphodiesterase 3B in TNF-α-stimulated adipocytes. *Int. J. Mol. Sci.* **2022**, *24*, 320. [CrossRef]
57. Salinas-Santander, M.A.; León-Cachón, R.B.; Cepeda-Nieto, A.C.; Sánchez-Domínguez, C.N.; González-Zavala, M.A.; Gallardo-Blanco, H.L.; Esparza-González, S.C.; González-Madrazo, M.Á. Assessment of biochemical parameters and characterization of TNFα-308G/A and PTPN22+ 1858C/T gene polymorphisms in the risk of obesity in adolescents. *Biomed. Rep.* **2016**, *4*, 107–111. [CrossRef]
58. Boumiza, S.; Chahed, K.; Tabka, Z.; Jacob, M.-P.; Norel, X.; Ozen, G. MMPs and TIMPs levels are correlated with anthropometric parameters, blood pressure, and endothelial function in obesity. *Sci. Rep.* **2021**, *11*, 20052. [CrossRef]
59. Wen, X.; Zhang, B.; Wu, B.; Xiao, H.; Li, Z.; Li, R.; Xu, X.; Li, T. Signaling pathways in obesity: Mechanisms and therapeutic interventions. *Signal Transduct. Target Ther.* **2022**, *7*, 298. [CrossRef]
60. Savova, M.S.; Mihaylova, L.V.; Tews, D.; Wabitsch, M.; Georgiev, M.I. Targeting PI3K/AKT signaling pathway in obesity. *Biomed. Pharmacother.* **2023**, *159*, 114244. [CrossRef]
61. Yochum, G.S. Multiple Wnt/ss-catenin responsive enhancers align with the MYC promoter through long-range chromatin loops. *PLoS ONE* **2011**, *6*, e18966. [CrossRef]
62. MacDougald, O.A.; Hwang, C.-S.; Fan, H.; Lane, M.D. Regulated expression of the obese gene product (leptin) in white adipose tissue and 3T3-L1 adipocytes. *Proc. Nat. Acad. Sci. USA* **1995**, *92*, 9034–9037. [CrossRef]
63. Jakab, J.; Miškić, B.; Mikšić, Š.; Juranić, B.; Ćosić, V.; Schwarz, D.; Včev, A. Adipogenesis as a potential anti-obesity target: A review of pharmacological treatment and natural products. *Diabetes Metab. Syndr. Obes. Targets Ther.* **2021**, *14*, 67–83. [CrossRef]
64. Prestwich, T.C.; MacDougald, O.A. Wnt/β-catenin signaling in adipogenesis and metabolism. *Curr. Opin. Cell Biol.* **2007**, *19*, 612–617. [CrossRef]

antioxidants

MDPI

Article

Massa Medicata Fermentata, a Functional Food for Improving the Metabolic Profile via Prominent Anti-Oxidative and Anti-Inflammatory Effects

Kyung-Mi Jung [1,†], Ga-Ram Yu [1,2,†], Da-Hoon Kim [1], Dong-Woo Lim [1,2,*] and Won-Hwan Park [1,*]

[1] Department of Diagnostics, College of Korean Medicine, Dongguk University, Goyang-si 10326, Republic of Korea; chaoskmj@hanmail.net (K.-M.J.); kalama2@dongguk.edu (G.-R.Y.); lilwayne224@gmail.com (D.-H.K.)

[2] Institute of Korean Medicine, Dongguk University, Goyang-si 10326, Republic of Korea

* Correspondence: greatwoodong@dongguk.edu (D.-W.L.); diapwh@dongguk.ac.kr (W.-H.P.)

† These authors contributed equally to this work.

Abstract: Massa Medicata Fermentata (MMF) is a naturally fermented product used to treat indigestion and increase stomach activity in traditional medicine. This study examined the ability of the hydrothermal extract of MMF to scavenge free radicals corresponding to biological oxidative stresses, further protecting essential biomolecules. The anti-inflammatory effects of MMF were evaluated in LPS-induced RAW264.7 macrophages and zebrafish. In addition, the effects of MMF on the body mass index (BMI) and cholesterol accumulation in adult zebrafish fed a high-cholesterol diet (HCD) for three weeks were examined. MMF prevented the DNA and lipid damage caused by oxidative stress, inhibited LDL oxidation, and reduced the expression of cytokines and related proteins (MAPK and NFκB), with prominent anti-oxidative pathway (NRF2-HO-1) activation properties. LPS-induced NO production was reduced, and the increase in BMI and TC caused by the HCD diet was suppressed by MMF in zebrafish embryos or adult zebrafish. The bioactive aglycone of quercetin may be contributing to the mechanisms of systemic effects. MMF has excellent antioxidant properties and is useful for improving inflammation status and metabolic profile, thus highlighting its potential as a healthy, functional food.

Keywords: anti-inflammation; anti-oxidant; metabolism; Massa Medicata Fermentata; nutraceutical; functional food

Citation: Jung, K.-M.; Yu, G.-R.; Kim, D.-H.; Lim, D.-W.; Park, W.-H. Massa Medicata Fermentata, a Functional Food for Improving the Metabolic Profile via Prominent Anti-Oxidative and Anti-Inflammatory Effects. *Antioxidants* **2024**, *13*, 1271. https://doi.org/10.3390/antiox13101271

Academic Editors: Baojun Xu and Bin Du

Received: 9 September 2024
Revised: 15 October 2024
Accepted: 18 October 2024
Published: 19 October 2024

1. Introduction

The interest in health-functional foods, such as nutritional supplements, herbal medicines, and health supplements, has increased steadily. They help maintain health, prevent disease, and improve the quality of life [1,2]. The health benefits of a plant-based diet are related to the antioxidant and anti-inflammatory mechanisms of various phytochemicals [3]. Recent studies have shown that both obesity and hypertension are associated with oxidative stress and inflammatory response. Oxidative stress and inflammatory processes adversely affect endothelial dysfunction in cardiovascular diseases and induce insulin resistance [4,5]. Insulin resistance plays a crucial role in the development and progression of metabolic diseases such as hypertension, diabetes, and non-alcoholic fatty liver disease [6].

The excessive production of reactive oxygen species (ROS), including superoxide radical anions, hydroxyl radicals, singlet oxygen, and hydrogen peroxide, induces oxidative stress and protein oxidation, contributing to the pathogenesis of various types of inflammatory diseases, such as atherosclerosis, rheumatoid arthritis, cancer, and allergies [7–9]. Phytochemicals with antioxidant effects have been studied extensively to treat inflammatory diseases [10]. Nrf2 (nuclear factor erythroid 2-related factor 2) is a major cellular sensor of oxidative stress. Under normal conditions, Nrf2 binds to Kelch-like ECH-associated protein 1 (Keap1) in the cytoplasm [11]. When the cells are exposed to oxidative stress, Nrf2 translocates to the nucleus. It binds to and activates antioxidant response elements (AREs), cis-acting enhancers in the promoter regions of a large and unique set of target genes, restoring redox homeostasis [12]. HO-1 (heme oxygenase 1) protects cells exposed to oxidizing agents. Increased HO-1 expression, which is regulated primarily by the Nrf2-ARE pathway, is associated with defense against oxidative stress [13–15].

Macrophages are immune cells that regulate the inflammatory process by releasing mediators such as cytokines and NO (nitric oxide). iNOS (inducible nitric oxide synthase) is a pro-inflammatory mediator that is key to assessing inflammation [16]. NO is generated in macrophages by iNOS following exposure to pro-inflammatory cytokines, including TNF-α, IL-1β, IL-6, and INF-γ, or microbial products, such as LPS [17,18]. The continuous and excessive production of NO can lead to a variety of diseases, such as atherosclerosis, diabetes, obesity, and cancer. The activation of NFκB (nuclear factor kappa-light-chain-enhancer of activated B cells) and MAPKs (mitogen-activated protein kinases) family members, including ERK (extracellular signal-regulated kinase), JNK (c-Jun N-terminal kinase), and p-p38 (p38 mitogen-activated protein kinase), are important for the production of pro-inflammatory factors and are triggered during inflammation under LPS induction in macrophages [19]. Thus, suppressing the NFκB and MAPK signaling pathways is a pharmacological approach to diseases such as cancer, arthritis, Alzheimer's disease, asthma allergies, and other inflammatory diseases [20,21].

MMF (Massa Medicata Fermentata) is used in oriental medicine to treat gastrointestinal disorders, such as indigestion, because it strengthens the spleen and stomach and promotes digestion [22,23]. In addition, MMF is widely used to treat vomiting, diarrhea, obesity, and related metabolic diseases. MMF is a natural fermentation product of bitter almond (the kernel of *Prunus armeniaca* L.), red bean (*Vigna umbellata* Thunb.), *Xanthium sibiricum* L., *Polygonum hydropiper* L., *Artemisia annua* L., and wheat bran or flour (*Triticum aestivum* L.) combined in a certain proportion and fermented by microorganisms [24–26].

Fermentation is a feasible strategy for improving the efficacy of natural products, which are widely used in food and drug processing [23,27]. For example, in Korean medicine, the raw materials of Samjunghwan, a traditional prescription, are subjected to a long-term fermentation process to improve bioactivity [27,28]. The compounds from natural products are metabolized into smaller ones, showing better absorption and bioactivities [29].

As previously reported, the composition of MMF changes during the fermentation process [30]. Many fermented herbal medicines were reported to benefit health, ultimately leading to improved metabolic profiles with multifaceted mechanisms [31,32]. Therefore, the MMF can be used as food or medicine to treat metabolic diseases.

This study evaluated the overall efficacy and mechanisms of MMF for combating the exogenous oxidative and endogenous inflammatory factors. Furthermore, based on these general activities, the therapeutic effects of MMF on imbalanced metabolic profiles in HCD-fed zebrafish were investigated to evaluate its potential as a functional food.

2. Materials and Methods

2.1. Chemicals

Fetal bovine serum (FBS) and penicillin/streptomycin solution were purchased from Invitrogen (Carlsbad, CA, USA), Dulbecco's Modified Eagle's Medium (DMEM), and Dulbecco's phosphate buffered saline (DPBS) were obtained from Gibco (Carlsbad, CA, USA). Lipopolysaccharide (LPS), 2,2-diphenyl-1-picrylhydrazyl (DPPH), human low-density lipoprotein (LDL), pBR322 plasmid DNA, DAF-FM DA (4-amino-5-methylamino-2′,7′-difluorofluorescein diacetate), and other reagents were acquired from Sigma Chemicals (St. Louis, MO, USA). The ELISA kits for mouse TNF-α, IL-1β, and IL-6 were supplied by R&D Systems (Minneapolis, MN, USA). The primary antibodies for iNOS (inducible nitric synthase), COX-2 (cyclooxygenase-2), Nrf2, HO-1, β-actin, and secondary antibodies were procured from Santa Cruz (Dallas, TX, USA). The primary antibodies against p-ERK, p-JNK, p-p38, p-IκB-α, and p-NF-κB (p-p65) were obtained from Cell Signaling Technology (Beverly, MA, USA). The total cholesterol (TC) assay kit was acquired from Biomax (Guri, Republic of Korea).

2.2. Preparation of MMF Extract

MMF was purchased from Humanherb (Chungcheongbuk-do, Republic of Korea). MMF (200 g) was ground to a powder and extracted in hot water (1 L) for 1 h at 95 °C. The crude extracts were filtered through Whatman filter paper, concentrated using a rotary evaporator (Buchi, Switzerland) at 95 °C, and freeze-dried to obtain a powder (12.5 g), which was eluted with DPBS and filtered through a 0.22 µm syringe filter before use.

2.3. Total Phenolic Content of MMF

The total phenolic content (TPC) was estimated using the Folin–Ciocalteu colorimetric method described previously with a slight modification. Briefly, 40 µL of MMF (10 mg/mL) was mixed with 200 µL of Folin–Ciocalteu reactant and 1.16 mL of distilled water and reacted for 3 min at room temperature and neutralized by adding 600 µL of 2% Na_2CO_3. The optical densities (ODs) of the reactants were measured at 765 nm using a microplate reader (Versamax, Molecular Devices, Silicon Valley, CA, USA) after incubation for 30 min [33]. Quantification was performed based on the standard curve of gallic acid. The results are expressed as grams of gallic acid equivalent (GAE) per 1 g of dry weight (DW). The total tannin content (TTC) was estimated using the Folin–Denis colorimetric method described previously, with a slight modification. Briefly, 15 µL of MMF (10 mg/mL) was mixed with 8.485 mL of distilled water, 0.5 mL of Folin–Denis reagent, and 1 mL of Na_2CO_3 and reacted at room temperature. The optical densities (ODs) of the reactants were measured at 700 nm using a microplate reader (Versamax, Molecular Devices, USA) [34]. Quantification was conducted based on the standard curve of tannic acid. The results were expressed as grams of tannic acid equivalent (TAE) per 1 g of dry weight (DW). The total flavonoid contents (TFC) were measured using a modified aluminum chloride colorimetric method. Briefly, 250 µL of MMF (10 mg/mL) was mixed with 1 mL of distilled water and 75 µL of $NaNO_2$ and reacted for 6 min at room temperature. Subsequently, 300 µL of 10% $AlCl_3$ (*w*/*v*) was added to the mixture. After six minutes, 500 µL of 1 M NaOH was added. The optical densities (ODs) of the reactants were measured at 510 nm using a microplate reader (Versamax, Molecular Devices, USA) [35]. Quantification was conducted based on the standard curve of catechin. The results are expressed as grams of catechin equivalent (CE) per 1 g dry weight (DW).

2.4. DPPH, ABTS Free Radical Scavenging Assay

The antioxidant capacity of the MMF, expressed as the donation of an electron or a hydrogen atom to the 2,2′-diphenyl-1-picrylhydrazyl (DPPH) free radical, was measured using a spectrophotometric method [36]. Briefly, 50 µL of MMF at various concentrations (50–800 µg/mL) was added to 1 mL of 0.1 mM DPPH in ethanol solution and 0.45 mL of 50 mM Tris-HCl (pH 7.4). The test tube was incubated in the dark for approximately 30 min at room temperature. Ascorbic acid was used as a positive control. The ODs of the reactants were measured at 517 nm using a microplate reader (Versamax, Molecular Devices, USA).

$$\text{DPPH inhibition (\%)} = (A_B - A_C)/A_B \times 100$$

where A_B is the recorded absorbance of the blank solution, and A_C is the recorded absorbance of the MMF solution.

On a similar basis, the free radical scavenging activity of MMF was measured with 2,2′-Azino-bis(3-ethylbenzothiazoline-6-sulfonic acid) (ABTS) free radical scavenging assay. Briefly, a preincubate mixture of 7.4 mM ABTS and 2.6 mM potassium persulfate was kept for 24 h in the dark. The reaction buffer was mixed with samples in a 10:1 ratio and reacted for 10 min. The optical density of the reactant was measured at 732 nm wavelength.

2.5. Superoxide-Free—Radical Scavenging Assay

This assay was performed to assess the antioxidant activity of MMF in scavenging superoxide free radicals. The superoxide radicals were generated in vitro by the hypoxanthine/xanthine oxidase system [37]. The scavenging potential of the extract was determined using the nitro-blue tetrazolium (NBT) reduction method. A reaction mixture was prepared using a 50 mM phosphate buffer (pH 7.5) containing EDTA (0.05 mM), hypoxanthine (0.2 mM), NBT (1 mM), aqueous or ethanolic extract (distilled water for the control), and xanthine oxidase. The xanthine oxidase (1.2 U/μL) was added last. The ODs of the reactants were measured at 560 nm using a microplate reader (Versamax, Molecular Devices, USA).

2.6. Ferric Thiocyanate (FTC) Assay

The level of lipid peroxidation inhibition was examined by the ferric thiocyanate (FTC) method [38,39]. A mixture of 0.4 mL MMF in ethanol, 0.2 mL of 2.5% linolenic acid in absolute ethanol, and 0.4 mL of 50 mM phosphate buffer (pH 7.0) was placed in a vial and placed in an oven at 40 °C in the dark. A 0.1 mL sample of this solution was added to 3 mL of 70% ethanol and 0.01 mL of aqueous ammonium thiocyanate (30%, *w/v*) and kept for three minutes after adding 0.01 mL of ferrous chloride (2.45 mg/mL in 3.5% HCl) to the reaction mixture. The ODs of the reactants were measured at 500 nm using a microplate reader (Versamax, Molecular Devices, USA).

2.7. DNA Nicking Assay

The DNA nicking assay was performed as described [40]. A 9 μL sample of MMF (5.5 mg/mL) was added to 1 μL of supercoiled pBR322 plasmid DNA (0.5 μL). The mixture solution was incubated at room temperature for 10 min, followed by the addition of 10 μL of Fenton's reagent (80 μmol/L $FeCl_3$, 50 μmol/L ascorbic acid, and 30 mmol/L H_2O_2). The mixture solution was incubated at 37 °C for 5 min. Finally, the DNA was analyzed on 1.0% agarose gel electrophoresis (80 V, 30 min) and visualized under an ultraviolet illuminator. One unit of catalase reactant was used as the positive control.

2.8. Relative Electrophoretic Mobility (REM) Assay

The relative electrophoretic mobility (REM) of native low-density lipoprotein (LDL) and oxidized LDL was performed to evaluate the protective efficacy of MMF against LDL oxidation, according to the procedures of references [41]. Before electrophoresis, the samples containing 10 μM of $CuSO_4$ and MMF were treated into LDL (120 μg/mL) and incubated for 12 h at 37 °C in the dark. The native LDL and pre-treated LDL samples loaded onto 0.7% agarose gel in TAE buffer (40 mM Tris-acetate with 1 mM EDTA, pH 8.0) and electrophoresed for 50 min at 85 V. After electrophoresis, the gel was fixed in 40% ethanol with 10% acetic acid for 30 min, stained with 0.15% coomassie brilliant blue R250, and LDL bands were visualized using the destaining solution.

2.9. Cell Culture and Treatment

RAW264.7 cells (a mouse macrophage cell line) were purchased from the Korea Cell Line Bank (KCLB, Seoul, Republic of Korea). The cells were cultured in DMEM medium supplemented with 10% FBS and 100 U/mL penicillin–streptomycin at 37 °C in a humidified atmosphere containing 5% CO_2 and maintained at ~70% confluence before being used in the experiments. The cells were seeded on cell culture dishes at 1.2×10^5 cells/mL and co-treated with LPS (1 μg/mL) and the MMF for 6–12 h (for immunofluorescence microscopy) or 12–24 h (for nitrite determination, Elisa, real-time PCR, and western blot).

2.10. Cell Viability Assay

The cell viabilities of RAW264.7 cells were determined by the WST (water-soluble tetrazolium salt) using an EZ-cytox assay kit according to the manufacturer's instructions. Briefly, the cells were seeded on cell culture dishes at 1.2×10^5 cells/mL and incubated at 37 °C in a humidified 5% CO_2 incubator for 24 h. The cells were treated with various concentrations of MMF extract in the absence or presence of LPS (1 μg/mL) for 24 h. The medium was replaced with DMEM containing 10% EZ-cytox and reacted for 30 min. The ODs of the reactants were measured at 450 nm.

2.11. Nitrite Assay

Griess reagent was used to examine the effects of MMF on LPS-induced nitrite levels [42]. Briefly, the cells were seeded on cell culture dishes at 1.2×10^5 cells/mL and incubated at 37 °C in a humidified 5% CO_2 incubator for 24 h. The cells were co-treated with LPS (1 μg/mL) and MMF at various concentrations for 24 h. Supernatants were collected and mixed with Griess reagent. The ODs were measured at 570 nm. The nitrite concentrations were calculated using a standard curve.

2.12. Preparation of Nuclear and Cytoplasm Fractions

Nuclear and cytosolic proteins were separated using an extraction reagents kit from Thermo Fisher Scientific (Rockford, IL, USA) [43]. Briefly, RAW264.7 macrophages were seeded on cell culture dishes at 1.2×10^5 cells/mL and incubated at 37 °C in a humidified 5% CO_2 incubator for 24 h. The cells were then co-treated with LPS (1 μg/mL) and MMF at various concentrations for 6 h. The nuclear translocation of NFκB from the cytoplasm was assessed by western blot.

2.13. Western Blot Analysis

The levels of proteins associated with oxidation and inflammation were determined by western blot. Briefly, the cells were lysed with RIPA (radioimmunoprecipitation assay) buffer (Thermo Fisher Scientific, Rockford, IL, USA) containing a protease and phosphatase inhibitor cocktail (Gendepot, Barker, TX, USA). The protein concentrations were measured using the BCA (bicinchoninic acid) protein assay kit (Thermo Fisher Scientific, Rockford, IL, USA). Protein lysates (35 μg) were loaded into 7.5–10% SDS-PAGE gels, electrophoresed, and transferred to the PVDF membranes at 100 V for 60 min using an electrophoretic transfer cell (Bio-rad, Hercules, CA, USA). Membranes were then blocked with 5% BSA in TBS/T (TBS containing 0.1% Tween 20) for 2 h at room temperature, incubated with the primary antibodies (1:1500 dilution in TBS/T) overnight with gentle agitation, rinsed with TBS/T, and incubated with secondary antibodies (1:3000 dilution in TBS/T) at room temperature for 2 h. The blots were detected using the Fusion Solo imaging system (Vilber Lourmat, Collegien, France), and the proteins were visualized using an ECL buffer (Super Signal West Pico, Thermo Fisher Scientific).

2.14. Quantitative Real-Time Polymerase Chain Reaction

The mRNA expression levels of the pro-inflammatory mediators associated genes were determined by qPCR. The total mRNA was isolated using the Trizol reagent (Invitrogen, Carlsbad, CA, USA) according to the manufacturer's instructions. Briefly, the isolated mRNA was subjected to cDNA synthesis using the AccuPower RT Premix kit (Bioneer, Daejeon, Republic of Korea) and oligo deoxythymine (dt) 18 primers (Invitrogen, Carlsbad, CA, USA). The primer-specific binding cDNA was amplified using a Light Cycler 480 PCR system (Roche, Basel, Switzerland). The PCR mix contained 10 μL of SYBR green master mixture (Roche, Switzerland), 8 μL of ultrapure water, 1 pmol/μL of gene primer, and 1 μL of template cDNA. Amplification was performed using the following schedule: initial denaturation at 95 °C for 10 min, followed by 45 cycles of denaturation at 95 °C for 10 s, annealing at 56–62 °C for 20 s, and extension at 72 °C for 20 s. Melting curve analysis was conducted at 95 °C for 5 min for a quality check. The threshold cycle (Ct) value was calculated to quantify the PCR results. The following primers were used: iNOS forward, 5′-GAGACAGGGAAGTCTGAAGCAC-3′, reverse, 5′-CCAGCAGTAGTTGCTCCTCTTC-3′; COX2 forward, 5′-GCGACATACTCAAGC AGGAGCA-3′, reverse, 5′-AGTGGTAACCGCTCAGGTGTTG-3′; TNFα forward, 5′-AAGCCTGTAGCCCACGTCGTA-3′, reverse, 5′-GGCACCACTAGTTGGTTGTCTTTG-3′; IL-1β forward, 5′-CTGAACTCAACTGTGAAATGCCA-3′, reverse, 5′-AAAGGTTTGGAAGCAG-CCCT-3′; IL-6 forward, 5′-CCACTTCACAAGTCGGAGGCTTA-3′, reverse, 5′-GCAAGTG-CATCATCGTTGTTCATAC-3′; β-actin forward, 5′-GCAAGTGCTTCTAGGCGGAC-3′, reverse, 5′-AAGAAAGGGTGTAAAACGCAGC-3′. The relative expression levels were calculated by dividing the gene Ct values by that of β-actin. All data were acquired using a LightCycler 480 instrument and software.

2.15. *Enzyme-Linked Immunosorbent Assay (ELISA)*

The concentrations of secretory inflammatory cytokines in cell culture supernatants were measured using Quantikine mouse ELISA kits (R&D Systems, Inc. Minneapolis, MN, USA). Briefly, the cells (1.2×10^5 cells/mL) were seeded in six-well plates and incubated at 37 °C in a humidified atmosphere containing 5% CO_2. Twenty-four hours after seeding, the cells were co-treated with LPS (1 μg/mL) and MMF for 24 h. The culture media were collected, and the TNFα, IL-1β, and IL-6 concentrations were measured according to the manufacturer's instructions. The ODs were measured at 450 nm using a microplate spectrophotometer.

2.16. *Immunofluorescence Microscopy*

The nuclear translocation of Nrf2 and NFκB was followed by growing the cells on Lab-Tek II chamber slides (Nalge Nunc, Naperville, IL, USA) using a slight modification of the previous study [44]. Briefly, the cells were fixed in 4% formaldehyde for 5 min, permeabilized with 0.1% Triton X-100 for 10 min at room temperature, blocked with 1% BSA for 1 h at room temperature, and labeled with 2 μg/mL of the primary antibody overnight at 4 °C. NFκB in cytoplasm and nuclei was detected by treating the cells with 2 μg/mL of FITC containing 0.2% BSA for 1 h at room temperature. The nuclei were stained using a mounting medium containing API (Vector Laboratories, Newark, CA, USA). The images were captured under a fluorescence microscope (BX50, Olympus, Tokyo, Japan).

2.17. *Zebrafish Maintenance*

All zebrafish (AB strain) were maintained at 28 °C with a 14/10 h light/dark cycle. The embryos were harvested from natural mating and cultured in the embryo media. The fertilized embryos were incubated in 2 mg/L methylene blue containing the E3 embryo media (1.6 g KCl, 34.8 g NaCl, 5.8 g $CaCl_2 \cdot 2H_2O$, and 9.78 g $MgCl_2 \cdot 6H_2O$ in 1 L of double-distilled water, pH 7.2 ± 0.1) at 28.5 °C. The effects of MMF on LPS-induced NO production in embryos were examined by stimulating the embryos with LPS (10 μg/mL) and MMF (500 μg/mL) for 24 h at 28.5 °C. After 7 h in the embryo medium E2 buffer, LPS-stimulated zebrafish larvae were transferred into 96-well plates and treated with a DAF-FM DA solution (2.5 μM) for 1 h in the dark at 28.5 °C, rinsed in fresh embryo medium, and fixed in 4% formaldehyde. The fluorescence intensities of individual larvae were assessed using a fluorescence microscope (BX50, Olympus, Japan). The effects of MMF on HCD-induced body mass index (BMI) and plasma total cholesterol in adult zebrafish were evaluated by selecting zebrafish with a standard length of approximately 3.8 cm and dividing them into four groups: control group; HCD 1 μg/mL group; HCD + MMF 10% group; and HCD + MMF 20% group. The HCD was prepared by mixing a diethyl ether solution of cholesterol with a normal feed to obtain 4 % (*w*/*w*) cholesterol in the diet after diethyl ether evaporation [45]. Each group was composed of ten fish and was fed an equal mass feed (30 mg/day, twice daily for 21 days) following the schedule, and the residual food was removed 1 h after feeding. The length and whole body weight of the zebrafish were measured. The BMI of the fish was obtained by dividing the body weight (g) of the fish by the square of the body length (cm). The zebrafish were sacrificed and analyzed individually, and the total cholesterol level was measured using commercial assay kits according to the manufacturer's instructions.

2.18. HPLC Analyses

Chromatographic analyses of MMF were performed using an HPLC system (Agilent 1260 infinity HPLC, Agilent, Santa Clara, CA, USA) equipped with a UV-Diode array detector, degasser, and auto-sampler. Quercetin standard (Thermo Fisher Scientific, Rockford, IL, USA) was diluted with DMSO to 10 mM, and MMF extract was diluted with DMSO to 10 mg/mL. The main chemical components of fractionated samples were analyzed using an Agilent Eclipse XDB-C18 chromatographic column (150 × 4.6 mm, 5 μm pore size) at a flow rate of 1 mL/min, a column temperature of 23.0 °C, and a detection wavelength of 300 nm by Isocratic elution (mobile phase A—water with 0.1 formic acid and mobile phase B—acetonitrile with 0.1 formic acid) as follows: 1–20 min, 70% A:30% B.

2.19. Statistical Analyses

The data were analyzed using Graph Pad Prism version 5.0 software (Graph Pad, La Jolla, CA, USA). The standard curves were estimated using Excel 2010 and PowerPoint 2010 (Microsoft, Redmond, WA, USA). An analysis of the variance (ANOVA) and One-Way ANOVA with Dunnett's multiple comparison test were used to determine the significance of the differences. The results are presented as means ± SDs, and p-values of <0.05 were considered significant.

3. Results

3.1. Total Phenolic Content and Radical Scavenging Activity

Figure 1A presents the total phenolic content (TPC), total tannin content (TTC), and total flavonoid content (TFC) of MMF. An aqueous extract of MMF had a TPC of 26.32 ± 1.05 μg GAE/mg, TTC of 28.84 ± 1.315 μg TAE/mg, and TFC of 1.92 ± 0.2 μg CE/mg. The antioxidant activity of MMF was determined by DPPH free radical scavenging assay, and their reducing ability was determined based on their concentration, providing the 50% inhibition (IC_{50}) value. The maximum inhibition value was observed at an MMF concentration of 400 μg/mL. As a result, the measured inhibition concentration 50% (IC_{50}) value was 213.36 ± 14.37 μg/mL (Figure 1B). In addition, an ABTS assay was performed, and the estimated IC_{50} of MMF was 833 μg/mL (Figure 1C). The superoxide anion radical scavenging activity of MMF was determined using the nitro-blue tetrazolium (NBT) reduction method. At 1 mg/mL, MMF showed a superoxide anion radical scavenging ability of approximately 30%. A slight increase was observed as the treatment concentration was increased (Figure 1D). The inhibition of the lipid peroxidation was examined using the ferric thiocyanate assay. Lipid peroxidation was prevented by 16.5% at an MMF concentration of 8 mg/mL. Lipid peroxidation has deleterious effects on cell physiology by disrupting cell signaling with reduced membrane protein activities, and MMF is believed to block this process. The Fenton reaction generates hydroxyl radicals (Figure 1E). The inhibitory effects of the MMF extract on DNA nicking are caused by hydroxyl radicals. Lanes 1, 2, 3, and 4 are pBR322 plasmid DNA, pBR322 plasmid DNA + Fenton's reagent, pBR322 plasmid DNA + MMF (5 mg/mL) + Fenton's reagent, and pBR322 plasmid DNA + catalase (1 unit) + Fenton's reagent, respectively. The MMF extracts had a protective effect against hydroxyl radical-mediated plasmid DNA damage. In particular, highly reactive hydroxyl radicals ultimately cause DNA damage, confirming that MMF reduces the formation of form II (single-strand nicked DNA) and form III (double-strand nicked DNA) to form I (double-strand DNA) in the nicking of DNA mediated by Fenton's reaction (Figure 1F). The protective effects of MMF on Cu^{2+}-induced human LDL oxidation. Lanes 1, 2, 3, 4, and 5 are native LDL, LDL + Cu^{2+}, LDL + Cu^{2+} + MMF (0.1 mg/mL), LDL + MMF (0.5 mg/mL) + Cu^{2+}, and LDL + ascorbic acid (0.1 mg/mL) + Cu^{2+}, respectively. In the aging process of blood vessels, the oxidation of low-density lipoprotein (LDL) is the most common vascular damage process known to mediate various inflammatory responses. MMF inhibited oxidized-LDL induced by Cu^{2+} in a concentration-dependent manner. The results were visualized by electrophoresis. The oxidation of apolipoprotein in LDL was demonstrated by the increased shift of bands. At 0.5 mg/mL, MMF notably protected Cu^{2+}-induced LDL oxidation (Figure 1G).

(A)

	Phenolic content		
	TPC (μg/mg)	TTC (μg/mg)	TFC (μg/mg)
MMF	26.32±1.05	28.84±1.315	1.92±0.2

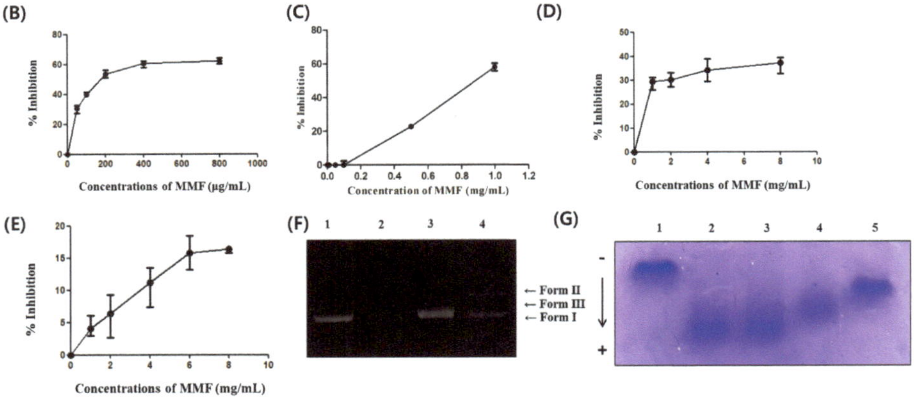

Figure 1. Phenolic contents and antioxidant activity of the MMF extract. (**A**) Total phenolic, tannin, and flavonoid contents of MMF extract. (**B**,**C**) Free-radical scavenging activity of MMF extract measured using a DPPH assay and ABTS assay. (**D**) The inhibitory effect of MMF extract was tested by NBT reduction system. (**E**) Lipid peroxidation inhibitory effect of MMF. (**F**) Showing a comparative electrophoretic pattern of pBR322 DNA nicking inhibition activity of MMF. Lane 1, pBR322 supercoil plasmid DNA; Lane 2, hydroxyl radical-mediated DNA nick form; Lane 3, pre-treatment of MMF (5 mg/mL, final concentration); Lane 4, pre-treatment of catalase (1 unit). (**G**) Effect of MMF on the Cu^{2+} mediated LDL oxidation by relative electrophoretic mobility (REM). Lane 1, native LDL; Lane 2, ox-LDL; Lane 3, pre-treatment of MMF (0.1 mg/mL); Lane 4, pre-treatment of MMF (0.5 mg/mL); Lane 5, pre-treatment of ascorbic acid (0.1 mg/mL). Dose-dependent effect of MMF against Cu^{2+}-mediated LDL oxidation by REM.

3.2. *Effects of MMF on the Viability of Murine Macrophage RAW264.7 Cells*

The cytotoxic effect of MMF on the viability of RAW264.7 cells was examined using the WST assay. The RAW264.7 cells were incubated with various concentrations (0–500 μg/mL) of MMF for 24 h. No significant decrease in cell viability was observed, even at the highest concentration of MMF (Figure 2A). On the other hand, the MMF treatment dose-dependently restored the cell viability reduced by LPS (1 μg/mL) from 71.95% to 109.75% (Figure 2B). Subsequent experiments were conducted using MMF at 300 or 500 μg/mL.

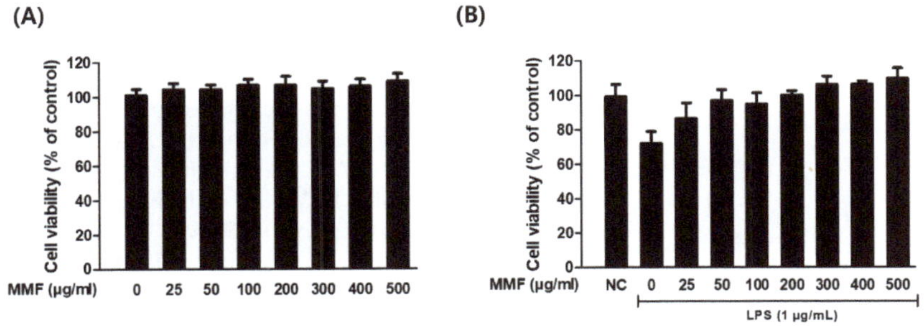

Figure 2. Effects of MMF on cell viability. The cells were treated with various concentrations of MMF extract in the absence (**A**) or presence (**B**) of LPS (1 μg/mL) for 24 h. The results are presented as the mean ± SDs of the percentages determined by three independent experiments versus non-treated controls. NC: non-treated negative control.

3.3. MMF Increased the Nuclear Translocation of Nrf2 in Macrophages

The antioxidant effects of MMF on the Nrf2/HO-1 signaling pathway were investigated by western blotting. MMF dose-dependently increased HO-1 expression, and significant nuclear translocation of Nrf2 was observed at 500 μg/mL (Figure 3A). The nuclear translocation of Nrf2 was reconfirmed through immunofluorescence microscopy. MMF dose-dependently increased the nuclear translocation of Nrf2 (Figure 3B).

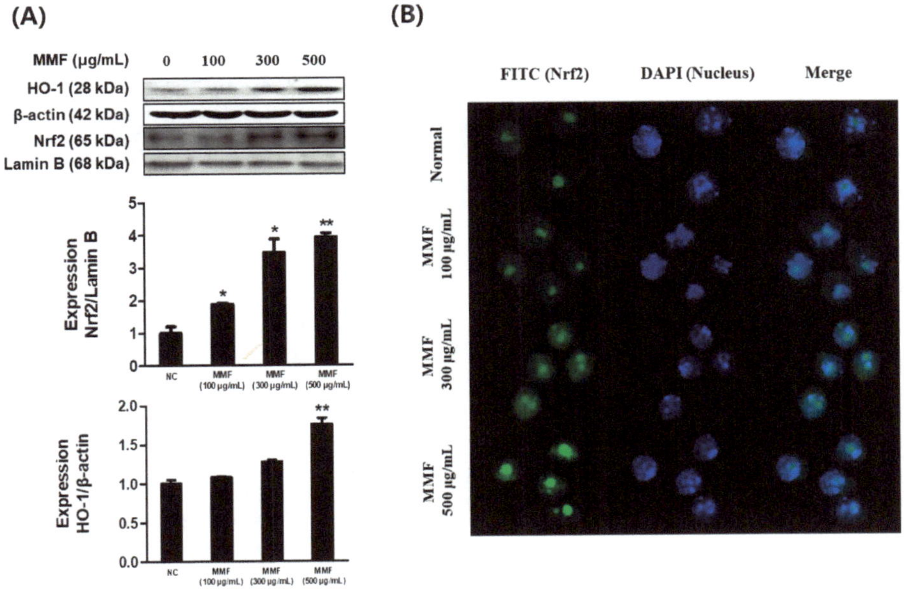

Figure 3. Effects of MMF on HO-1 expression and the nuclear translocation of Nrf2. (**A**) Raw 264.7 cells were incubated with various concentrations of MMF for 6 h and HO-1 expression was examined by western blot analysis. The Nrf2 expression levels in cytoplasm and nuclear were assessed by western blot analysis. (**B**) Nuclear Nrf2 translocation was visualized under an immunofluorescence microscope. The band intensities were measured by densitometry and normalized to the intensities of the total forms and β-actin. The results are presented as the means ± SDs of three independent experiments. * $p < 0.05$, ** $p < 0.01$ versus non-treated negative control. NC: non-treated negative control.

3.4. MMF Inhibited iNOS and COX-2 Expressions and the Phosphorylation of MAPKs in LPS-Stimulated Macrophages

The effects of MMF on NO production in LPS-stimulated RAW264.7 cells were evaluated. The NO levels in the medium significantly increased by LPS at 1 μg/mL for 24 h. MMF significantly and dose-dependently reduced NO production in LPS-stimulated RAW264.7 cells. At 500 μg/mL, MMF significantly decreased LPS-induced NO production from 107.66 to 17.84 μM in RAW264.7 cells (Figure 4A). The effects of MMF on the mRNA and protein expression levels of iNOS and COX2 in LPS-stimulated RAW264.7 cells were examined. The change in iNOS protein in relation to NO production was observed. MMF notably and dose-dependently inhibited the LPS-induced mRNA and protein expression levels of iNOS and COX2 in RAW264.7 cells (Figure 4B,C). The inhibitory effects of MMF on the MAPK signaling pathway in LPS-stimulated RAW264.7 cells were investigated by western blotting. LPS markedly elevated the phosphorylation level of ERK, JNK, and p38. On the other hand, high MMF concentrations significantly inhibited the phosphorylation of ERK, JNK, and p38 in LPS-stimulated RAW264.7 cells (Figure 4D).

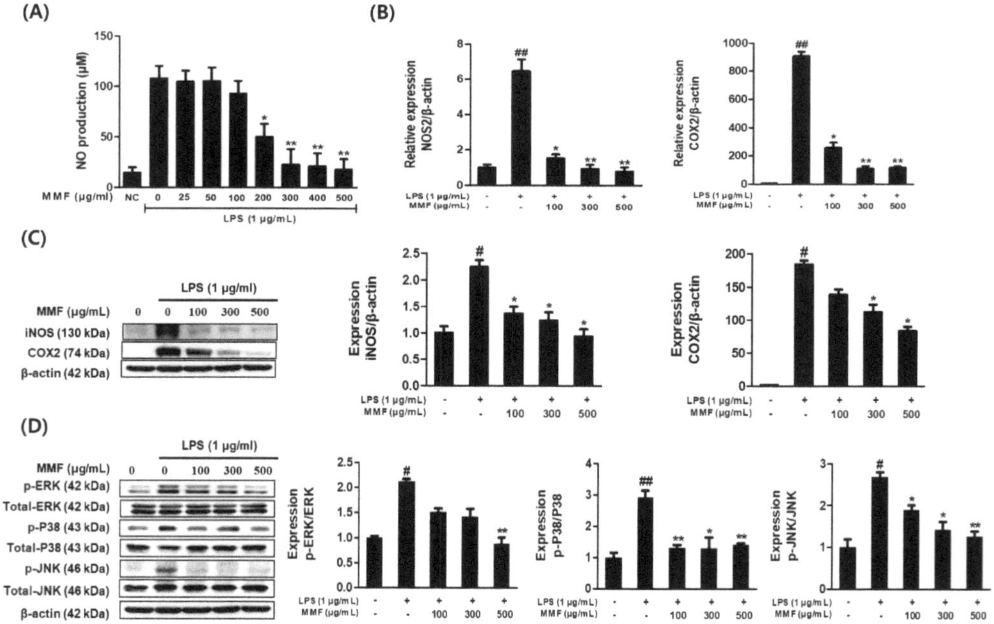

Figure 4. MMF inhibited the up-regulation of iNOS and COX2 and the phosphorylation of MAPKs in LPS-treated RAW264.7 macrophages. RAW264.7 macrophages were co-treated with MMF and LPS (1 μg/mL) for 12–24 h. (**A**) NO concentrations were estimated using a Griess reaction. (**B**) Gene expression levels of NOS2 and COX2 obtained by Real-time PCR (**C**) Western blot analysis revealed the effects on the relative iNOS and COX2 protein levels. (**D**) Relative MAPK expression determined by western blot analysis. The band intensities were measured by densitometric analysis and normalized versus the intensities of the total forms and β-actin. The results are presented as the means ± SDs of three independent experiments. # $p < 0.05$, ## $p < 0.01$ versus LPS-treated controls, and * $p < 0.05$, ** $p < 0.01$ versus LPS-treated RAW264.7 cells. NC: non-treated negative control.

3.5. MMF Inhibited the Nuclear Translocation of NFκB in LPS-Stimulated Macrophages

The effects of MMF on NFκB family members in LPS-stimulated RAW264.7 cells were investigated by analyzing the phosphorylated protein levels. Western blot images showed that MMF significantly inhibited IκB and NFκB phosphorylation by LPS (Figure 5A).

Immunofluorescence microscopy showed that MMF reduced NFκB translocation to the nucleus in LPS-stimulated RAW264.7 cells compared to LPS alone (Figure 5B). Moreover, cytosolic NFκB activation was also increased. These results suggested that the MMF treatment partially affected NFκB activation in response to LPS induction through NFκB phosphorylation.

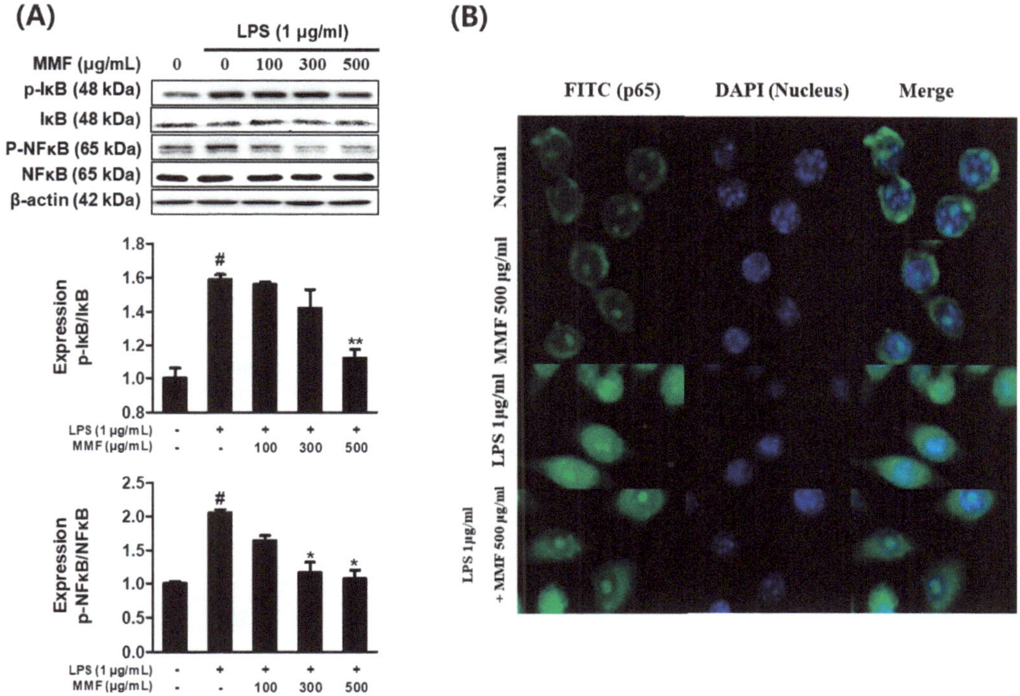

Figure 5. MMF inhibited NFκB translocation by reducing the phosphorylation of IκB in RAW264.7 macrophages. RAW264.7 macrophages were co-treated with MMF and LPS (1 μg/mL) for 12 h. (**A**) Relative phosphorylation of the IκB and NFκB protein levels as determined by western blot analysis. (**B**) visualized by immunofluorescence microscopy. The results are presented as the means ± SDs of three independent experiments. # $p < 0.05$ versus LPS-treated controls, and * $p < 0.05$, ** $p < 0.01$ versus LPS-treated RAW264.7 cells.

3.6. MMF Reduced Pro-Inflammatory Cytokine Levels in LPS-Treated Macrophages

ELISA and qPCR were performed to examine the effects of MMF on pro-inflammatory cytokine production. LPS increased TNFα, IL-1β, and IL-6 production in RAW264.7 cells. ELISA showed that LPS stimulation significantly upregulated the production of the pro-inflammatory cytokines (TNFα, IL-1β, and IL-6) in RAW264.7 cells. On the other hand, at 500 μg/mL, MMF significantly downregulated the TNFα, IL-1β, and IL-6 levels induced by LPS (Figure 6A). The qPCR results also showed that MMF dose-dependently decreased the mRNA expression levels of TNFα, IL-1β, and IL-6 (Figure 6B). These data suggest that MMF exerts anti-inflammatory activity by suppressing pro-inflammatory cytokines in LPS-stimulated RAW264.7 cells.

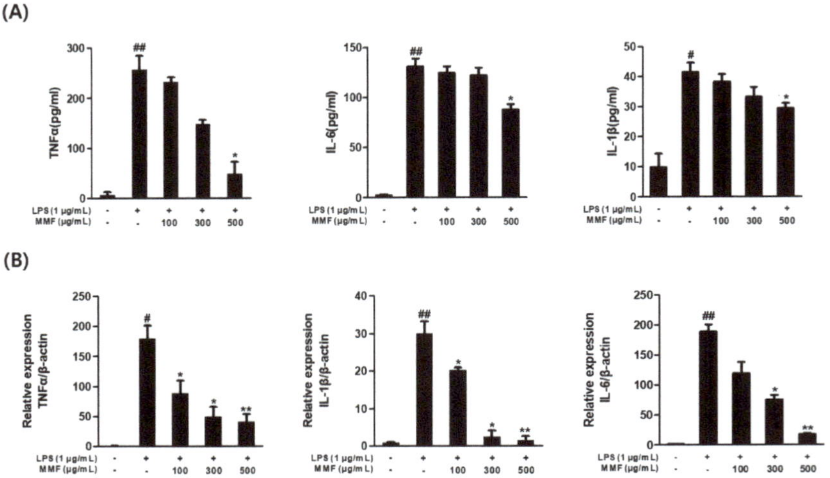

Figure 6. MMF significantly reduced LPS-induced pro-inflammatory cytokine levels in RAW264.7 macrophages. RAW264.7 macrophages were simultaneously treated with MMF and LPS (1 μg/mL) for 24 h. (**A**) Relative expression of TNFα, IL-1β, and IL-6, as determined by ELISA. (**B**) Relative expression of TNFα, IL-1β, and IL-6, as determined by qPCR. The results are presented as the means ± SDs of three independent experiments. # $p < 0.05$; ## $p < 0.01$ versus LPS-treated controls, and * $p < 0.05$; ** $p < 0.01$ versus LPS-treated RAW264.7 cells.

3.7. MMF Decreased NO Production in LPS-Stimulated Zebrafish

The inhibitory effects of MMF on NO production in LPS-induced zebrafish larvae were evaluated using the fluorescent probe DAF-FM DA. The LPS treatment induced NO production in zebrafish larvae, and MMF reduced this LPS-induced increase in NO production (Figure 7).

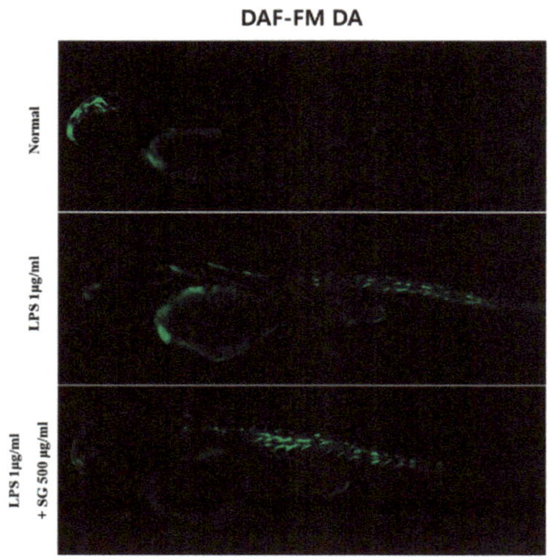

Figure 7. Protective effects of MMF on LPS-induced NO generation in zebrafish larvae. The zebrafish larvae were pre-treated with MMF (500 μg/mL) for 1 h and then exposed to LPS (10 μg/mL) for 24 h. The levels of NO generation were observed under a fluorescence microscope after staining with DAF-FM DA.

3.8. MMF Improved Body Weight and Serum Lipid Levels

The zebrafish study evaluated the benefits of MMF on high-cholesterol diet-induced obesity in adult zebrafish. There was no difference between the HCD-fed groups (HCD, HCD + MMF 10%, and HCD + MMF 20%) until the first week, but a lower BMI was observed in the MMF intake group after the second week. On the sacrifice day, a difference was observed between mean body weights in the three groups (NCD, HCD + MMF 10%, and HCD + MMF 20%) and the HCD group (Figure 8A). The plasma total cholesterol levels in adult zebrafish fed on the HCD showed similar metabolic features to human obesity. MMF significantly and dose-dependently improved the plasma total cholesterol in high-cholesterol diet-induced obesity in adult zebrafish (Figure 8B).

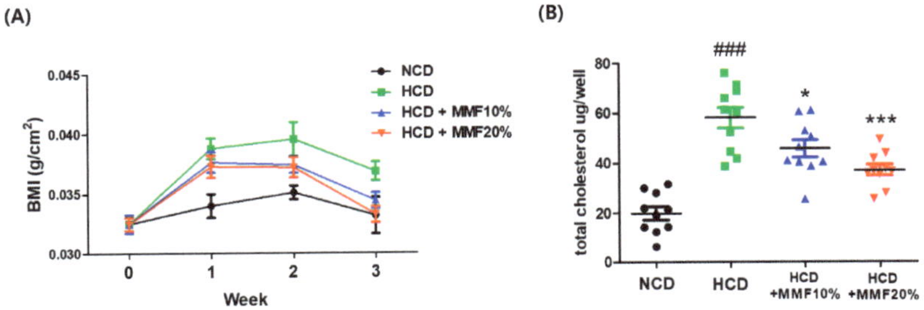

Figure 8. Assessment of the BMI and plasma total cholesterol in adult zebrafish fed MMF. (**A**) Change in BMI (g/cm^2) during the experimental period for three weeks. (**B**) TC levels in HCD-induced obese zebrafish at three weeks fed 10% and 20% MMF + HCD. NCD: normal cholesterol diet; HCD: high-cholesterol diet. * $p < 0.05$, *** $p < 0.001$, ### $p < 0.001$.

3.9. HPLC Analysis Indicates the Potential Presence of Quercetin

The HPLC analysis was performed with MMF extract and previously known ingredients of MMF. HPLC chromatogram showed a peak of an identified standard compound of quercetin at a retention time of 2.711 min (Figure 9A). This peak corresponded to the peak in MMF used in this study at a retention time of 2.747 min (Figure 9B). We estimated the concentration of quercetin contents in MMF extract as 891.9 μg/g (*w*/*w*) based on the standard curve quantification.

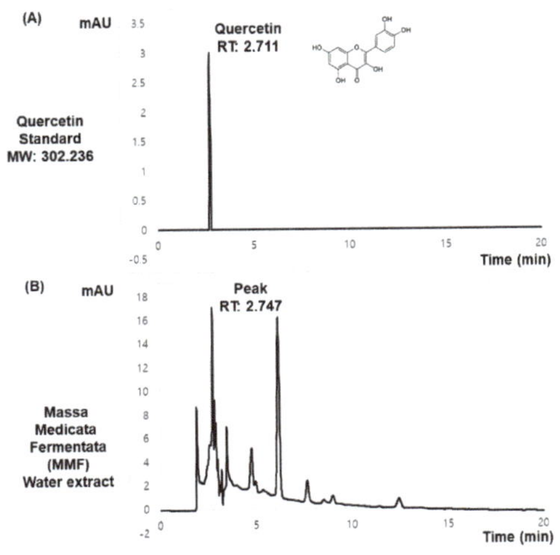

Figure 9. HPLC fingerprinting analysis of MMF extract. (**A**) HPLC chromatogram analysis of Quercetin, a standard compound of MMF. (**B**) HPLC analysis of hydrothermal extract of MMF used in this study.

4. Discussion

Metabolic disorders are complex conditions characterized by abnormal metabolic profiles and are closely related to diabetes, obesity, and hyperlipidemia [46]. Hyperglycemia, a hyperlipidemia condition of the host, directly damages the systemic oxidant/antioxidant balance in vascular epithelial tissue or immune cells and aggravates the inflammatory status [47]. A closer look showed that a low-grade inflammatory status in adipose tissue stimulates preadipocyte hyperplasia and hypertrophy [48]. On the other hand, excessive ROS generated from beta-oxidation mechanisms of adipocytes underlies the cellular pathophysiology of these metabolic diseases. ROS generation activates multifarious inflammatory signaling pathways and promotes inflammatory response [49]. Inflammation and oxidative stress play a significant role in the development of metabolic comorbidities such as hyperlipidemia, hypertension, and glucose intolerance, which lead to metabolic dysfunction [50,51]. The trio of oxidative stress, inflammation, and adiposity make it tricky to stop the vicious cycle of metabolic disorders. A functional food or medicine with this multifaceted potential might efficiently impact patients with metabolic disorders Health-functional foods derived from natural resources have multifaceted potentials, such as nutritional value and pharmacological properties [52].

In this study, the potential of MMF as a functional food was evaluated using a series of experiments, including chemical assays and in vitro and in vivo system-based studies MMF was highly expected to have antioxidant ability by evaluating the phenol, tannin, and flavonoid contents. The free radical scavenging ability was evaluated. The IC_{50} value obtained in the DPPH assay revealed no significant effect in the nitroblue tetrazolium (NBT) assay. Therefore, MMF is believed to scavenge cation radicals rather than anion radicals. Reactive oxygen species are a major source of oxidative and nitrosative stress for organisms, damaging macromolecules, including lipids and DNA [53]. The antioxidant activity of MMF during linoleic acid oxidation, as measured by the thiocyanate method, did not increase by more than 16.5%. On the other hand, the antioxidant activity of MMF during LDL oxidation was significant, suggesting that the antioxidant activity of natural antioxidants is system-dependent and that a wide range of activities can be observed according to the lipid systems used as substrates. This study evaluated the potential of MMF as an antioxidant defense against hydroxyl radical-mediated plasmid DNA damage using agarose gel electrophoresis. The results show that MMF water extract protects against DNA damage induced by metal-catalyzed Fenton reactions. The antioxidants protect DNA from oxidative damage, changing the three-dimensional structure of DNA and altering DNA mobility in electric fields. The plasmid DNA appears in three distinct forms on an agarose gel. Form I is a supercoiled circular form that migrates faster than the other forms. A nicked circular form (form II) is formed when the supercoiled DNA form is broken. This form moves much more slowly than the other forms. Another form is Form III, a linear form between Forms I and II. MMF significantly minimized this cleavage.

The inflammatory response is a complex biological mechanism of the host's immune system that defends the body against harmful stimuli. On the other hand, persistent inflammation can lead to diseases and tumorigenesis, such as atherosclerosis, rheumatoid arthritis, and cancer [54]. During an inflammatory immune response, macrophages release pro-inflammatory mediators, such as NO, iNOS, COX2, and pro-inflammatory cytokines and chemokines. MMF reduced iNOS and COX-2 expression and the release of cytokines, such as TNFα, IL-1β, and IL-6, in LPS-stimulated macrophages [55,56]. Inhibiting the production of inflammatory mediators might be a useful therapeutic strategy in inflammatory diseases. NFκB regulates the transcription of inflammatory factors, and its activation is regulated by a MAPK (mitogen-activated protein kinase) and is involved in the inflammatory process through nuclear translocation [57,58]. In this study, using RAW264.7 cells, MMF attenuated LPS-induced phosphorylation of the NFκB pathway proteins and the nuclear translocation of p65. The MAPK pathway regulates various metabolic processes of cells and is strongly involved in the immune defense and inflammatory response. MMF significantly decreased the phosphorylation of LPS-stimulated MAPK pathway protein expression.

This study applied an inflammation model using LPS and an obesity model using an HCD diet in zebrafish. The zebrafish model is used widely in studies to screen drugs with anti-inflammatory properties because it acts as an in vivo inflammation model exhibiting similar inflammatory responses to mammals [59]. In addition, zebrafish are increasingly utilized as models of human obesity and metabolic diseases, including visceral adiposity, hepatic steatosis, atherosclerosis, and type 2 diabetes [60]. MMF inhibited LPS-induced NO production in vivo embryos, as shown by DAF-FM DA fluorescence analysis. Moreover, the BMI and TC of zebrafish were reduced dose-dependently by the MMF treatment. In addition to the prominent anti-inflammatory properties in vitro, the MMF was assumed to exert weight reduction and hypocholesterolemic effects by controlling the systemic oxidative and inflammatory status.

As fermentation progresses, crude phytochemical components are metabolized into bioactive substances by bacteria and fungi, which benefit human health. For example, MMF is reported to be habituated by numerous resident bacteria, and the fermentation progressed by dominant bacteria genera of Enterobacter, Pediococcus, Pseudomonas, Mucor, and Saccharomyces [61]. Lactic acid was produced by fermentation [30], and many studies reported the benefits of lactic acid or lactic acid bacteria-derived metabolites against obesity or metabolic diseases [62].

Quercetin, which is universally found in numerous herbs, can be produced from its various glycoside forms during the fermentation process [29]. Several studies have shown that MMF contains quercetin as a major compound [63]. This bioactive aglycone of quercetin has been reported to exert potent antioxidant activity, thereby providing multifaceted benefits for metabolic disorders [64]. As presented in the HPLC analysis, our MMF extract contains quercetin, which is equivalent to around 1.49 μM, which may be contributing to the mechanisms of systemic effects.

The standardization of MMF is indeed a significant issue for researchers. Consequently, many studies have investigated the major compounds of MMF, particularly following the fermentation process. Major composition profiles of MMFs from different batches have been investigated by researchers. A recent study investigated the differences among nine MMFs produced from various batches in China and Korea [65]. While it is not a simple issue, we may assert that many studies have been conducted and are currently underway by researchers to address the standardization issues of MMF.

This study has several limitations in revealing the detailed mechanisms for the anti-obesity effects, e.g., appetite suppression and enhancement of metabolic rate or inhibition of nutrient digestion, because of the technical limitation of the zebrafish model. On the other hand, MMF may exert systemic benefits on HCD-fed zebrafish, as shown by fluorescence imaging. Nevertheless, no zebrafish in the MMF group (in 10% and 20%) expired, indicating the low toxic profile of MMF in vivo. A low toxicity profile is a critical factor for a new food material to be developed as a functional food [66]. We can reasonably assume that at least quercetin, as detected in our MMF sample, may help mitigate HCD or LPS-induced harmful conditions in zebrafish due to its notable antioxidant activities.

MMF, a fermented food used in traditional medicine, was evaluated as a potential modern health functional food. It showed notable potential for metabolic disorders and reliable background effects on treating oxidative and inflammatory conditions. Thus, MMF may have high value as a healthy, functional food, but additional research is needed.

5. Conclusions

MMF can scavenge free radicals, activate the Nrf2/HO-1 pathway in RAW264.7 cells, and show anti-inflammatory effects by attenuating the production of the pro-inflammatory mediators by inhibiting NFκB and MAPK signaling activation. In a metabolic disorder model using zebrafish, the BMI and TC of MMF groups were reduced significantly compared to the HCD group. These results highlight the potential of MMF as a healthy, functional food.

Author Contributions: Conceptualization, K.-M.J.; methodology, G.-R.Y. and D.-H.K.; software, D.-H.K.; validation, D.-W.L.; data curation, G.-R.Y. and D.-W.L.; writing—original draft preparation, G.-R.Y.; writing—review and editing, D.-W.L.; supervision, D.-W.L. and W.-H.P.; funding acquisition, G.-R.Y. and D.-W.L. All authors have read and agreed to the published version of the manuscript.

Funding: This work was supported by the Basic Science Research Program of the Korean National Research Foundation, funded by the Ministry of Education (Grant no. 2022R1A6A3A01087626).

Institutional Review Board Statement: Not applicable.

Informed Consent Statement: Not applicable.

Data Availability Statement: Data are contained within the article.

Conflicts of Interest: The authors declare no conflicts of interest.

References

1. Cencic, A.; Chingwaru, W. The role of functional foods, nutraceuticals, and food supplements in intestinal health. *Nutrients* **2010**, *2*, 611–625. [CrossRef] [PubMed]
2. Hasler, C.M. Functional foods: Benefits, concerns and challenges—A position paper from the American Council on Science and Health. *J. Nutr.* **2002**, *132*, 3772–3781. [CrossRef] [PubMed]
3. Rodríguez-Negrete, E.V.; Morales-González, Á.; Madrigal-Santillán, E.O.; Sánchez-Reyes, K.; Álvarez-González, I.; Madrigal-Bujaidar, E.; Valadez-Vega, C.; Chamorro-Cevallos, G.; Garcia-Melo, L.F.; Morales-González, J.A. Phytochemicals and Their Usefulness in the Maintenance of Health. *Plants* **2024**, *13*, 523. [CrossRef] [PubMed]
4. Ndisang, J.F.; Vannacci, A.; Rastogi, S. Oxidative stress and inflammation in obesity, diabetes, hypertension, and related cardiometabolic complications. *Oxidative Med. Cell. Longev.* **2014**, *2014*, 506948. [CrossRef]
5. Matsuda, M.; Shimomura, I. Increased oxidative stress in obesity: Implications for metabolic syndrome, diabetes, hypertension, dyslipidemia, atherosclerosis, and cancer. *Obes. Res. Clin. Pract.* **2013**, *7*, e330–e341. [CrossRef]
6. Fujii, H.; Kawada, N.; Japan Study Group of NAFLD (JSG-NAFLD). The role of insulin resistance and diabetes in nonalcoholic fatty liver disease. *Int. J. Mol. Sci.* **2020**, *21*, 3863. [CrossRef]
7. Chapple, I. Reactive oxygen species and antioxidants in inflammatory diseases. *J. Clin. Periodontol.* **1997**, *24*, 287–296. [CrossRef]
8. Di Rosanna, P.; Salvatore, C. Reactive oxygen species, inflammation, and lung diseases. *Curr. Pharm. Des.* **2012**, *18*, 3889–3900. [CrossRef]
9. Ahmed, O.M.; Mohammed, M.T. Oxidative stress: The role of reactive oxygen species (ROS) and antioxidants in human diseases. *Plant Arch* **2020**, *20*, 4089–4095.
10. Yatoo, M.; Gopalakrishnan, A.; Saxena, A.; Parray, O.R.; Tufani, N.A.; Chakraborty, S.; Tiwari, R.; Dhama, K.; Iqbal, H. Anti-inflammatory drugs and herbs with special emphasis on herbal medicines for countering inflammatory diseases and disorders-a review. *Recent Pat. Inflamm. Allergy Drug Discov.* **2018**, *12*, 39–58. [CrossRef]
11. Motohashi, H.; Yamamoto, M. Nrf2–Keap1 defines a physiologically important stress response mechanism. *Trends Mol. Med.* **2004**, *10*, 549–557. [CrossRef] [PubMed]
12. Itoh, K.; Wakabayashi, N.; Katoh, Y.; Ishii, T.; O'Connor, T.; Yamamoto, M. Keap1 regulates both cytoplasmic-nuclear shuttling and degradation of Nrf2 in response to electrophiles. *Genes Cells* **2003**, *8*, 379–391. [CrossRef] [PubMed]
13. Loboda, A.; Damulewicz, M.; Pyza, E.; Jozkowicz, A.; Dulak, J. Role of Nrf2/HO-1 system in development, oxidative stress response and diseases: An evolutionarily conserved mechanism. *Cell. Mol. Life Sci.* **2016**, *73*, 3221–3247. [CrossRef] [PubMed]
14. Wang, Y.; Gao, L.; Chen, J.; Li, Q.; Huo, L.; Wang, Y.; Wang, H.; Du, J. Pharmacological modulation of Nrf2/HO-1 signaling pathway as a therapeutic target of Parkinson's disease. *Front. Pharmacol.* **2021**, *12*, 757161. [CrossRef]
15. Lee, J.-M.; Johnson, J.A. An important role of Nrf2-ARE pathway in the cellular defense mechanism. *BMB Rep.* **2004**, *37*, 139–143. [CrossRef]
16. Zamora, R.; Vodovotz, Y.; Billiar, T.R. Inducible nitric oxide synthase and inflammatory diseases. *Mol. Med.* **2000**, *6*, 347–373. [CrossRef]
17. Weinberg, J.; Misukonis, M.; Shami, P.; Mason, S.; Sauls, D.; Dittman, W.; Wood, E.; Smith, G.; McDonald, B.; Bachus, K. Human mononuclear phagocyte inducible nitric oxide synthase (iNOS): Analysis of iNOS mRNA, iNOS protein, biopterin, and nitric oxide production by blood monocytes and peritoneal macrophages. *Blood* **1995**, *86*, 1184–1195. [CrossRef]
18. Chi, D.S.; Qui, M.; Krishnaswamy, G.; Li, C.; Stone, W. Regulation of nitric oxide production from macrophages by lipopolysaccharide and catecholamines. *Nitric Oxide* **2003**, *8*, 127–132. [CrossRef] [PubMed]
19. Kim, S.H.; Kim, J.; Sharma, R.P. Inhibition of p38 and ERK MAP kinases blocks endotoxin-induced nitric oxide production and differentially modulates cytokine expression. *Pharmacol. Res.* **2004**, *49*, 433–439. [CrossRef]
20. Yong, H.-Y.; Koh, M.-S.; Moon, A. The p38 MAPK inhibitors for the treatment of inflammatory diseases and cancer. *Expert Opin. Investig. Drugs* **2009**, *18*, 1893–1905. [CrossRef]
21. Kunnumakkara, A.B.; Sailo, B.L.; Banik, K.; Harsha, C.; Prasad, S.; Gupta, S.C.; Bharti, A.C.; Aggarwal, B.B. Chronic diseases, inflammation, and spices: How are they linked? *J. Transl. Med.* **2018**, *16*, 1–25. [CrossRef] [PubMed]

22. Gao, S.; Shan, L.; Shi, Y.; Zhao, Y.; Mu, Q.; Cui, Y.; Chai, X.; Wang, Y. Exploration of the variations of amino acids in Massa Medicata Fermentata and their effects on gastrointestinal diseases. *LWT* **2023**, *173*, 114309. [CrossRef]
23. Hussain, A.; Bose, S.; Wang, J.-H.; Yadav, M.K.; Mahajan, G.B.; Kim, H. Fermentation, a feasible strategy for enhancing bioactivity of herbal medicines. *Food Res. Int.* **2016**, *81*, 1–16. [CrossRef]
24. Sun, Y.N.; Yang, S.Y.; Koh, Y.-S.; Kim, Y.H.; Li, W. Isolation and identification of benzochroman and acylglycerols from massa medicata fermentata and their inhibitory effects on LPS-stimulated cytokine production in bone marrow-derived dendritic cells. *Molecules* **2018**, *23*, 2400. [CrossRef]
25. Fu, F.Q.; Xu, M.; Wei, Z.; Li, W. Biostudy on traditional Chinese medicine massa medicata fermentata. *ACS Omega* **2020**, *5*, 10987–10994. [CrossRef]
26. Wang, Z.; Okutsu, K.; Yoshizaki, Y.; Futagami, T.; Tamaki, H.; Maruyama, T.; Toume, K.; Komatsu, K.; Hashimoto, F.; Takamine, K. Effects of microbial fermentation on enzyme activity and volatile properties of Massa Medicata Fermentata. *Tradit. Kampo Med.* **2022**, *9*, 10–17. [CrossRef]
27. Lim, D.-W.; Bose, S.; Wang, J.-H.; Choi, H.S.; Kim, Y.-M.; Chin, Y.-W.; Jeon, S.-H.; Kim, J.-E.; Kim, H. Modified SJH alleviates FFAs-induced hepatic steatosis through leptin signaling pathways. *Sci. Rep.* **2017**, *7*, 45425. [CrossRef]
28. Ansari, A.; Bose, S.; Patra, J.K.; Shin, N.R.; Lim, D.-W.; Kim, K.-W.; Wang, J.-H.; Kim, Y.-M.; Chin, Y.-W.; Kim, H. A controlled fermented samjunghwan herbal formula ameliorates non-alcoholic hepatosteatosis in HepG2 Cells and OLETF rats. *Front. Pharmacol.* **2018**, *9*, 596. [CrossRef]
29. Lim, D.-W.; Wang, J.-H. Gut microbiome: The interplay of an "invisible organ" with herbal medicine and its derived compounds in chronic metabolic disorders. *Int. J. Environ. Res. Public Health* **2022**, *19*, 13076. [CrossRef]
30. Zhang, H.; Gao, S.; Zhang, X.; Meng, N.; Chai, X.; Wang, Y. Fermentation characteristics and the dynamic trend of chemical components during fermentation of Massa Medicata Fermentata. *Arab. J. Chem.* **2022**, *15*, 103472. [CrossRef]
31. Chen, G.; Li, Z.; Liu, S.; Tang, T.; Chen, Q.; Yan, Z.; Peng, J.; Yang, Z.; Zhang, G.; Liu, Y. Fermented Chinese herbal medicine promoted growth performance, intestinal health, and regulated bacterial microbiota of weaned piglets. *Animals* **2023**, *13*, 476. [CrossRef] [PubMed]
32. Zhang, Y.; Zhang, J.; Yan, J.; Qi, X.; Wang, Y.; Zheng, Z.; Liang, J.; Ling, J.; Chen, Y.; Tang, X. Application of fermented Chinese herbal medicines in food and medicine field: From an antioxidant perspective. *Trends Food Sci. Technol.* **2024**, *148*, 104410. [CrossRef]
33. Nikolaeva, T.; Lapshin, P.; Zagoskina, N. Method for determining the total content of phenolic compounds in plant extracts with Folin–Denis reagent and Folin–Ciocalteu reagent: Modification and comparison. *Russ. J. Bioorganic Chem.* **2022**, *48*, 1519–1525. [CrossRef]
34. Negi, A.; Dobhal, K.; Ghildiyal, P. Antioxidant potential and effect of extraction solvent on total phenol content, flavonoids content and tannin content of Ficus palmata Forssk. *Int. J. Pharm. Sci. Rev. Res.* **2018**, *49*, 19–24.
35. Shraim, A.M.; Ahmed, T.A.; Rahman, M.M.; Hijji, Y.M. Determination of total flavonoid content by aluminum chloride assay: A critical evaluation. *LWT* **2021**, *150*, 111932. [CrossRef]
36. Mensor, L.L.; Menezes, F.S.; Leitão, G.G.; Reis, A.S.; Santos, T.C.d.; Coube, C.S.; Leitão, S.G. Screening of Brazilian plant extracts for antioxidant activity by the use of DPPH free radical method. *Phytother. Res.* **2001**, *15*, 127–130. [CrossRef]
37. Parejo, I.; Viladomat, F.; Bastida, J.; Rosas-Romero, A.; Saavedra, G.; Murcia, M.A.; Jiménez, A.M.; Codina, C. Investigation of Bolivian plant extracts for their radical scavenging activity and antioxidant activity. *Life Sci.* **2003**, *73*, 1667–1681. [CrossRef] [PubMed]
38. Lee, J.-C.; Kim, H.-R.; Kim, J.; Jang, Y.-S. Antioxidant property of an ethanol extract of the stem of Opuntia ficus-indica var. saboten. *J. Agric. Food Chem.* **2002**, *50*, 6490–6496. [CrossRef]
39. Kim, J.-S. Evaluation of in vitro antioxidant activity of the water extract obtained from dried pine needle (Pinus densiflora). *Prev. Nutr. Food Sci.* **2018**, *23*, 134. [CrossRef]
40. Lim, D.-W.; Kim, H.; Park, J.-Y.; Kim, J.-E.; Moon, J.-Y.; Park, S.-D.; Park, W.-H. Amomum cardamomum L. ethyl acetate fraction protects against carbon tetrachloride-induced liver injury via an antioxidant mechanism in rats. *BMC Complement. Altern. Med.* **2016**, *16*, 155. [CrossRef]
41. Yoon, M.-A.; Jeong, T.-S.; Park, D.-S.; Xu, M.-Z.; Oh, H.-W.; Song, K.-B.; Lee, W.S.; Park, H.-Y. Antioxidant effects of quinoline alkaloids and 2, 4-di-tert-butylphenol isolated from Scolopendra subspinipes. *Biol. Pharm. Bull.* **2006**, *29*, 735–739. [CrossRef] [PubMed]
42. Yu, G.R.; Lim, D.W.; Karunarathne, W.A.H.M.; Kim, G.Y.; Kim, H.; Kim, J.E.; Park, W.H. A non-polar fraction of Saponaria officinalis L. acted as a TLR4/MD2 complex antagonist and inhibited TLR4/MyD88 signaling in vitro and in vivo. *FASEB J.* **2022**, *36*, e22387. [CrossRef]
43. Liu, Z.; Zhang, J.; Huang, X.; Huang, L.; Li, S.; Wang, Z. Magnesium sulfate inhibits the secretion of high mobility group box 1 from lipopolysaccharide-activated RAW264. 7 macrophages in vitro. *J. Surg. Res.* **2013**, *179*, e189–e195. [CrossRef]
44. Yu, G.R.; Kim, J.E.; Lim, D.W.; Park, W.H. The combination of Ephedrae herba and coixol from Coicis semen attenuate adiposity via glucocorticoid receptor regulation. *Sci. Rep.* **2023**, *13*, 20324. [CrossRef] [PubMed]
45. Luo, H.; Li, Q.-Q.; Wu, N.; Shen, Y.-G.; Liao, W.-T.; Yang, Y.; Dong, E.; Zhang, G.-M.; Liu, B.-R.; Yue, X.-Z.; et al. Chronological in vivo imaging reveals endothelial inflammation prior to neutrophils accumulation and lipid deposition in HCD-fed zebrafish. *Atherosclerosis* **2019**, *290*, 125–135. [CrossRef]

46. Heindel, J.J.; Blumberg, B.; Cave, M.; Machtinger, R.; Mantovani, A.; Mendez, M.A.; Nadal, A.; Palanza, P.; Panzica, G.; Sargis, R Metabolism disrupting chemicals and metabolic disorders. *Reprod. Toxicol.* **2017**, *68*, 3–33. [CrossRef] [PubMed]
47. El-Shinnawi, U.; Soory, M. Actions of adjunctive nutritional antioxidants in periodontitis and prevalent systemic inflammatory diseases. *Endocr. Metab. Immune Disord.-Drug Targets (Former. Curr. Drug Targets-Immune Endocr. Metab. Disord.)* **2015**, *15*, 261–276 [CrossRef]
48. Henninger, A.J.; Eliasson, B.; Jenndahl, L.E.; Hammarstedt, A. Adipocyte hypertrophy, inflammation and fibrosis characterize subcutaneous adipose tissue of healthy, non-obese subjects predisposed to type 2 diabetes. *PLoS ONE* **2014**, *9*, e105262. [CrossRef]
49. Carrillo, J.L.M.; Campo, J.; Coronado, O.G.; Gutiérrez, P.T.V.; Cordero, J.F.C.; Juárez, J.V. Adipose tissue and inflammation. In *Adipose Tissue*; InTech: London, UK, 2018; Volume 93.
50. Veloso, C.D.; Belew, G.D.; Ferreira, L.L.; Grilo, L.F.; Jones, J.G.; Portincasa, P.; Sardão, V.A.; Oliveira, P.J. A mitochondrial approach to cardiovascular risk and disease. *Curr. Pharm. Des.* **2019**, *25*, 3175–3194. [CrossRef]
51. Raut, S.K.; Khullar, M. Oxidative stress in metabolic diseases: Current scenario and therapeutic relevance. *Mol. Cell. Biochem.* **2023**, *478*, 185–196. [CrossRef]
52. Dama, A.; Shpati, K.; Daliu, P.; Dumur, S.; Gorica, E.; Santini, A. Targeting metabolic diseases: The role of nutraceuticals in modulating oxidative stress and inflammation. *Nutrients* **2024**, *16*, 507. [CrossRef] [PubMed]
53. Song, P.; Zou, M.H. Roles of reactive oxygen species in physiology and pathology. In *Atherosclerosis: Risks, Mechanisms, and Therapies*; John Wiley & Sons: Hoboken, NJ, USA, 2015; pp. 379–392.
54. Costa, C.; Incio, J.; Soares, R. Angiogenesis and chronic inflammation: Cause or consequence? *Angiogenesis* **2007**, *10*, 149–166. [CrossRef]
55. Zhang, X.; Cao, J.; Zhong, L. Hydroxytyrosol inhibits pro-inflammatory cytokines, iNOS, and COX-2 expression in human monocytic cells. *Naunyn-Schmiedeberg's Arch. Pharmacol.* **2009**, *379*, 581–586. [CrossRef]
56. Cho, H.; Yun, C.-W.; Park, W.-K.; Kong, J.-Y.; Kim, K.S.; Park, Y.; Lee, S.; Kim, B.-K. Modulation of the activity of pro-inflammatory enzymes, COX-2 and iNOS, by chrysin derivatives. *Pharmacol. Res.* **2004**, *49*, 37–43. [CrossRef] [PubMed]
57. Zhang, N.; Ahsan, M.H.; Zhu, L.; Sambucetti, L.C.; Purchio, A.F.; West, D.B. Regulation of IκBα expression involves both NF-κB and the MAP kinase signaling pathways. *J. Inflamm.* **2005**, *2*, 10. [CrossRef]
58. Jijon, H.; Allard, B.; Jobin, C. NF-κB inducing kinase activates NF-κB transcriptional activity independently of IκB kinase γ through a p38 MAPK-dependent RelA phosphorylation pathway. *Cell. Signal.* **2004**, *16*, 1023–1032. [CrossRef] [PubMed]
59. Zanandrea, R.; Bonan, C.D.; Campos, M.M. Zebrafish as a model for inflammation and drug discovery. *Drug Discov. Today* **2020**, *25*, 2201–2211. [CrossRef]
60. Schlegel, A. Zebrafish models for dyslipidemia and atherosclerosis research. *Front. Endocrinol.* **2016**, *7*, 159. [CrossRef]
61. Xu, Y.; Xie, Y.-B.; Zhang, X.-R.; Chen, C.; Xiang, H.; Xie, Q. Monitoring of the bacterial and fungal biodiversity and dynamics during Massa Medicata Fermentata fermentation. *Appl. Microbiol. Biotechnol.* **2013**, *97*, 9647–9655. [CrossRef]
62. Shen, Y.L.; Zhang, L.Q.; Yang, Y.; Yin, B.C.; Ye, B.C.; Zhou, Y. Advances in the role and mechanism of lactic acid bacteria in treating obesity. *Food Bioeng.* **2022**, *1*, 101–115. [CrossRef]
63. Zhang, X.; Zhang, H.; Huang, Q.; Sun, J.; Yao, R.; Wang, J. Effect of Massa Medicata Fermentata on the gut microbiota of dyspepsia mice based on 16S rRNA technique. *Evid.-Based Complement. Altern. Med.* **2020**, *2020*, 7643528. [CrossRef] [PubMed]
64. Hosseini, A.; Razavi, B.M.; Banach, M.; Hosseinzadeh, H. Quercetin and metabolic syndrome: A review. *Phytother. Res.* **2021**, *35*, 5352–5364. [CrossRef] [PubMed]
65. Wang, Z.; Okutsu, K.; Futagami, T.; Yoshizaki, Y.; Tamaki, H.; Maruyama, T.; Toume, K.; Komatsu, K.; Hashimoto, F.; Takamine, K. Microbial community structure and chemical constituents in Shinkiku, a fermented crude drug used in Kampo medicine. *Front. Nutr.* **2020**, *7*, 115. [CrossRef] [PubMed]
66. Huggett, A.; Schilter, B. Research needs for establishing the safety of functional foods. *Nutr. Rev.* **1996**, *54*, S143. [CrossRef]

antioxidants

Article

Protective Effect of Ergothioneine against Oxidative Stress-Induced Chondrocyte Death

Shuzo Sakata [1], Ryo Kunimatsu [2,*] and Kotaro Tanimoto [2]

1 Department of Orthodontics, Division of Oral Health and Development, Hiroshima University Hospital, 1-2-3 Kasumi, Minami-ku, Hiroshima 734-8553, Japan; shuzosakata@hiroshima-u.ac.jp

2 Department of Orthodontics and Craniofacial Developmental Biology, Graduate School of Biomedical and Health Sciences, Hiroshima University, 1-2-3 Kasumi, Minami-ku, Hiroshima 734-8553, Japan; tkotaro@hiroshima-u.ac.jp

* Correspondence: ryoukunimatu@hiroshima-u.ac.jp; Tel.: +81-82-257-5686

Abstract: Reactive oxygen species (ROS) induce oxidative stress in cells and are associated with various diseases, including autoimmune diseases. Ergothioneine (EGT) is a natural amino acid derivative derived from the ergot fungus and has been reported to exhibit an effective antioxidant function in many models of oxidative stress-related diseases. Recently, mutations in OCTN1, a membrane transporter of EGT, have been reported to be associated with rheumatoid arthritis. Therefore, we investigated the chondrocyte-protective function of EGT using a model of oxidative stress-induced injury of chondrocytes by hydrogen peroxide (H_2O_2). Human chondrocytes were subjected to oxidative stress induced by H_2O_2 treatment, and cell viability, the activity of lactate dehydrogenase (LDH) released into the medium, dead cell ratio, intracellular ROS production, and mitochondrial morphology were assessed. EGT improved chondrocyte viability and LDH activity in the medium and strongly suppressed the dead cell ratio. EGT also exerted protective effects on intracellular ROS production and mitochondrial morphology. These results provide evidence to support the protective effects of EGT on chondrocytes induced by oxidative stress.

Keywords: reactive oxygen species; ergothioneine; chondrocyte protection; rheumatoid arthritis; oxidative stress

Citation: Sakata, S.; Kunimatsu, R.; Tanimoto, K. Protective Effect of Ergothioneine against Oxidative Stress-Induced Chondrocyte Death. *Antioxidants* **2024**, *13*, 800. https://doi.org/10.3390/antiox13070800

Academic Editors: Bin Du and Baojun Xu

Received: 4 June 2024
Revised: 21 June 2024
Accepted: 25 June 2024
Published: 1 July 2024

1. Introduction

Reactive oxygen species (ROS) are a group of highly oxidatively active molecules that are produced during oxidative phosphorylation in mitochondria, and ROS are associated with the various physiological functions of cells [1]. ROS can be produced by many factors such as ultraviolet (UV) light, radiation, smoking, aging, food, and inflammation. Cells have ROS scavenging systems such as superoxide dismutase, catalase, and glutathione peroxidase, which maintain the homeostasis of intracellular ROS concentrations. However, when the production of ROS exceeds the antioxidant capacity of the cell and the homeostasis of intracellular ROS concentration is disrupted, intracellular lipids, enzymes, and nucleic acids are peroxidized, disrupting the maintenance of vital functions [2]. This condition is called oxidative stress, and oxidative stress caused by the excessive accumulation of ROS has been reported to be involved in many diseases, including cancer [3], hypertension [4], atherosclerosis [5], neurodegenerative diseases [6], diabetes [7], and autoimmune diseases [8].

Recently, oxidative stress has been reported to be involved in rheumatoid arthritis (RA) and osteoarthritis (OA) [9,10]. These diseases are caused by the chronic inflammation of the synovial membrane, which leads to the degeneration of bone, cartilage, and other joint tissues [11]. The excessive accumulation of ROS in chondrocytes induces mitochondrial dysfunction associated with membrane lipid peroxidation, leading to chondrocyte death [12]. Since articular cartilage has low proliferative capacity and lacks the ability to self-repair, chondrocyte death is considered an important factor in the pathological progression of joint-destructive diseases [13]. Therefore, the reduction of excess intracellular ROS accumulation and the protection of cellular physiological functions by supplementation with antioxidants could be an effective therapeutic strategy against oxidative stress-related diseases. The standard treatment for OA and RA is the oral administration of anti-rheumatic drugs, non-steroidal anti-inflammatory drugs, and steroids, or the intra-articular injection of hyaluronic acids; however, these treatments have certain side effects [14,15]. Therefore, the development of a safe supplement therapy that complements the patient's antioxidant capacity through the intake of dietary antioxidants is necessary [16].

Ergothioneine (EGT) is a unique amino acid derivative derived from the ergot fungus and has a strong antioxidant capacity and high chemical stability [17]. The potential efficacy of EGT as a therapeutic agent is becoming clear, with many reports that EGT is effective in neurodegenerative diseases such as Parkinson's disease caused by neuronal oxidative stress [18,19]. Recently, OCTN1, a plasma membrane transporter of EGT, has been reported to be associated with RA [20]. These reports suggest that investigating the effects of EGT on oxidative stress in chondrocytes is important for establishing new therapeutic strategies for OA and RA. Therefore, this study aimed to examine the potential effectiveness of EGT for degenerative diseases of articular cartilage by evaluating the protective effect of EGT against the cytotoxicity induced by oxidative stress.

2. Materials and Methods

2.1. Cell Culture

Primary normal human knee articular chondrocytes (NHAC-Kn) were obtained from Lonza (Walkersville, MD, USA). Chondrocytes were cultured in chondrocyte basal medium (Lonza) supplemented with 5% fetal bovine serum (FBS), 0.2% R3-insulin-like-growth factor-1, 0.5% human recombinant fibroblast growth factor-beta, 1% transferrin, 0.1% insulin, and 0.1% gentamicin/amphotericin-B according to the manufacturer's instructions. NHAC-Kn cells were maintained in a 37 °C humidified atmosphere of 5% CO_2 and used for all experiments at passage 5.

2.2. Quantitative Polymerase Chain Reaction (qPCR)

Total RNA was extracted from chondrocytes treated with 1 mM EGT for 24 h using an RNeasy Mini Kit (Qiagen, Valencia, CA, USA) according to the manufacturer's instructions. The control group included chondrocytes treated with phosphate-buffered saline (PBS), the vehicle for EGT. Complementary DNA (cDNA) was synthesized from 1 µg of total RNA using a ReverTra Ace qPCR Master Mix (Toyobo, Osaka, Japan) and SimpliAmp Thermal Cycler (Thermo Fisher Scientific, Waltham, MA, USA). qPCR was performed using a Real-time PCR Master Mix (Toyobo) and a Thermal Cycler Dice Real Time System (Takara, Shiga, Japan). The following primers were used for qPCR: *collagen type 2 alpha 1 (COL2A1)*, forward 5′-GCTCCCAGAACATCACCTACCA-3′, reverse 5′-AACAGTCTTGCCCCACTTACCG-3′ and actin beta (*ACTB*); forward 5′-AGAGCTACGAGCTGCCTGAC-3′ and reverse 5′-AGCACTGTGTTGGCGTACAG-3′. *COL2A1* mRNA levels were normalized to *ACTB* mRNA levels, and the relative gene expression of *COL2A1* was calculated with the $2^{-\Delta\Delta CT}$ method.

2.3. Cell Viability and Cytotoxicity Analysis

Cell viability was examined using a Cell Counting Kit-8 (CCK-8, Dojindo, Kumamoto, Japan). The chondrocytes were seeded with 5000 cells in a 96-well plate, and after 24 h incubation, the cells were treated with EGT (0, 0.1, 0.25, 0.5, and 1 mM) or hydrogen peroxide (H_2O_2; 0, 0.1, 0.25, 0.5, and 1 mM). After 24 h incubation, CCK-8 solution was added to the medium for 1 h at 37 °C. The absorbance at 450 nm was measured by using a microplate reader Multiskan FC (Themo scientific, Waltham, MA, USA).

Cell cytotoxicity was examined using a Cytotoxicity LDH Assay Kit-WST (Dojindo). Chondrocytes were seeded with 10,000 cells in a 24-well plate, and after 24 h incubation, the cells were treated with H_2O_2 (1 mM) or with the addition of EGT (0.1, 0.25, 0.5, and 1 mM). After an additional 24 h incubation, the activity of lactate dehydrogenase (LDH) in the supernatant was measured following the manufacturer's instructions. Cell morphology and confluency were analyzed by using CELLCYTE X (CYTENA, Freiburg, Germany).

2.4. Live/Dead-Cell Assay

To detect H_2O_2-induced chondrocyte death, Diyo-1 (AAT Bioquest, Sunnyvale, CA, USA) and SYTO-59 (Invitrogen, Carlsbad, CA, USA) double-staining was performed. Diyo-1 is a membrane-impermeable nuclear-staining dye that penetrates dead cell membranes and emits green fluorescence. By contrast, SYTO-59 is a membrane-permeable nuclear-staining dye that can penetrate the cell membrane of both dead and living cells and emits red fluorescence. Therefore, the ratio of dead cells to total cells can be calculated by dividing the number of green fluorescence-stained cells by the number of red fluorescence-stained cells. These fluorescent dyes were dissolved at a concentration of 1 μM each in chondrocyte basal medium containing FBS and supplements. Chondrocytes were seeded with 10,000 cells in a 24-well plate, and after 24 h incubation, the cells were treated with H_2O_2 (1 mM) or with the addition of EGT (1 mM) at 37 °C for 3, 6, and 12 h. The control group was treated with PBS. Following treatments, fluorescence images were obtained by using CELLCYTE X (CYTENA).

2.5. ROS Assay

Intracellular ROS production was evaluated using dichloro-dihydro-fluorescein diacetate (DCFH-DA) dye (ROS Assay Kit Photo-oxidation Resistant DCFH-DA, Dojindo). The chondrocytes were treated with H_2O_2 (1 mM) and EGT (1 mM) at 37 °C for 3, 6, and 12 h in the presence of DCFH-DA dye according to the manufacturer's instructions. The control group was treated with PBS. Live-cell fluorescence images were obtained and analyzed by using CELLCYTE X (CYTENA).

2.6. Mitochondrial Staining

Mitochondrial visualization was performed using MitoView Green (Biotium, San Francisco, CA, USA). This probe is independent of mitochondrial membrane potential and fluoresces when fractionated into the mitochondrial membrane. Therefore, mitochondrial mass can be evaluated. Chondrocytes were treated with H_2O_2 (1 mM) and EGT (1 mM) at 37 °C for 2 h in the presence of MitoView Green (200 nM), and fluorescence images were obtained using CELLCYTE X (CYTENA).

2.7. Statistical Analysis

The results are shown as mean ± standard deviation (SD). Statistical analysis was performed using JMP Pro version 17 software (JMP Statistical Discovery LlC, Cary, NC, USA) and statistically significant differences between the groups were determined using one-way analysis of variance followed by Dunnett's test or Tukey's test. The statistical significance was set at $p < 0.05$.

3. Results

3.1. Effect of EGT and H_2O_2 on Collagen Gene Expression and Cell Viability

An increase in the gene expression of COL2A1 in chondrocytes treated with EGT was observed compared to untreated chondrocytes (Figure 1B). When treating the chondrocytes with 0.1, 0.25, 0.5, and 1 mM of EGT for 24 h, no significant difference was observed (Figure 1C). On the other hand, on application of 0.1, 0.25, 0.5, and 1 mM of H_2O_2 to chondrocytes for 24 h, a significant decrease in cell viability of approximately 20% was observed at 0.5 mM H_2O_2, and a significant decrease of approximately 95% was observed at 1 mM (Figure 1D).

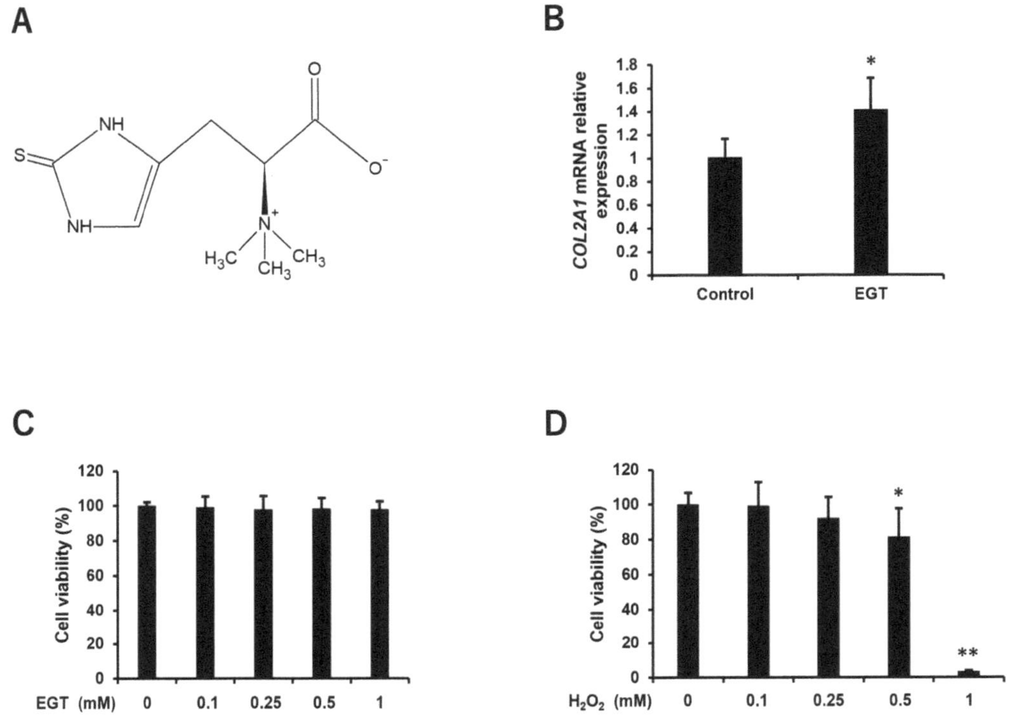

Figure 1. Effect of EGT and H_2O_2 on cell viability of human chondrocytes. (**A**) Structural formula of EGT. (**B**) Relative gene expression of *COL2A1* in chondrocytes treated with EGT (1 mM) for 24 h. *COL2A1* mRNA levels were normalized to ACTB mRNA levels. The values are presented as mean ± SD (n = 4). (**C**) Cell viability of chondrocytes treated with EGT at various concentrations (0–1 mM) for 24 h. (**D**) Cell viability of chondrocytes treated with H_2O_2 at various concentrations (0–1 mM) for 24 h. Cell viability was determined by CCK-8 assay. The values of cell viability are presented as mean ± SD (n = 8). * $p < 0.05$, ** $p < 0.01$ vs. control group. EGT, ergothioneine; H_2O_2, hydrogen peroxide; *COL2A1*, collagen type 2 alpha 1; ACTB, actin beta; CCK-8, Cell Counting Kit-8; SD, standard deviation.

3.2. EGT Ameliorates H_2O_2-Induced Chondrocyte Damage

In comparison, when treating the chondrocytes with 0.1, 0.25, 0.5, and 1 mM of EGT in addition to 1 mM H_2O_2, the addition of 0.5 and 1 mM of EGT improved the morphological changes in chondrocytes and decreased the viability caused by H_2O_2 (Figure 2A,B). Similarly, the activity of extracellularly released LDH, which was increased by H_2O_2 treatment, was significantly decreased by the addition of 0.5 and 1 mM of EGT (Figure 2C).

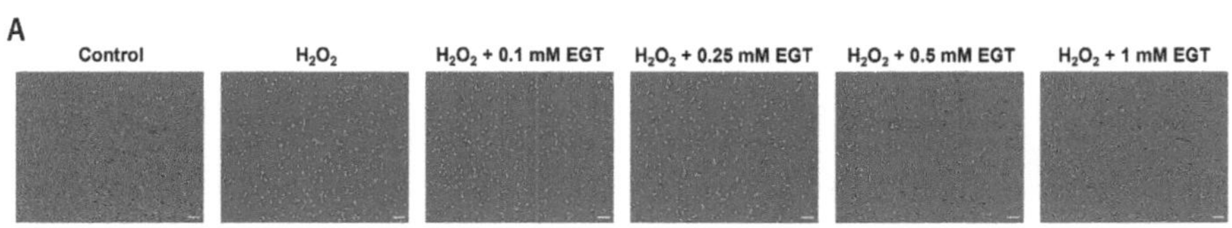

Figure 2. *Cont.*

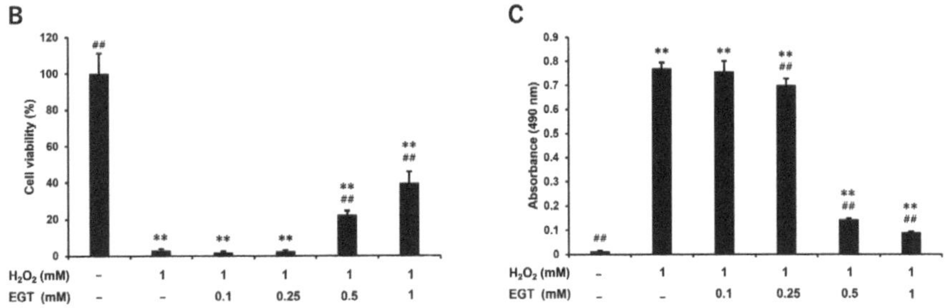

Figure 2. EGT ameliorates H_2O_2-induced cell damage in chondrocytes. (**A**) Morphology of chondrocytes treated with H_2O_2 (1 mM) and EGT (0–1 mM) for 24 h (scale bar: 100 μm). (**B**) Cell viability quantified by CCK-8 assay. Values are presented as mean ± SD (n = 6). (**C**) LDH activity of supernatant of chondrocytes treated with H_2O_2 (1 mM) and EGT (0–1 mM) for 24 h. LDH activity was measured using Cytotoxicity LDH Assay Kit-WST. Values are presented as mean ± SD (n = 6). ** $p < 0.01$ vs. control group; ## $p < 0.01$ vs. H_2O_2-treated group. EGT, ergothioneine; H_2O_2, hydrogen peroxide; CCK-8, Cell Counting Kit-8; LDH, lactate dehydrogenase; SD, standard deviation.

3.3. Inhibitory Effect of EGT on H_2O_2-Induced Chondrocyte Death

The essential mechanism of cell death by treatment with H_2O_2 is the peroxidation of polyunsaturated fatty acids by excess intracellular ROS accumulation and subsequent disruption of cell membrane continuity [21,22]. Therefore, the double-staining assay for live and dead cells using the difference in cell membrane permeability of nuclear-staining dyes was performed. No dead cells were observed at 3 h after the addition of 1 mM H_2O_2 and 1 mM EGT, but a significant increase in the dead cell ratio was observed in the H_2O_2-treated group at 6 h after treatment (Figure 3A,B). After 12 h of treatment, most of the cells were identified as dead cells in the H_2O_2-treated group. On the other hand, in the H_2O_2 + EGT-treated group, these cell deaths were strongly inhibited and no significant increase in the dead cell ratio was observed at 12 h after treatment. No significant change in the dead cell ratio was observed in the group treated with EGT compared to the control group.

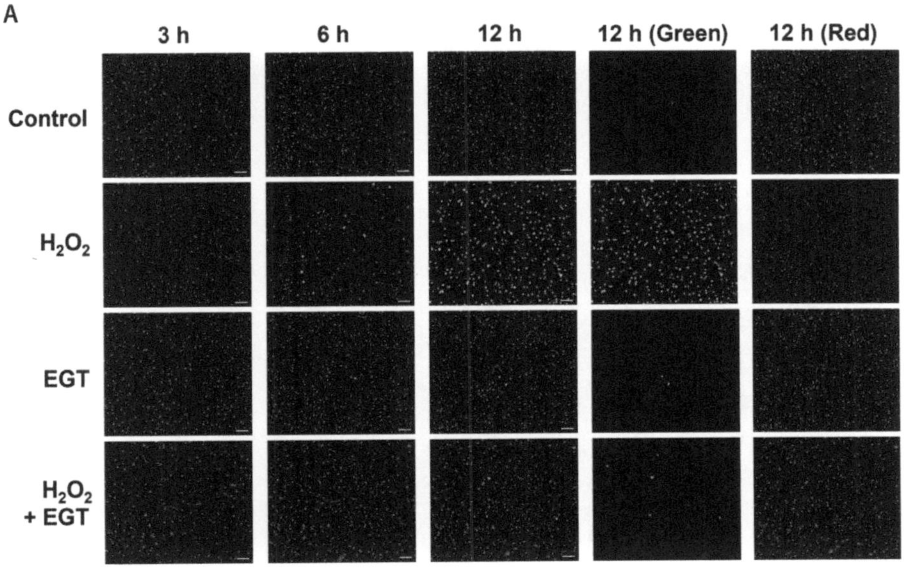

Figure 3. *Cont.*

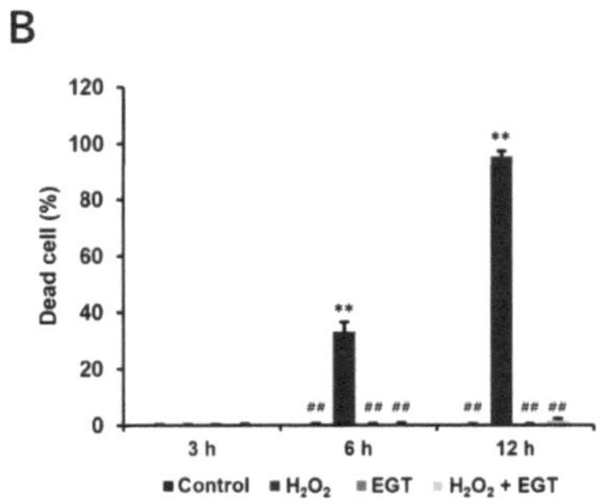

Figure 3. EGT inhibits H_2O_2-induced cell death in chondrocytes. (**A**) Fluorescence image of chondrocytes treated with H_2O_2 (1 mM) and EGT (1 mM) for 3, 6, and 12 h (scale bar: 100 μm). Dead cells (green) and total cells (red) were double-stained with Diyo-1 and SYTO59 nuclear staining dyes. (**B**) Quantitative data on the ratio of dead cells to total cells calculated by double-staining. Values are presented as mean ± SD (n = 3). ** $p < 0.01$ vs. control group; ## $p < 0.01$ vs. H_2O_2-treated group. EGT, ergothioneine; H_2O_2, hydrogen peroxide; SD, standard deviation.

3.4. *Effect of EGT on ROS Production in H_2O_2-Treated Chondrocytes*

Cells treated with H_2O_2 have increased intracellular ROS concentrations. Therefore, intracellular ROS accumulation was monitored using DCFH-DA dye, a fluorescent probe for intracellular ROS. Consequently, significant increases in intracellular ROS concentrations were observed at 3, 6, and 12 h after H_2O_2 treatment, but the addition of EGT significantly suppressed these increases (Figure 4A,B). No significant changes in intracellular ROS concentrations were observed after treatment with EGT. No changes in cell counts were observed (Figure 4C).

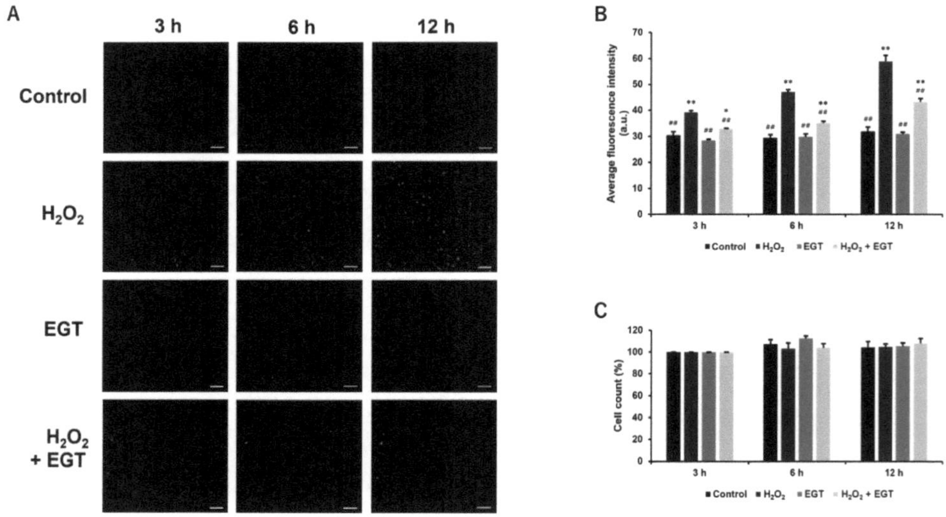

Figure 4. EGT inhibits H_2O_2-induced intracellular ROS production in chondrocytes. (**A**) Fluorescence images of chondrocytes treated with H_2O_2 (1 mM) and EGT (1 mM) for 3, 6, and 12 h in the presence of DCFH-DA dye (scale bar: 100 μm). (**B**) Quantitative data on the average fluorescence intensity of chondrocytes obtained from fluorescence images. Values are presented as mean ± SD (n = 3). (**C**) Changes in cell count over time. The values are expressed as 100% of the cell count in each group

at 3 h. * $p < 0.05$, ** $p < 0.01$ vs. control group; ## $p < 0.01$ vs. H_2O_2-treated group. EGT, ergothioneine; H_2O_2, hydrogen peroxide; ROS, reactive oxygen species; SD, standard deviation; DCFH-DA, dichloro-dihydro-fluorescein diacetate.

3.5. Effect of EGT on Mitochondrial Morphology of H_2O_2-Treated Chondrocytes

Cell death due to oxidative stress causes mitochondrial dysfunction in its early stages [23]. Therefore, mitochondrial staining using MitoView Green dye was performed to observe mitochondrial morphology in the pre-membrane disruption phase caused by H_2O_2. In chondrocytes treated with H_2O_2 for 2 h, mitochondrial condensation was observed prior to plasma membrane damage, with a significant increase in fluorescence intensity (Figure 5A,B). However, no significant change in fluorescence intensity was observed in the H_2O_2 + EGT- and EGT-treated groups compared to the control group.

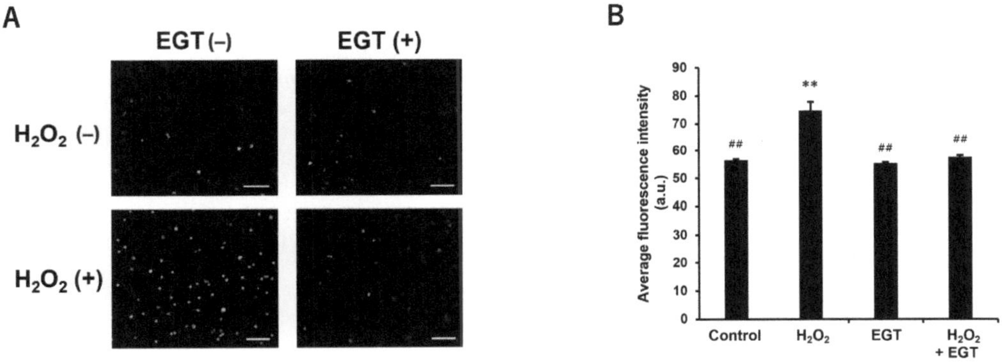

Figure 5. EGT inhibits H_2O_2-induced mitochondrial condensation in chondrocytes. (**A**) Fluorescence images of chondrocytes treated with H_2O_2 (1 mM) and EGT (1 mM) for 2 h in the presence of MitoView Green dye (scale bar: 100 μm). (**B**) Quantitative data on the average fluorescence intensity of chondrocytes obtained from fluorescence images. Values are presented as mean ± SD (n = 3). ** $p < 0.01$ vs. control group; ## $p < 0.01$ vs. H_2O_2-treated group. EGT, ergothioneine; H_2O_2, hydrogen peroxide; SD, standard deviation.

4. Discussion

EGT is a naturally occurring amino acid derivative with safe and potent antioxidant properties that may be a potential therapeutic agent for oxidative stress-related diseases. In this study, to examine the chondrocyte-protective effects of EGT, an oxidative stress-induced chondrocyte injury model was generated using H_2O_2. The results showed that 1 mM EGT potently inhibited H_2O_2-induced chondrocyte death, demonstrating its efficacy in a model of oxidative stress-induced cartilage damage in vitro.

Firstly, the cytotoxicity of EGT on chondrocytes was investigated. The results of the CCK-8 assay showed that EGT has no cytotoxicity to chondrocytes up to a concentration of 1 mM. Previous studies have reported no cytotoxicity after 24 h of exposure to 1 mM EGT [24–26], which is consistent with our results. However, some reports have suggested that even concentrations not reaching 1 mM may be cytotoxic [27,28], possibly due to differences in the expression of the EGT transporter (OCTN1) in different cell types. OCTN1 has been found to be widely distributed in many mammalian organs and cells [29]. Accordingly, EGT, the substrate of OCTN1, is absorbed from the small intestine and then transported to all tissues in the body and retained for a long period of time [30]. Therefore, when administrating EGT to the human body, the appropriate EGT-application concentration should be carefully considered.

The protective effect of EGT on chondrocytes against H_2O_2-induced oxidative stress was evaluated using the LDH-release assay and the cell nuclear double-staining assay. In this study, chondrocyte damage caused by 1 mM H_2O_2 was inhibited in a dose-dependent manner by the addition of EGT. In addition, 1 mM EGT markedly inhibited chondrocyte death.

H_2O_2 is most commonly used as an oxidant for oxidative stress-induced cellular damage [31–34]. Several studies have reported on the protective effects of EGT against H_2O_2-induced oxidative stress, including reports on phaeochromocytoma [35] and hippocampal neural cells [36]. Similar to these studies, the results of this study showed that EGT had a cytoprotective effect against H_2O_2-induced oxidative stress.

On monitoring intracellular ROS using a fluorescence probe, EGT significantly inhibited the H_2O_2-induced increase in intracellular ROS concentration. This is similar to previous studies [37]. EGT did not reduce intracellular ROS concentration exclusively.

H_2O_2, a metabolite of superoxide, is produced mainly by leukocytes and macrophages and has harmful effects on normal tissue in autoimmune diseases such as RA [38]. Under physiological conditions, the extracellular concentration of H_2O_2, the most stable ROS, is up to approximately 10 μM, whereas under pathological conditions, the concentration of ROS reaches as high as 1 mM, nearly 100 times higher than that in natural conditions [39,40].

ROS are not only harmful to the organism but are also involved in physiological activities [41]. EGT reacts strongly with free radicals, including hydroxyl radicals, whereas it is less directly reactive with H_2O_2 [42,43]. This property implies that only the most dangerous hydroxyl radicals can be scavenged without excessively scavenging the ROS necessary for physiological activity. In the present study, EGT treatment alone was able to potently inhibit cell death without showing cytotoxicity, which may be due to this property. Further, this property may be responsible for EGT not reducing the intracellular ROS production exclusively.

Finally, mitochondrial morphology was observed using MitoView Green. Hydroxyl radicals produced in mitochondria peroxidize mitochondrial membranes, leading to mitochondrial dysfunction [44,45]. In the present study, EGT suppressed the changes in mitochondrial morphology caused by H_2O_2. This indicates that EGT protects mitochondria from oxidative stress damage caused by H_2O_2.

Several studies have tested the efficacy of antioxidant therapy for RA [46–48]. EGT, similar to these antioxidants, is possibly safer because it does not violate the body's natural ROS homeostasis, which is expected to be effective against RA. Further, drug therapy for RA may elevate oxidative stress in patients [49,50]. EGT is also expected to serve as an adjunctive supplement to reduce these side effects of conventional drug therapy.

The present study investigated H_2O_2-induced chondrocyte death, and the findings are consistent with previous findings conducted with the naturally occurring antioxidants vitamin C and vitamin E [51,52]. Compared to these antioxidants, however, EGT has pharmacokinetic properties more suitable for supplement therapy as its blood levels are maintained for long periods of time after a single administration. Hence, EGT could be a better antioxidant supplement.

However, the future application of antioxidant therapy with EGT for RA requires the further evaluation of the efficacy and safety of EGT in vivo. In addition, since EGT is transported to organs throughout the body via the ubiquitously expressed OCTN1, its systemic effects must also be carefully investigated.

This study has a few limitations. First, the results of this study did not directly demonstrate efficacy against RA. Further studies in disease models using rheumatoid cells from patients are needed to further evaluate efficacy. Furthermore, the pathogenesis of RA is complex, and, in addition to oxidative stress, many factors contribute, including chronic inflammatory conditions and cartilage matrix destruction. Therefore, further detailed studies using cellular and animal models of inflammation and matrix destruction are needed to explore the potential therapeutic effects of EGT for RA.

5. Conclusions

In this study, EGT potently inhibited chondrocyte death induced by H_2O_2 without harming chondrocytes. EGT also improved ROS production and the mitochondrial morphological changes caused by H_2O_2. This result suggests the efficacy of EGT in oxidative stress-induced chondrocyte damage. Thus, EGT may be a candidate therapeutic agent for degenerative diseases caused by oxidative stress.

Author Contributions: Conceptualization, S.S. and R.K.; Methodology, S.S.; Software, S.S.; Validation, S.S. and R.K.; Formal analysis, S.S.; Investigation, S.S. and R.K.; Resources, S.S., R.K. and K.T.; Data curation, S.S.; Writing—original draft, S.S.; Writing—review and editing, R.K. and K.T.; Visualization, S.S.; Supervision, K.T.; Project administration, R.K. and K.T.; Funding acquisition, S.S. All authors have read and agreed to the published version of the manuscript.

Funding: This work was funded by grants from the JSPS KAKENHI (grant number: 23K16203).

Institutional Review Board Statement: Not applicable.

Informed Consent Statement: Not applicable.

Data Availability Statement: The data supporting the findings of this study are available from the corresponding author upon reasonable request.

Conflicts of Interest: The authors declare no conflicts of interest.

References

1. Sies, H.; Jones, D.P. Reactive oxygen species (ROS) as pleiotropic physiological signalling agents. *Nat. Rev. Mol. Cell Biol.* **2020**, *21*, 363–383. [CrossRef]
2. Forman, H.J.; Zhang, H. Targeting oxidative stress in disease: Promise and limitations of antioxidant therapy. *Nat. Rev. Drug Discov.* **2021**, *20*, 689–709. [CrossRef] [PubMed]
3. Hayes, J.D.; Dinkova-Kostova, A.T.; Tew, K.D. Oxidative stress in cancer. *Cancer Cell* **2020**, *38*, 167–197. [CrossRef] [PubMed]
4. Touyz, R.M.; Rios, F.J.; Alves-Lopes, R.; Neves, K.B.; Camargo, L.L.; Montezano, A.C. Oxidative stress: A unifying paradigm in hypertension. *Can. J. Cardiol.* **2020**, *36*, 659–670. [CrossRef]
5. Yang, X.; Li, Y.; Li, Y.; Ren, X.; Zhang, X.; Hu, D.; Gao, Y.; Xing, Y.; Shang, H. Oxidative stress-mediated atherosclerosis: Mechanisms and therapies. *Front. Physiol.* **2017**, *8*, 600. [CrossRef]
6. Butterfield, D.A.; Halliwell, B. Oxidative stress, dysfunctional glucose metabolism and Alzheimer disease. *Nat. Rev. Neurosci.* **2019**, *20*, 148–160. [CrossRef]
7. Gray, S.P.; Di Marco, E.; Okabe, J.; Szyndralewiez, C.; Heitz, F.; Montezano, A.C.; de Haan, J.B.; Koulis, C.; El-Osta, A.; Andrews, K.L.; et al. NADPH oxidase 1 plays a key role in diabetes mellitus–Accelerated atherosclerosis. *Circulation* **2013**, *127*, 1888–1902. [CrossRef]
8. Smallwood, M.J.; Nissim, A.; Knight, A.R.; Whiteman, M.; Haigh, R.; Winyard, P.G. Oxidative stress in autoimmune rheumatic diseases. *Free. Radic. Biol. Med.* **2018**, *125*, 3–14. [CrossRef]
9. Zamudio-Cuevas, Y.; Martínez-Flores, K.; Martínez-Nava, G.A.; Clavijo-Cornejo, D.; Fernández-Torres, J.; Sánchez-Sánchez, R. Rheumatoid arthritis and oxidative stress. *Cell. Mol. Biol.* **2022**, *68*, 174–184. [CrossRef]
10. Lepetsos, P.; Papavassiliou, A.G. ROS/oxidative stress signaling in osteoarthritis. *Biochim. Et Biophys. Acta* **2016**, *1862*, 576–591. [CrossRef]
11. Loeser, R.F.; Goldring, S.R.; Scanzello, C.R.; Goldring, M.B. Osteoarthritis: A disease of the joint as an organ. *Arthritis Rheum.* **2012**, *64*, 1697–1707. [CrossRef] [PubMed]
12. Henrotin, Y.E.; Bruckner, P.; Pujol, J.-P. The role of reactive oxygen species in homeostasis and degradation of cartilage. *Osteoarthr. Cartil.* **2003**, *11*, 747–755. [CrossRef] [PubMed]
13. Salucci, S.; Falcieri, E.; Battistelli, M. Chondrocyte death involvement in osteoarthritis. *Cell Tissue Res.* **2022**, *389*, 159–170. [CrossRef] [PubMed]
14. Wang, W.; Zhou, H.; Liu, L. Side effects of methotrexate therapy for rheumatoid arthritis: A systematic review. *Eur. J. Med. Chem.* **2018**, *158*, 502–516. [CrossRef] [PubMed]
15. Solomon, D.H.; Husni, M.E.; Wolski, K.E.; Wisniewski, L.M.; Borer, J.S.; Graham, D.Y.; Libby, P.; Lincoff, A.M.; Lüscher, T.F.; Menon, V.; et al. Differences in safety of non-steroidal antiinflammatory drugs in patients with osteoarthritis and patients with rheumatoid arthritis: A randomized clinical trial. *Arthritis Rheumatol.* **2018**, *70*, 537–546. [CrossRef]
16. Li, Y.; Schellhorn, H.E. New developments and novel therapeutic perspectives for vitamin C. *J. Nutr.* **2007**, *137*, 2171–2184. [CrossRef]
17. Cheah, I.K.; Halliwell, B. Ergothioneine; antioxidant potential, physiological function and role in disease. *Biochim. Biophys. Acta* **2012**, *1822*, 784–793. [CrossRef] [PubMed]

18. Song, T.-Y.; Chen, C.-L.; Liao, J.-W.; Ou, H.-C.; Tsai, M.-S. Ergothioneine protects against neuronal injury induced by cisplatin both in vitro and in vivo. *Food Chem. Toxicol.* **2010**, *48*, 3492–3499. [CrossRef]
19. Yang, N.C.; Lin, H.C.; Wu, J.H.; Ou, H.C.; Chai, Y.C.; Tseng, C.Y.; Liao, J.W.; Song, T.Y. Ergothioneine protects against neuronal injury induced by beta-amyloid in mice. *Food Chem. Toxicol.* **2012**, *50*, 3902–3911. [CrossRef]
20. Tokuhiro, S.; Yamada, R.; Chang, X.; Suzuki, A.; Kochi, Y.; Sawada, T.; Suzuki, M.; Nagasaki, M.; Ohtsuki, M.; Ono, M.; et al. An intronic SNP in a RUNX1 binding site of SLC22A4, encoding an organic cation transporter, is associated with rheumatoid arthritis. *Nat. Genet.* **2003**, *35*, 341–348. [CrossRef]
21. Stockwell, B.R.; Angeli, J.P.F.; Bayir, H.; Bush, A.I.; Conrad, M.; Dixon, S.J.; Fulda, S.; Gascón, S.; Hatzios, S.K.; Kagan, V.E.; et al. Ferroptosis: A regulated cell death nexus linking metabolism, redox biology, and disease. *Cell* **2017**, *171*, 273–285. [CrossRef] [PubMed]
22. Dixon, S.J.; Lemberg, K.M.; Lamprecht, M.R.; Skouta, R.; Zaitsev, E.M.; Gleason, C.E.; Patel, D.N.; Bauer, A.J.; Cantley, A.M.; Yang, W.S.; et al. Ferroptosis: An iron-dependent form of nonapoptotic cell death. *Cell* **2012**, *149*, 1060–1072. [CrossRef] [PubMed]
23. Park, C.; Lee, H.; Noh, J.S.; Jin, C.Y.; Kim, G.Y.; Hyun, J.W.; Leem, S.H.; Choi, Y.H. Hemistepsin A protects human keratinocytes against hy-drogen peroxide-induced oxidative stress through activation of the Nrf2/HO-1 signaling pathway. *Arch. Biochem. Biophys* **2020**, *691*, 108512. [CrossRef] [PubMed]
24. D'Onofrio, N.; Martino, E.; Balestrieri, A.; Mele, L.; Cautela, D.; Castaldo, D.; Balestrieri, M.L. Diet-derived ergothioneine induces necroptosis in colorectal cancer cells by activating the SIRT3/MLKL pathway. *FEBS Lett.* **2022**, *596*, 1313–1329. [CrossRef] [PubMed]
25. D'onofrio, N.; Servillo, L.; Giovane, A.; Casale, R.; Vitiello, M.; Marfella, R.; Paolisso, G.; Balestrieri, M.L. Ergothioneine oxidation in the protection against high-glucose induced endothelial senescence: Involvement of SIRT1 and SIRT6. *Free. Radic. Biol. Med.* **2016**, *96*, 211–222. [CrossRef] [PubMed]
26. Ko, H.J.; Kim, J.; Ahn, M.; Kim, J.H.; Lee, G.S.; Shin, T. Ergothioneine alleviates senescence of fibroblasts induced by UVB damage of keratinocytes via activation of the Nrf2/HO-1 pathway and HSP70 in keratinocytes. *Exp. Cell Res.* **2021**, *400*, 112516. [CrossRef] [PubMed]
27. Hseu, Y.C.; Lo, H.W.; Korivi, M.; Tsai, Y.C.; Tang, M.J.; Yang, H.L. Dermato-protective properties of ergothioneine through induction of Nrf2/ARE-mediated antioxidant genes in UVA-irradiated Human keratinocytes. *Free Radic. Biol. Med.* **2015**, *86*, 102–117. [CrossRef]
28. Wang, Z.; Ma, J.; Miao, Z.; Sun, Y.; Dong, M.; Lin, Y.; Wu, Y.; Sun, Z. Ergothioneine inhibits the progression of osteoarthritis via the Sirt6/NF-κB axis both in vitro and in vivo. *Int. Immunopharmacol.* **2023**, *119*, 110211. [CrossRef]
29. Kato, Y.; Kubo, Y.; Iwata, D.; Kato, S.; Sudo, T.; Sugiura, T.; Kagaya, T.; Wakayama, T.; Hirayama, A.; Sugimoto, M.; et al. Gene knockout and metabolome analysis of carnitine/organic cation transporter OCTN1. *Pharm. Res.* **2010**, *27*, 832–840. [CrossRef]
30. Paul, B.D. Ergothioneine: A stress vitamin with antiaging, vascular, and neuroprotective roles? *Antioxid Redox Signal* **2022**, *36*, 1306–1317. [CrossRef]
31. Wang, L.; Chen, Q.; Zhuang, S.; Wen, Y.; Cheng, W.; Zeng, Z.; Jiang, T.; Tang, C. Effect of Anoectochilus roxburghii flavonoids extract on H_2O_2—Induced oxidative stress in LO2 cells and D-gal induced aging mice model. *J. Ethnopharmacol.* **2020**, *254*, 112670. [CrossRef] [PubMed]
32. Siu, P.M.; Wang, Y.; Alway, S.E. Apoptotic signaling induced by H_2O_2-mediated oxidative stress in differentiated C_2C1_2 myotubes. *Life Sci.* **2009**, *84*, 468–481. [CrossRef] [PubMed]
33. Ou, X.; Lee, M.R.; Huang, X.; Messina-Graham, S.; Broxmeyer, H.E. SIRT1 Positively regulates autophagy and mitochondria function in embryonic stem cells under oxidative stress. *Stem Cells* **2014**, *32*, 1183–1194. [CrossRef] [PubMed]
34. Hübner, A.A.; Demarque, D.P.; Lourenço, F.R.; Rosado, C.; Baby, A.R.; Kikuchi, I.S.; Bacchi, E.M. Phytocompounds recovered from the waste of cabernet sauvignon (*Vitis vinifera* L.) vinification: Cytotoxicity (in normal and stressful conditions) and in vitro photoprotection efficacy in a sunscreen system. *Cosmetics* **2022**, *10*, 2. [CrossRef]
35. Colognato, R.; Laurenza, I.; Fontana, I.; Coppede, F.; Siciliano, G.; Coecke, S.; Aruoma, O.; Benzi, L.; Migliore, L. Modulation of hydrogen peroxide-induced DNA damage, MAPKs activation and cell death in PC12 by ergothioneine. *Clin. Nutr.* **2006**, *25*, 135–145. [CrossRef]
36. Kushairi, N.; Phan, C.W.; Sabaratnam, V.; Naidu, M.; David, P. Dietary amino acid ergothioneine protects HT22 hippocampal neurons against H_2O_2-induced neurotoxicity via antioxidative mechanism. *PharmaNutrition* **2020**, *13*, 100214. [CrossRef]
37. Servillo, L.; D'Onofrio, N.; Casale, R.; Cautela, D.; Giovane, A.; Castaldo, D.; Balestrieri, M.L. Ergothioneine products derived by super-oxide oxidation in endothelial cells exposed to high-glucose. *Free Radic Biol. Med.* **2017**, *108*, 8–18. [CrossRef] [PubMed]
38. Mittal, M.; Siddiqui, M.R.; Tran, K.; Reddy, S.P.; Malik, A.B. Reactive oxygen species in inflammation and tissue injury. *Antioxid. Redox Signal.* **2014**, *20*, 1126–1167. [CrossRef] [PubMed]
39. Kumar, R.; Han, J.; Lim, H.-J.; Ren, W.X.; Lim, J.-Y.; Kim, J.S. Mitochondrial induced and self-monitored intrinsic apoptosis by antitumor theranostic prodrug: In vivo imaging and precise cancer treatment. *J. Am. Chem. Soc.* **2014**, *136*, 17836–17843. [CrossRef]
40. Cadahía, J.P.; Bondebjerg, J.; Hansen, C.A.; Previtali, V.; Hansen, A.E.; Andresen, T.L.; Clausen, M.H. Synthesis and evaluation of hydrogen peroxide sensitive prodrugs of methotrexate and aminopterin for the treatment of rheumatoid arthritis. *J. Med. Chem.* **2018**, *61*, 3503–3515. [CrossRef]

41. Lennicke, C.; Cochemé, H.M. Redox metabolism: ROS as specific molecular regulators of cell signaling and function. *Mol. Cell* **2021**, *81*, 3691–3707. [CrossRef]
42. Franzoni, F.; Colognato, R.; Galetta, F.; Laurenza, I.; Barsotti, M.; Di Stefano, R.; Bocchetti, R.; Regoli, F.; Carpi, A.; Balbarini, A.; et al. An in vitro study on the free radical scavenging capacity of ergothioneine: Comparison with reduced glutathione, uric acid and trolox. *Biomed. Pharmacother.* **2006**, *60*, 453–457. [CrossRef] [PubMed]
43. Akanmu, D.; Cecchini, R.; Aruoma, O.I.; Halliwell, B. The antioxidant action of ergothioneine. *Arch. Biochem. Biophys.* **1991**, *288*, 10–16. [CrossRef]
44. Takashi, Y.; Tomita, K.; Kuwahara, Y.; Roudkenar, M.H.; Roushandeh, A.M.; Igarashi, K.; Nagasawa, T.; Nishitani, Y.; Sato, T. Mitochondrial dysfunction promotes aquaporin expression that controls hydrogen peroxide permeability and ferroptosis. *Free. Radic. Biol. Med.* **2020**, *161*, 60–70. [CrossRef]
45. Ransy, C.; Vaz, C.; Lombès, A.; Bouillaud, F. Use of H_2O_2 to cause oxidative stress, the catalase issue. *Int. J. Mol. Sci.* **2020**, *21*, 9149. [CrossRef] [PubMed]
46. Zeng, L.; Yu, G.; Yang, K.; Li, J.; Hao, W.; Chen, H. The efficacy of antioxidative stress therapy on oxidative stress levels in rheumatoid arthritis: A systematic review and meta-analysis of randomized controlled trials. *Oxidative Med. Cell. Longev.* **2021**, *2021*, 3302886. [CrossRef]
47. Qamar, N.; John, P.; Bhatti, A. Emerging role of selenium in treatment of rheumatoid arthritis: An insight on its antioxidant properties. *J. Trace Elements Med. Biol.* **2021**, *66*, 126737. [CrossRef]
48. Islam, M.T.; Sarkar, C.; Hossain, R.; Bhuia, S.; Mardare, I.; Kulbayeva, M.; Ydyrys, A.; Calina, D.; Habtemariam, S.; Kieliszek, M.; et al. Therapeutic strategies for rheumatic diseases and disorders: Targeting redox imbalance and oxidative stress. *Biomed. Pharmacother.* **2023**, *164*, 114900. [CrossRef] [PubMed]
49. Nawaz, H.; Ali, A.; Rehman, T.; Aslam, A. Chronological effects of non-steroidal anti-inflammatory drug therapy on oxidative stress and antioxidant status in patients with rheumatoid arthritis. *Clin. Rheumatol.* **2021**, *40*, 1767–1778. [CrossRef]
50. Ezhilarasan, D. Hepatotoxic potentials of methotrexate: Understanding the possible toxicological molecular mechanisms. *Toxicology* **2021**, *458*, 152840. [CrossRef]
51. Chang, Z.; Huo, L.; Li, P.; Wu, Y.; Zhang, P. Ascorbic acid provides protection for human chondrocytes against oxidative stress. *Mol. Med. Rep.* **2015**, *12*, 7086–7092. [CrossRef] [PubMed]
52. Bhatti, F.-U.; Mehmood, A.; Wajid, N.; Rauf, M.; Khan, S.N.; Riazuddin, S. Vitamin E protects chondrocytes against hydrogen peroxide-induced oxidative stress in vitro. *Inflamm. Res.* **2013**, *62*, 781–789. [CrossRef] [PubMed]

antioxidants

MDPI

Article

Antioxidant and Stress Resistance Properties of Flavonoids from Chinese Sea Buckthorn Leaves from the Qinghai–Tibet Plateau

Jinmei Zhao, Yumei Jiang, Yang Bi * and Juan Wei *

College of Food Science and Engineering, Gansu Agricultural University, Lanzhou 730070, China; zhaojinmei01085@st.gsau.edu.cn (J.Z.); jym316@gsau.edu.cn (Y.J.)
* Correspondence: biyang@gsau.edu.cn (Y.B.); weijuan@gsau.edu.cn (J.W.)

Abstract: The unique ecological environment of the Qinghai–Tibetan Plateau has endowed Chinese sea buckthorn leaves with rich bioactivities. In this study, we investigated the bioactivity and stress resistance mechanisms of flavonoids derived from Chinese sea buckthorn leaves (FCL) native to the Qinghai–Tibet Plateau. Our analysis identified a total of 57 flavonoids, mainly flavonol glycosides, from FCL, of which 6 were novel flavonoids. Isorhamnetin glycosides, quercetin glycosides and kaempferol glycosides were the three most dominant classes of compounds in FCL. In particular, isorhamnetin-3-*O*-glucoside-7-*O*-rhamnoside emerged as the most abundant compound. Our results showed that FCL possesses potent antioxidant properties, as evidenced by its ability to effectively scavenge DPPH free radicals and demonstrate ferric reducing antioxidant power (FRAP) and oxygen radical absorbance capacity (ORAC) levels comparable to Trolox, a well-known antioxidant standard. Furthermore, FCL showed remarkable efficacy in reducing reactive oxygen species (ROS) levels and malondialdehyde (MDA) levels while enhancing the activities of key antioxidant enzymes, namely superoxide dismutase (SOD) and catalase (CAT), in *Caenorhabditis elegans*, a widely used model organism. Mechanistically, we elucidated that FCL exerts its stress resistance effects by modulating of transcription factors DAF-16 and HSF-1 within the insulin/insulin-like growth factor-1 signaling pathway (IIS). Activation of these transcription factors orchestrates the expression of downstream target genes including *sod-3*, *ctl-1*, *hsp16.2*, and *hsp12.6*, thus enhancing the organism's ability to cope with stressors. Overall, our study highlights the rich reservoir of flavonoids in Chinese sea buckthorn leaves as promising candidates for natural medicines, due to their robust antioxidant properties and ability to enhance stress resistance.

Keywords: Chinese sea buckthorn; leaves; flavonoids; antioxidant activity; stress resistance

Citation: Zhao, J.; Jiang, Y.; Bi, Y.; Wei, J. Antioxidant and Stress Resistance Properties of Flavonoids from Chinese Sea Buckthorn Leaves from the Qinghai–Tibet Plateau. *Antioxidants* **2024**, *13*, 763. https://doi.org/10.3390/antiox13070763

Academic Editor: Baojun Xu

Received: 15 May 2024
Revised: 15 June 2024
Accepted: 18 June 2024
Published: 25 June 2024

1. Introduction

Chinese sea buckthorn (*Hippophae rhamnoides* subsp. *sinensis* Rousi) is an important subspecies of *Hippophae rhamnoides* L., which accounts for about 85% of all sea buckthorn in China. Chinese sea buckthorn, including wild forests and artificial forests, is a pioneering species for windbreak and sand fixation and has good economic value and ecological efficiency [1,2]. It is mainly distributed on the Loess Plateau, Inner Mongolia Plateau, and Qinghai–Tibet Plateau of China [2]. The Tibetan Plateau, which serves as the primary source of sea buckthorn in China, provides unique conditions for the secondary metabolism of sea buckthorn. These conditions result from its high altitude, low latitude, and intense ultraviolet radiation environment [1,3].

Flavonoids are the major active constituents of Chinese sea buckthorn leaves. They are mainly present in the form of flavonoid aglycones and flavonoid glycosides, which include flavones, flavonols, flavanols, flavanones, dihydroflavonol, chalcones, and procyanidins [4]. Studies by Wei et al. (2022) and Kang et al. (2019) have identified different flavonoids in sea buckthorn leaves [5,6]. Flavonols and their glycoside derivatives are particularly abundant and are formed by combining of flavonoid aglycones such as quercetin, isorhamnetin, and kaempferol with glycosyl groups such as glucose and rhamnose via glycosidic bonds [7,8]. Major flavonols in Chinese sea buckthorn leaves include isorhamnetin-3-glucoside, myricitrin, rutin, isorhamnetin-rhamnoside, kaempferol, quercetin, and hyperin [5,9]. Other flavonol derivatives such as isorhamnetin-3-glucoside-rhamnoside, narcissoside, nicotiflorin, rutin, and isoquercitrin have also been reported [2].

Sea buckthorn leaf extract exhibits potent antioxidant activity. Research indicates that the ethyl acetate extract of sea buckthorn leaves effectively scavenges superoxide anions ($O_2^{-}\cdot$), showing a significant dose–response relationship [10]. Similarly, the methanol extract of sea buckthorn leaves demonstrates effective scavenging of DPPH free radicals, with an activity reaching 54.17 mg TE·g^{-1} [11]. Moreover, Chinese sea buckthorn leaf extract not only exhibits high cellular antioxidant capacity but also possesses anti-HepG2 cell proliferation properties [5]. Furthermore, it reduces oxidative stress in diabetic kidneys by decreasing the accumulation of advanced glycation end products (AGE), thereby ameliorating kidney damage [12]. It also shows cytoprotective and antioxidant properties against oxidative stress in mouse macrophages [13].

Despite previous characterizations of the flavonoid composition in Chinese sea buckthorn leaves, these studies predominantly used traditional solvent extraction and ultrasonic-assisted extraction methods, often resulting in incomplete identification of flavonoid types. Additionally, investigations into the antioxidant activity of flavonoids from Chinese sea buckthorn leaves sourced from the Tibetan Plateau, together with their anti-stress effects and underlying molecular mechanisms, remain largely unexplored. Therefore, this study aims to address these gaps by preparing a flavonoid extract (FCL) from Chinese sea buckthorn leaves using ultra-high pressure-assisted technology and AB-8 macroporous resin purification methods. Subsequently, the flavonoids in FCL will be systematically characterized using UPLC-ESI-QTOF-MS/MS and UPLC-PAD techniques. Moreover, the antioxidant activity of FCL will be evaluated to elucidate its anti-stress effects and mechanisms in a *C. elegans* model.

2. Material and Methods

2.1. Materials

Chinese sea buckthorn leaves were harvested from 10 different sampling sites in Lexiu Town, Hezuo City, Gannan Tibetan Autonomous Prefecture, Gansu Province, China in September 2021 (longitude 102.92′ E, local latitude 34.89′ N, about 3028 m above sea level), and the samples were collected by the S-shape sampling method, mixed, and then sampled according to the quartering method. After natural drying (28 ± 2 °C, RH 38 ± 2%), the leaves were crushed with a high-speed multifunctional pulverizer (H2489, Hebei Zhizhong Machinery Technology Co., Ltd., Xingtai, China) and sieved with a 40-mesh sieve, after which the sieved powder was treated with petroleum ether by Soxhlet extraction for 2 h. The sea buckthorn leaf powder was collected and placed at room temperature to evaporate the petroleum ether to dryness and then packed in sealed packages and refrigerated for further study.

We purchased *Escherichia coli* OP50 and wild-type N2 *Caenorhabditis elegans* from Fujian Shangyuan Biological Science & Technology Co., Ltd. (Fuzhou, China). TJ375, TJ356, CF1553, CL2166, and LD1 were purchased from the official website of Caenorhabditis Genetics Center (CGC). Without any specific instructions, *C. elegans* were maintained on agar plates at 20 °C using *E. coli* OP50 as the food source. Simultaneous treatment of adult *C. elegans* using a sodium perchlorate bleaching procedure to ensure that the growth period of each *C. elegans* remained constant during the experiment [14].

2.2. Preparation of Leaf Flavonoids

Based on our previous studies, the defatted sea buckthorn leaf powder was extracted with 50% ethanol by ultra-high pressure-assisted extraction under the conditions of pressure 143 MPa, temperature 44 °C, pressure hold time 3 min, and liquid–solid ratio 41:1. AB-8 macroporous resin was used to purify the crude extract, and the eluate was concentrated and lyophilized to obtain the purified sea buckthorn leaf flavonoids (FCL).

2.3. Identification of FCL Flavonoids

Identification and confirmation of chromatographic peaks by QTOF mass spectrometer (1290-6546, Agilent Technologies Co., Ltd., Santa Clara, CA, USA). Separation by chromatography was performed on a ZORBAX Eclipse Plus C18 column (3.0 × 100 mm, 1.8 μm). A binary gradient elution system consisting of water–formic acid (100:0.1, v/v) (A) and acetonitrile (B) was used as follows: 0–2.5 min, 14% B; 2.5–5 min, 14–18% B; 5–10 min, 18–19% B; 10–17.5 min, 19–28% B; 17.5–22.5 min, 28–32% B; 22.5–28.5 min, 32–45% B; 28.5–30 min, 45–98% B. The sample had an injection volume of 2 μL and a flow rate of 0.3 mL·min^{-1}. The ionization source employed in this study was ESI, and the MS spectra were scanned with the scanning range set to m/z 100–1100 [9].

2.4. Quantification of the Major Flavonoid Components of FCL

The determination was made using ACQUITY UPLC (Waters, Milford, MA, USA), and all separations were carried out on a ZORBAX SB-C18 column (4.6 × 250 mm, 5 μm). Acetonitrile (B) and water–formic acid (100:0.1, v/v) (A) were utilized in a binary gradient elution system. 0–5 min, 86% A; 5–10 min, 86–82% A; 10–20 min, 82–81% A; 20–35 min, 81–72% A; 35–45 min, 72–68% A; 45–55 min, 68–55% A; 55–58 min, 55–86% A; 58–60 min, 86% A were the steps in the gradient elution process. Full-wavelength scanning was performed using the photodiode array detector (PAD), and 360 nm was ultimately chosen as the detection wavelength. The temperature of the column was kept at 30 °C, and the volume of injection was 10 μL, while the flow rate was 1 mL·min^{-1} [9].

2.5. Determination of Antioxidant Activity In Vitro

The DPPH free radical scavenging ability follows the method proposed by Zhang et al. (2010) [15]. FCL and Trolox reference standards were prepared in mixtures with 50% ethanol to create solutions ranging from 0–0.5 mg·mL^{-1}. A 10 μL quantity of sample solution and 190 μL DPPH (50 mg·L^{-1}) solution were added to ELISA plates. After reacting in dark for 30 min, the absorbance was measured at a wavelength of 517 nm. Results were expressed using DPPH free radical scavenging (%), with Trolox reference standards as positive controls.

The ferric reducing antioxidant power (FRAP) was determined using the kit. FCL was prepared as a 0.01–0.05 mg·mL^{-1} solution with 50% ethanol. The 950 μL of FRAP working solution and 50 μL of sample solution were mixed completely and the reaction was carried out for 20 min and then the absorbance was measured at 593 nm. FRAP values were expressed as mg Trolox (TE)·mL^{-1}.

The oxygen radical absorbance capacity (ORAC) was based on the method proposed by Zhang et al. (2010) [15]. Briefly, 20 μL Trolox standard solution, 20 μL phosphate buffer (blank), and 20 μL sample solution were added to ELISA plates and incubated in a fluorescence microplate reader at 37 °C for 10 min. Then, 200 μL 6.0 μmol·L^{-1} fluorescein sodium working solution was added to ELISA plates, and after shaking incubation for 20 min, 20 μL 119 mmol·L^{-1} AAPH solution was added to each well. The fluorescence values of each well were measured every 5 min for a total of 31 measurements. The net area under the curve (AUC) was determined by subtracting the AUC of the blank from the AUC of the sample/Trolox fluorescence intensity. The ORAC values of the samples were computed by plotting the standard curve derived by the net AUC of Trolox; results were expressed in mmol TE·g^{-1} DW.

The method proposed by Adom and Liu (2005) was used as the basis for peroxide radical scavenging capacity (PSC) determination [16]. Briefly, 100 μL sample solution, 100 μL Trolox standard solution, and 100 μL phosphate buffer (blank) were added to the ELISA plate, and then 100 μL DCFH-DA solution and 80 μL 200 mmol·L^{-1} AAPH were added to each well, respectively. Fluorescence values were measured every 2 min at

37 °C and recorded for 40 min. Construct a standard curve based on the curve area of the Trolox dynamic curve to calculate the PSC value of the sample. The result is expressed as mmol TE·g^{-1} DW.

2.6. Determination of FCL Stress Resistance

2.6.1. Heat and Oxidative Stress Assays

Synchronized L4-stage nematodes were selected and placed on NGM plates containing three different mass concentrations of FCL (50, 200, and 400 µg·mL^{-1} in *E. coli* OP50 bacterial solution, respectively), all of which contained 150 µM of 5-fluorouracil to inhibit the reproduction of nematode offspring. M9 buffer was used in place of FCL solution in the control group. Each group was set up with 3 NGM plates, and 40 nematodes were transferred to each plate and incubated at 20 °C. In the heat stress experiment, nematodes were treated with FCL for 96 h, and the culture temperature was changed from 20 °C to 35 °C, and then the culture was continued at constant temperature. Numbers of surviving nematodes in each group were monitored and counted at 2-h intervals until all nematodes died, for a total of three independent experiments. In the oxidative stress test, the nematodes were fed with FCL for 96 h and placed on NGM plates with 10 mM H_2O_2. The nematodes were observed and counted at 1 h intervals until all nematodes had died, for a total of three independent experiments [17].

2.6.2. Determination of Reactive Oxygen Species (ROS) Levels

The nematodes were treated with FCL for 96 h, and then they were exposed to heat stress for 6 h at 35 °C to cause oxidative damage. In contrast, the control group did not receive either heat shock or FCL treatment. After three rounds of washing in M9 buffer, the nematodes on NGM plates were moved to an opaque 96-well plate that was filled with 50 µL of 100 µmol·L^{-1} DCFH-DA and 50 µL of M9 buffer solution. Following two hours of incubation at 20 °C, the nematodes were removed from the plate and placed on 2% agarose slides. Next, fluorescence intensity was measured with a fluorescence microscope [18]. Images under bright field and fluorescent field were recorded separately for each nematode and superimposed using Image Pro Plus 6.0 software.

2.6.3. Determination of Malondialdehyde (MDA) Content and Superoxide Dismutase (SOD) and Catalase (CAT) Activities

The nematodes were treated with FCL at 20 °C for 96 h and then were exposed for 6 h to heat stress at 35 °C to induce oxidative damage. After three rounds of washing in M9 buffer, the nematodes on NGM plates were moved to centrifuge tubes, with approximately 800 nematodes per group. The nematodes were centrifuged at 5000 rpm·min^{-1} for 1 min, discarding the supernatant, and washing three times with M9 buffer solution. The nematodes were then crushed with a pestle provided in the kit, and the precipitate was adjusted with 1 mL of M9 buffer. Centrifuge the homogenate at 5000 rpm at 4 °C for 10 min and set the supernatant aside for testing. According to the instructions of the reagent kit (Soraibao Technology Co., Ltd., Beijing, China), MDA content and CAT, SOD activities were determined in the homogenate of nematodes, and the enzyme activities were expressed as U·mg pro^{-1}.

2.6.4. RNA Extraction and Determination of Gene Expression

Transfer of approximately 2000 synchronized nematodes to plates with or without 200 µg·mL^{-1} FCL was incubated at 20 °C for 48 h, then heat stressed at 35 °C for 6 h to induce oxidative damage. Total RNA was extracted using TRIzol reagent and reverse transcribed into cDNA using UnionScript First-stand cDNA Synthesis Mix for qPCR kit (Genesand Biotech Co., Ltd., Jinan, China). The transcript levels of *daf-2*, *akt-2*, gcs-1, *daf-16*, *gst-4*, *sod-3*, *hsf-1*, *hsp12.6*, *ctl-1*, *hsp16.2*, and *skn-1* genes were detected using GS AntiQ qPCR SYBR Green Fast Mix real-time fluorescence quantitative PCR kit (Genesand Biotech

Co., Ltd.). The $2^{-\Delta\Delta CT}$ method was used to assess relative gene expression levels and normalize them to the expression of the gene *β-actin*.

2.6.5. Detection of Nuclear Translocation of DAF-16::GFP and SKN-1::GFP

LD1 (SKN-1::GFP) and TJ356 (DAF-16::GFP), synchronized transgenic strains of the L4 stage, were placed on NGM plates and treated at 20 °C for 48 h with or without 200 μg·mL^{-1} FCL, and groups of nematodes were divided into two parts: one part was treated with heat stress for 6 h at 35 °C, and the other was untreated. Nematodes were washed with M9 buffer solution and transferred to 2% agarose slides for observation under a fluorescence microscope. DAF-16 localization in each nematode was divided into 3 types (nucleus, cytoplasm, and intermediates), taking into account the primary location of DAF-16::GFP [19], with a count of at least 60 nematodes per group.

2.6.6. Determination of GFP Fluorescence Intensity in Transgenic Strains

Synchronized transgenic nematodes TJ375 (HSP-16.2::GFP), CF1553 (SOD-3::GFP) and CL2166 (GST-4::GFP), and were cultured for 48 h in NGM plates with or without 200 μg·mL^{-1} FCL. After heat stress treatment for 6 h at 35 °C, nematodes were washed with M9 buffer solution and transferred to 2% agarose slides for observation under a fluorescence microscope.

2.7. Data Statistics

All tests were repeated three times, and data are presented as mean ± standard deviation. All data were analyzed by analysis of variance (ANOVA) with Duncan's test of significance of differences using SPSS 20.0 statistical software. ChemDraw20.0 was used to draw the chemical structural formula of flavonoids, and Origin 9.0 was used to draw the graph. To determine the lifespan of *C. elegans*, Kaplan–Meier survival analysis was performed and plotted using Graphpad Pism9.5. The fluorescence images of *C. elegans* were processed and analyzed using Image Pro Plus 6.0 software, and the relative fluorescence values were statistically analyzed, with at least 15 images in each group.

3. Results

3.1. Identification of Flavonoids in FCL

A total of 60 flavonoids were found in FCL (Supplementary Figure S1), of which 57 were finally identified. Although there were characteristic fragments of flavonoid aglycones present in the MS/MS fragments of compounds **37**, **54** and **57**, they were not identified due to insufficient structural information (Table 1).

Table 1. Flavonoids in the leaves of Chinese sea buckthorn.

Compounds	Rt/min	Formula	[M-H]$^-$ *m/z*	Fragment Ion *m/z*	Identification
			Flavonols		
3	2.806	$C_{33}H_{40}O_{21}$	771.1972	609.1447, 463.0854, 301.0349, 178.9999, 151.0043	Quercetin-3-*O*-rutinoside-7-*O*-glucoside
4	2.865	$C_{27}H_{30}O_{17}$	625.1401	463.0851, 301.0346, 299.0196, 178.9962, 151.0035	Quercetin-3,7-*O*-diglucoside
5	3.221	$C_{33}H_{40}O_{21}$	771.1958	625.1390, 447.0879, 301.0337, 300.0259, 271.0252, 151.0034	Quercetin-3-*O*-sophoroside-7-*O*-rhamnoside
6	4.118	$C_{33}H_{40}O_{20}$	755.2014	609.1433, 285.0393, 431.0911, 255.0302	Kaempferol-3-*O*-sophoroside-7-*O*-rhamnoside
8	4.473	$C_{28}H_{32}O_{17}$	639.1553	477.1011, 315.0500, 301.0300, 300.0262, 271.0249	Isorhamnetin-3,7-*O*-diglucoside
9	4.499	$C_{34}H_{42}O_{21}$	785.2126	623.1602, 477.0988, 315.0507, 300.0267, 151.0052	Isorhamnetin-3-*O*-rutinoside-7-*O*-glucoside
10	4.845	$C_{34}H_{42}O_{21}$	785.2111	639.1535, 461.1051, 315.0500, 300.0283, 151.0039	Isorhamnetin-3-*O*-sophoroside-7-*O*-rhamnoside
11	7.443	$C_{27}H_{30}O_{16}$	609.1443	447.0865, 301.0344, 300.0233, 299.0186	Quercetin-3-*O*-glucoside-7-*O*-rhamnoside

Table 1. *Cont.*

Compounds	Rt/min	Formula	[M-H]$^-$ *m/z*	Fragment Ion *m/z*	Identification
12	7.597	$C_{33}H_{40}O_{20}$	755.2022	447.0865, 301.0344, 300.0233, 271.0248, 151.0039	Quercetin-3-*O*-rutinoside-7-*O*-rhamnoside
13	7.604	$C_{27}H_{30}O_{17}$	625.1401	301.0302, 300.0271, 271.0239, 255.0305, 151.0032	Quercetin-3-*O*-sophoroside
14	8.069	$C_{27}H_{30}O_{17}$	625.1401	317.0271, 287.0191, 271.0237, 151.0034, 125.0231	Myricetin-3-*O*-rutinoside
15	8.230	$C_{33}H_{40}O_{20}$	755.2022	593.1499, 447.0921, 285.0376, 271.0198	Kaempferol-3-*O*-rutinoside-7-*O*-glucoside
16	8.493	$C_{44}H_{50}O_{25}$	977.2535	832.2069, 771.1926, 625.1458, 545.0402, 447.0858, 301.0302, 299.0186	Quercetin-3-*O*-(-6-*O*-sinapoyl)-sophoroside-7-*O*-rhamnoside
17	8.603	$C_{21}H_{20}O_{13}$	479.0819	317.0246, 316.0197, 287.0184, 271.0246, 214.0274	Myricetin-3-*O*-galactoside
18	8.802	$C_{43}H_{48}O_{24}$	947.2414	801.1856, 771.2014, 625.1439, 447.0885, 301.0365, 271.0212	Quercetin-3-*O*-sophoroside-O-glucuronide-7-*O*-rhamnoside
19	9.017	$C_{33}H_{40}O_{19}$	739.2073	593.1483, 285.0378, 284.0321, 125.0245	Kaempferol-3-*O*-rutinoside-7-*O*-rhamnoside
20	9.068	$C_{28}H_{32}O_{16}$	623.1603	477.0993, 461.1008, 315.0495, 285.0375, 275.0158	Isorhamnetin-glucoside-rhamnoside
21	9.136	$C_{27}H_{30}O_{15}$	593.1511	447.0914, 431.0927, 285.0400, 255.0299	Kaempferol-3-*O*-glucoside-7-*O*-rhamnoside
22	9.212	$C_{34}H_{42}O_{20}$	769.2179	623.1594, 315.0474, 299.011, 285.0354	Isorhamnetin-3-*O*-rutinoside-7-*O*-rhamnoside
23	9.254	$C_{27}H_{30}O_{16}$	609.1443	429.0860, 285.0367, 255.0325, 227.0374, 151.0038	Kaempferol-3-*O*-sophoroside
24	9.423	$C_{34}H_{42}O_{20}$	769.2179	315.0490, 299.0187	7-Methylquercetin-3-galactoside-6″-rhamnoside-3‴-rhamnoside
25	9.559	$C_{45}H_{52}O_{25}$	991.2686	845.2133, 639.1541, 477.1049, 315.0484, 300.0202, 275.0174	Isorhamnetin-3-*O*-(-6-*O*-sinapoyl)-sophoroside-7-*O*-rhamnoside
26	9.601	$C_{28}H_{32}O_{16}$	623.1598	477.1057, 461.1088, 315.0502, 271.0226, 183.0470	Isorhamnetin 3-*O*-glucoside-7-*O*-rhamnoside
27	10.016	$C_{43}H_{48}O_{23}$	931.2484	785.1874, 609.1463, 465.0655, 285.0388, 125.0253	Kaempferol-3-*O*-sophoroside-*O*-glucuronide-7-*O*-rhamnoside
28	10.024	$C_{44}H_{50}O_{24}$	961.2598	815.2039, 639.1500, 461.1050, 485.01212, 315.0480, 300.0268, 135.0301	Isorhamnetin-3-*O*-sophoroside-*O*-glucuronide-7-*O*-rhamnoside
29	10.253	$C_{27}H_{30}O_{15}$	593.1500	461.1059, 447.0907, 315.0506, 300.0260, 285.0360, 270.0138	Isorhamnetin-3-*O*-arabinoside-7-*O*-rhamnoside
30	10.295	$C_{27}H_{30}O_{16}$	609.1441	301.034, 300.0274, 271.0250, 178.9991, 151.0035	Quercetin-3-*O*-rutinoside
31	10.870	$C_{27}H_{30}O_{15}$	593.1512	447.0885, 315.0537, 300.0289, 285.0412, 227.0363	Kaempferol-3-*O*-galactoside-7-*O*-rhamnoside
32	10.956	$C_{28}H_{32}O_{17}$	639.1553	331.0440, 330.0378, 317.0239, 315.0140, 287.0204, 271.0237, 215.0364, 178.9997, 151.0032	Myricetin-3′-methyl-3-*O*-rutinoside
33	11.051	$C_{21}H_{20}O_{12}$	463.0876	301.0299, 272.0293, 255.0310, 217.0233, 151.0035	Quercetin-3-*O*-galactoside
34	11.522	$C_{21}H_{20}O_{12}$	463.0876	301.0324, 271.0248, 255.0299, 227.0355, 151.0042	Quercetin-3-*O*-glucoside
35	13.519	$C_{27}H_{30}O_{15}$	593.1495	285.0394, 255.0296, 227.0355	Kaempferol-3-*O*-rutinoside
36	14.314	$C_{28}H_{32}O_{16}$	623.1598	315.0502, 301.0308, 285.0406, 271.0252, 255.0290	Isorhamnetin-3-*O*-rutinoside
38	14.957	$C_{21}H_{20}O_{11}$	447.0918	285.0380, 255.0290, 227.0347	Kaempferol-3-*O*-glucoside
39	15.550	$C_{23}H_{24}O_{13}$	507.1131	344.0537, 301.0355, 287.0582, 273.0399, 258.0168	Syringetin-3-*O*-galactoside
40	15.609	$C_{22}H_{22}O_{12}$	477.1024	315.0502, 300.0266, 285.0392, 271.0251, 151.0042	Isorhamnetin-3-*O*-glucoside
41	16.278	$C_{22}H_{22}O_{12}$	477.1028	315.0503, 300.0208, 285.0241, 299.0242	Isorhamnetin-3-*O*-galactoside
42	17.466	$C_{15}H_{10}O_8$	317.0296	178.09979, 151.0038, 137.0243, 107.0138, 83.0145	Myricetin
44	18.63	$C_{21}H_{20}O_{11}$	447.0921	301.034, 300.0262, 151.0029, 121.0295, 107.0133	Quercetin-7-*O*-rhamnoside
48	20.672	$C_{43}H_{54}O_{22}$	921.2998	759.2463, 722.0562, 447.0300, 286.0375, 285.0362	3-*O*-β-D-sophorosyl-kaempferol-7-*O*-{3-*O*-[2(*E*)-2,6-dimethyl-6-hydroxy-2,7-octadienoyl]}-α-L-rhamnoside
50	22.049	$C_{21}H_{20}O_{10}$	431.0980	344.0179, 285.0403, 284.0321, 257.0436, 254.9958	Kaempferol-rhamnoside
51	22.442	$C_{27}H_{30}O_{15}$	593.1291	487.0871, 447.0906, 285.0390, 284.0328, 255.0297, 151.0014	Kaempferol-3-*O*-glucoside-7-*O*-rhamnoside

Table 1. *Cont.*

Compounds	Rt/min	Formula	[M-H]⁻ *m/z*	Fragment Ion *m/z*	Identification
52	22.624	$C_{22}H_{22}O_{11}$	461.1089	431.0113, 315.0536, 299.0167, 287.0528, 271.0239, 107.0142	Isorhamnetin-rhamnoside
53	22.692	$C_{15}H_{10}O_7$	301.0350	178.9981, 163.0049, 151.0039, 107.0139, 65.0032	Quercetin
55	25.436	$C_{37}H_{44}O_{17}$	759.2483	597.2049, 596.1804, 447.0926, 358.8980, 285.0427	3-*O*-β-D-glucosyl-kaempferol-7-*O*-{2-*O*-[2(*E*)-2,6-dimethyl-6-hydroxy-2,7-octadienoyl]}-α-L-rhamnoside
59	28.803	$C_{15}H_{10}O_6$	285.0399	241.0512, 229.0497, 154.0426, 107.0141, 93.0356, 63.0365	Kaempferol
60	28.201	$C_{16}H_{12}O_7$	315.0503	301.0300, 300.0272, 271.0249, 227.0398, 151.0044, 107.0134	Isorhamnetin
				New Flavonoids	
43	17.64	$C_{43}H_{54}O_{23}$	937.2957	773.8043, 629.1837, 609.1457, 457.7284, 447.1069, 301.0346,	Quercetin-3-*O*-rutinoside-7-*O*-[2(*E*)-2,6-dimethyl-6-hydroxy-2,7-octadienoyl(1→6)]-glucoside
45	19.129	$C_{44}H_{56}O_{23}$	951.3107	767.2035, 623.1589, 459.0922, 315.0515, 313.0312	Isorhamnetin-3-*O*-rutinoside-7-*O*-[2(*E*)-2,6-dimethyl-6-hydroxy-2,7-octadienoyl(1→6)]-glucoside
46	19.519	$C_{38}H_{46}O_{19}$	805.2537	643.1941, 477.1037, 458.5028, 315.0494, 313.0336, 299.0067,	Isorhamnetin-3-*O*-glucoside-7-*O*-[2(*E*)-2,6-dimethyl-6-hydroxy-2,7-octadienoyl(1→6)]-glucoside
47	19.578	$C_{43}H_{54}O_{23}$	937.2957	775.2385, 668.9043, 609.1446, 463.0873, 447.0838, 301.0346,	Quercetin-3-*O*-sophoroside-7-*O*-[2(*E*)-2,6-dimethyl-6-hydroxy-2,7-octadienoyl(1→2)]-rhamnoside
49	20.911	$C_{44}H_{56}O_{23}$	951.3107	789.2574, 623.1507, 477.1026, 315.0499, 299.0190	Isorhamnetin-3-*O*-sophoroside-7-*O*-[2(*E*)-2,6-dimethyl-6-hydroxy-2,7-octadienoyl(1→2)]-rhamnoside
56	25.552	$C_{38}H_{46}O_{18}$	789.2582	627.2021, 477.0975, 315.0502, 300.0324, 299.0218, 286.0425	Isorhamnetin-3-*O*-glucoside-7-*O*-[2(*E*)-2,6-dimethyl-6-hydroxy-2,7-octadienoyl(1→2)]-rhamnoside
				Flavanols	
1	2.595	$C_{15}H_{14}O_7$	305.0667	219.0669, 164.0113, 139.0417, 125.0251	Gallocatechin
2	2.647	$C_{15}H_{14}O_7$	305.0667	219.0658, 164.0116, 139.0404, 125.0250	Epigallocatechin
7	4.348	$C_{15}H_{14}O_6$	289.0715	245.0813, 219.0279, 203.0713, 125.0239, 109.0298	Epicatechin
				Flavane	
58	25.863	$C_{15}H_{12}O_5$	271.0606	151.0035, 119.0497, 107.0144, 83.0138, 63.0246	Naringenin
				Unknown	
37	14.907	$C_{43}H_{48}O_{24}$	947.2414	783.6926, 639.1375, 609.1426, 486.9109, 301.0346, 285.3524,	Unidentified
54	23.800	$C_{37}H_{44}O_{18}$	775.244	612.1838, 510.0540, 463.0849, 301.0333, 299.0196	Unidentified
57	25.713	$C_{38}H_{46}O_{18}$	789.2582	627.1983, 477.1037, 315.0500, 299.0164	Unidentified

According to the MS/MS fragment information of the flavonoid aglycone reference standard (Supplementary Figure S2), it can be inferred that the major flavonoid components of FCL are flavonol glycosides derived from isorhamnosin, quercetin, kaempferol, and myricetin. In MS/MS spectrometry, glycosidic bond cleavage is a characteristic cleavage mode of flavonoid glycosides [9,20]. As an example, compound **9** had an $[M-H]^-$ ion at *m/z* 785.2126, and its molecular formula was speculated to be $C_{34}H_{42}O_{21}$. The four major fragment ions were observed at *m/z* 623.16, 477.10, 315.05, and 300.03. The fragment ion at *m/z* 623.16 was raised as a result of the loss of glucose, and the loss of glucose further produced the ion at *m/z* 477.10. The fragment ion *m/z* 315.05 was generated by the loss of glucose at *m/z* 477.10, which was further cleaved to produce the same characteristic ion peaks 300.02 and 151.01 as the reference standard of isorhamnetin. Combined with Chemspider, Massbank information and MS/MS spectral data preliminary identification of compound **9** as isorhamnetin-3-*O*-rutin-7-*O*-glucoside, and its possible cleavage pathway was shown in Supplementary Figure S3. Based on the above fragmentation pattern, a total of 45 flavonols were identified from FCL (Figure 1) [20,21]. The 8 compounds **43**, **45–49**, **55** and **56** all contain an ionic fragment at m/z 166.0978, of which compounds **48** and **55** are identified according to the reference as 3-*O*-β-D-sophorosyl-kaempferol-7-*O*-{3-*O*-[2(*E*)-2,6-dimethyl-6-hydroxy-2,7-octadienoyl]}-α-L-rhamnoside and 3-*O*-β-D-glucosyl-kaempferol-7-*O*-{2-*O*-[2(*E*)-2,6-dimethyl-6-hydroxy-2,7-octadienoyl]}-α-L-rhamnoside [22], and deduced m/z 166.0978 as [2(*E*)-2,6-dimethyl-6-hydroxy-2,7-octadienoyl]$^-$. Based on the ion fragmentation information, we identified the remaining 6 compounds. Take compound **47** as an example: its quasimolecular ion was m/z 937.2957 [M-H]$^-$ with molecular formula $C_{43}H_{54}O_{23}$, and five major ion fragments were observed at m/z 775.24, 609.14, 463.09, 447.08, and 301.03. The attributions of the fragment ions are: m/z 775.24 for [M-H-Glc]$^-$, m/z 609.14 for [M-H-Glc-S]$^-$, m/z 447.08 for [M-H-Glc-S-Glc]$^-$, m/z 463.08 for [M-H-Glc-S-Rha]$^-$, and m/z 315.05 for [M-H-Glc-S-Rha-Glc]$^-$, from which compound **47** was presumed to be Quercetin-3-*O*-sophoroside-7-*O*-[2(*E*)-2,6-dimethyl-6-hydroxy-2,7-octadienoyl(1→2)]-rhamnoside, whose possible cleavage pathway is shown in Figure S4. In addition, three flavanols (compounds **1**, **2**, and **7**) and one flavane (**58**) were identified. Compounds **1** and **2** (m/z 305.0667) were speculated to be isomers based on their similar fragment paths, and fragment information was compared with a reference and identified as gallocatechin and epigallocatechin [23]. Based on Chemspider and Massbank databases, compound **7** (m/z 289.0715) was characterized as epicatechin, while compound **58** (m/z 271.0606) was identified as naringenin by comparison of MS and fragmentation information with the reference standard of naringenin.

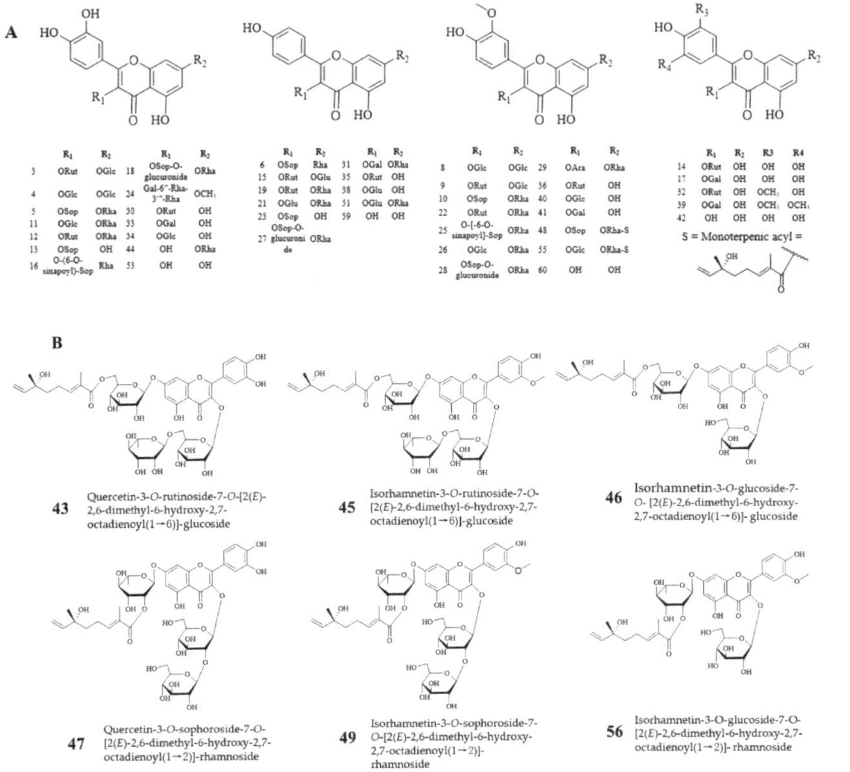

Figure 1. Flavonols and 6 new flavonoids in Chinese sea buckthorn leaves. (**A**) Flavonols, (**B**) 6 new flavonoids.

The flavonoid components of FCL were mainly flavonol glycosides formed by the combination of flavonoid aglycones and glycosyl groups. The glycoconjugates mainly include glucose, rhamnose, rutinose, sophorose, arabinose and galactose. According to the glycosylation method, the flavonoid components of FCL are mainly classified as *O*-glycosides, and the glycosidic linkages were mainly formed at the 3-position of the C-ring and the 7-position of the A-ring (Figure 1). Among the 57 identified flavonol glycosides, there were 18 types of isorhamnetin glycosides, 16 types of quercetin glycosides, and 14 types of kaempferol glycosides, among which 6 compounds were new flavonoid compounds (**43**, **45**, **46**, **47**, **49**, and **56**), and 7 compounds (compounds **16**, **24**, **25**, **32**, **39**, **48**, and **55**) were detected for the first time in sea buckthorn leaves. The 3 unidentified flavonoids (compounds **37**, **54**, and **57**) were found for the first time in sea buckthorn.

3.2. *Quantification of Major Flavonoids*

FCL contained abundant flavonoids (64.35 ± 0.79 mg·g^{-1} DW) (Supplementary Table S1, Figure S5), more than olive leaves (4.92–18.29 mg·g^{-1} DW) and mulberry leaves (31.212 mg·g^{-1} DW) [24,25]. The flavonoid fraction of FCL could be classified into three classes according to the difference in the number of glycosidic chains, namely flavonoid disaccharide chain glycosides, flavonoid monosaccharide chain glycoside, and flavonoid aglycone. Among them, flavonoid disaccharide chain glycosides were the dominant component, followed by flavonoid monosaccharide chain glycoside, with flavonoid aglycone content accounting for only 13.78% of the total content. It supports the findings of Liu et al. [4] and suggests that the flavonoids abundant in Chinese sea buckthorn leaves were primarily flavonoid glycosides [19]. Its flavonoid glycosides were mainly derivatives of kaempferol, isorhamnetin, and quercetin. Among them, quercetin glycosides and isorhamnetin glycosides accounted for the largest portion (78.13%) of the total flavonoids, and their species were more diverse, suggesting that sea buckthorn leaves are a good source of isorhamnetin and quercetin derivatives. Compared to sea buckthorn berry flavonol profiles [8], kaempferol glycosides (totaling 17.39% of total flavonoids) were more abundant in sea buckthorn leaves. Among the 16 flavonoids quantified, the most abundant was isorhamnetin-3-*O*-glucoside-7-*O*-rhamnoside (17.74% of the total flavonoids), followed by quercetin-3-*O*-rutinoside (13.17%), kaempferol-3-*O*-sophoroside-7-*O*-rhamnoside (13.38%) and quercetin-3-*O*-galactoside (12.74%).

3.3. *Antioxidant Activity of FCL In Vitro*

A dose-dependent DPPH radical scavenging ability was observed for both FCL and Trolox (Figure 2A), with IC_{50} values of 0.123 mg·mL^{-1} and 0.157 mg·mL^{-1}, respectively, suggesting that FCL scavenges DPPH radicals 1.28 times more than Trolox. The FRAP of FCL increased with the increasing concentration (Figure 2B), and in the concentration range of 0.01–0.05 mg·mL^{-1}, the iron-reducing antioxidant power of FCL was equivalent to that of Trolox. Figure 2C, D plot the temporal kinetic curves of the relative fluorescence intensity of Trolox standard solution and FCL decaying and increasing with time. The ORAC value of FCL was 3.74 ± 0.22 mmol TE·g^{-1}, which was 0.935 times higher than that of Trolox. The fluorescence intensity decay trend of 12.5 μg·mL^{-1} FCL was similar to that of Trolox (Figure 2C). Trolox and FCL showed a clear dynamic trend of fluorescence increase, and the fluorescence intensity increase trend of 25 μg·mL^{-1} FCL and 6.25 μg·mL^{-1} Trolox was similar (Figure 2D). Taken together, this indicated that FCL can effectively scavenge DPPH radicals and has iron-reducing antioxidant power and ORAC comparable to that of Trolox.

The radical scavenging reaction of DPPH and the FRAP reaction belong to the single electron transfer reaction, which mainly reflects the reducing ability to high valence ions [26] The antioxidant reaction of ORAC and PSC belongs to the hydrogen atom transfer mechanism, which mainly reflects the ability of substrates and antioxidants to compete for peroxy radicals [15]. Flavonoids can provide hydrogen atoms to free radicals, acting as good electron donors to inhibit the development of peroxide chain reactions [15,27]. Thus, we could infer that the strong antioxidant capacity of FCL is mainly due to its abundant flavonoid components.

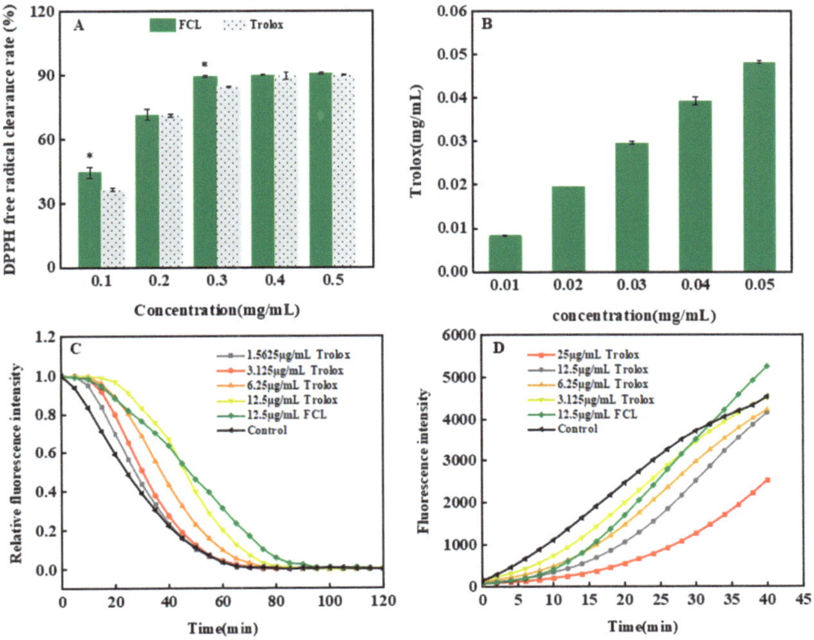

Figure 2. Effects of FCL on DPPH free radical scavenging ability (**A**), FRAP (**B**), ORAC (**C**), and PSC (**D**). Compared with the Trolox group, * indicates significant difference ($p < 0.05$).

3.4. The Effect of FCL on the Stress Resistance in *C. elegans*

3.4.1. Effects of FCL Treatment on the Ability of *C. elegans* to Resist Oxidative and Heat Stress

Under oxidative stress treatment, the mean survival time of nematodes groups fed with three different concentrations of FCL was significantly higher than that of the control group (Figure 3A). The mean survival time of nematodes in the 200 μg·mL^{-1} FCL group was the longest and 29.58% higher than that of the control group. The survival curves of the FCL-treated group were generally shifted to the right compared with the control group (Figure 3B), suggesting that FCL could effectively enhance the tolerance to oxidative stress in nematode. High temperature can lead to metabolic disorders and enzyme inactivation in the body, generating large amounts of ROS and causing oxidative stress [28]. Under the heat stress treatment, the mean survival time of nematodes in all three FCL groups with different concentrations was extremely significantly higher than that of the control group ($p < 0.01$) (Figure 3C). Among them, the 400 μg·mL^{-1} FCL-treated group showed the longest survival time of nematodes. Survival curves were generally shifted to the right in the FCL-treated group compared with the control group (Figure 3D), indicating that FCL could effectively enhance the heat stress resistance of nematodes and showed strong antioxidant capacity. Since flavonoids such as quercetin and rutin could enhance the resistance of *C. elegans* to heat stress [29], and isorhamnetin and its derivatives had potential protective effects against oxidative stress in human RPE cells [30], it was suggested that the abundant isorhamnetin and quercetin derivatives in FCL may be mainly responsible for the enhanced resistance of *C. elegans* under stress conditions.

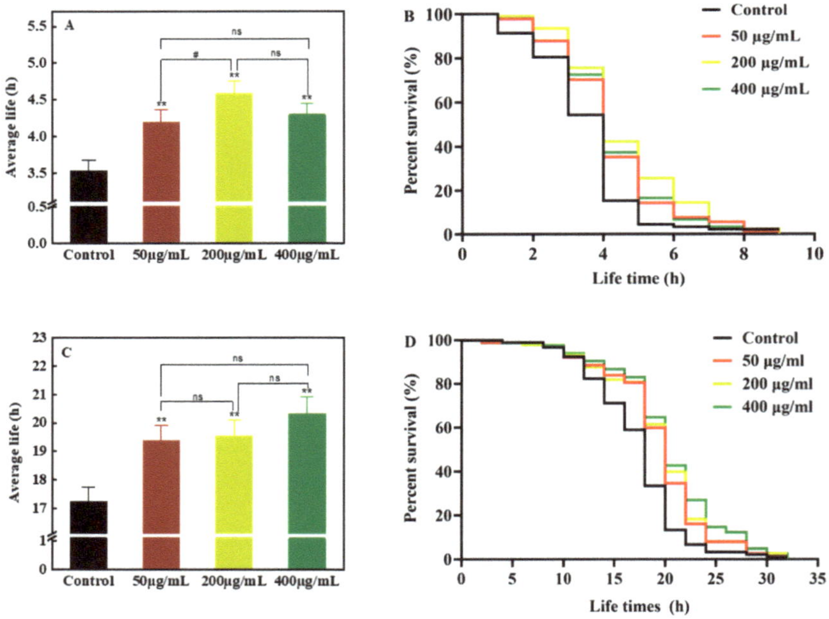

Figure 3. Effects of FCL on average lifespan (**A**), life curve (**B**) under oxidative stress, average lifespan (**C**) and life curve (**D**) under heat stress in *C. elegans*. The symbol ** indicates an extremely significant difference compared to the control group ($p < 0.01$), the symbol # indicates the comparison between different treatment groups (# $p < 0.05$), and ns indicates no significant difference.

3.4.2. Effects of FCL Treatment on ROS, MDA Levels, and SOD and CAT Activities

The ROS level in the heat stress-treated model group increased sharply (Figure 4A,B), which was 2.64 times that of the control group (without heat stress treatment), indicating that stress treatment could significantly increase the ROS level in *C. elegans*. Under heat stress, FCL treatment significantly reduced ROS levels, 50 $\mu g \cdot mL^{-1}$ FCL reduced ROS levels by 18.76%, and at high concentrations of 200 $\mu g \cdot mL^{-1}$ and 400 $\mu g \cdot mL^{-1}$, ROS levels in *C. elegans* were reduced by 49.62% and 44.08%, respectively. This indicates that FCL treatment could effectively remove ROS levels in *C. elegans*, with the high concentration of FCL being particularly effective in ROS removal. Similarly, the FCL-treated group resulted in *C. elegans* with lower MDA levels (Figure 4C), which were reduced by 14.22%, 32.0%, and 37.07% in the 50 $\mu g \cdot mL^{-1}$, 200 $\mu g \cdot mL^{-1}$ and 400 $\mu g \cdot mL^{-1}$ treatment groups, respectively, compared with the control. Peroxidation and oxidative stress in the organism are directly caused by excessive ROS [31]. The level of lipid peroxidation in the body can be indicated by the amount of MDA present, which is created by oxygen free radicals in cell membranes and the peroxidation of unsaturated fatty acids [30]. The findings of this study demonstrated that the FCL significantly reduced the amount of ROS and lipid peroxidation in nematodes under acute stress. Consequently, we deduced that FCL would improve oxidative stress resistance of *C. elegans* by inhibiting the accumulation of excess ROS and the onset of lipid peroxidation.

The SOD activity of nematodes fed with different concentrations of FCL were significantly higher than the control ($p < 0.01$). The SOD activities of the 50 $\mu g \cdot mL^{-1}$, 200 $\mu g \cdot mL^{-1}$, and 400 $\mu g \cdot mL^{-1}$ FCL groups were 50.92%, 65.86%, and 72.80% higher than those of the control group, respectively (Figure 4D). In the same way, *C. elegans* treated by various FCL doses exhibited significantly increased CAT activity compared to the control group ($p < 0.05$). The 200 $\mu g \cdot mL^{-1}$ and 400 $\mu g \cdot mL^{-1}$ FCL groups were 106.08% and 96.86% higher than the control group, respectively. This indicated that, under situations of heat stress,

FCL might considerably increase the SOD and CAT activities and eliminate excess ROS, extending the survival rate of *C. elegans*.

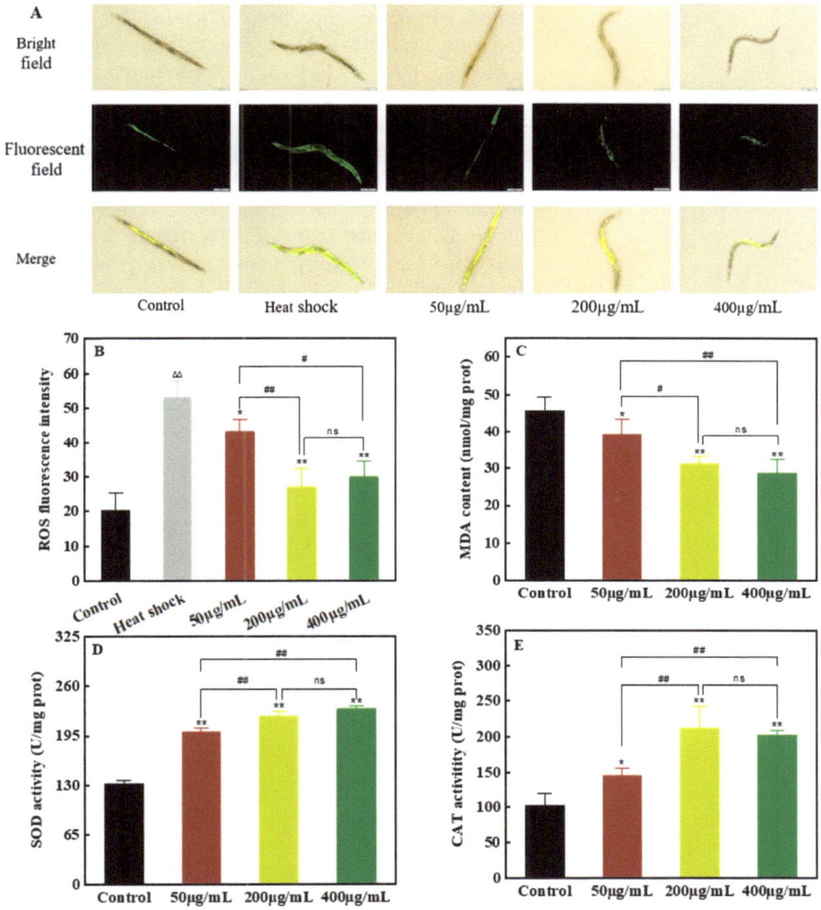

Figure 4. Effects of FCL on the levels of ROS and MDA and the activities of SOD and CAT in *C. elegans*. (**A**) Comparison of the fluorescence levels of ROS (scale 200 μm). (**B**) Quantitative determination of ROS fluorescence intensity. The symbol ΔΔ indicates that the difference was highly significant compared with the control group ($p < 0.01$), the symbol * indicates the comparison with the heat stress group (* $p < 0.05$, ** $p < 0.01$), and the symbol # indicates the comparison between different treatment groups (# $p < 0.05$, ## $p < 0.01$), and ns indicates no significant difference. (**C**) MDA content; (**D**) SOD activity; (**E**) CAT activity.

3.5. The Molecular Mechanism of FCL Regulation of Stress Resistance in C. elegans

The insulin/insulin-like growth factor-1 signaling pathway (IIS) is a classical and conserved signaling pathway for aging regulation, which can regulate the activities of three transcription factors (DAF-16, SKN-1, and HSF-1), thereby inducing the expression of a series of genes related to stress response, homeostatic regulation, and metabolism [32]. To investigate if the IIS pathway has a role in FCL-mediated stress resistance in nematodes, we detected the expression levels of genes and proteins related to the IIS pathway.

The 200 $\mu g \cdot mL^{-1}$ FCL treatment significantly decreased the transcript levels of *daf-2* and *akt-2* in the IIS pathway compared with the control ($p < 0.05$) and significantly increased the transcript levels of *daf-16* and its downstream genes, *ctl-1* and *sod-3* ($p < 0.05$), with a

1.02-fold increase in the expression of the *sod-3* gene (Figure 5A). In the DAF-16 nuclear translocation assay (Figure 5B), FCL treatment decreased the DAF-16::GFP expression in the cytoplasm and increased the proportion of *C. elegans* with its localization in the nucleus. In the control group, the proportion of DAF-16::GFP in the nucleus was 8.3%, while in the FCL-treated group, it increased to 22.5%; after the stress treatment, FCL increased the nuclear localization of DAF-16::GFP to 75.0% compared to 67.9% in the control group (Figure 5C). DAF-16 is an important transcription factor that influences the ability of *C. elegans* to resist stress [33]. The genes upstream of the IIS pathway, *daf-2* and *akt-2* are repressed in a stress environment, which prompts DAF-16 to dephosphorylate and translocate from the cytoplasm to the nucleus. There, it binds to the DNA promoter-binding region, inducing the expression of target genes and increasing the resistance of *C. elegans* to stress [30]. Results of this study indicated that FCL activated the transcription factor DAF-16, promoted its nuclear translocation, up-regulated the expression of the downstream target genes *sod-3* (sod gene) and *ctl-1* (CAT gene), and enhanced the resistance of *C. elegans*. This result was also verified in the CF1553 transgenic strain, in which FCL treatment dramatically up-regulated the SOD-3 protein expression level (Figure 5F,G).

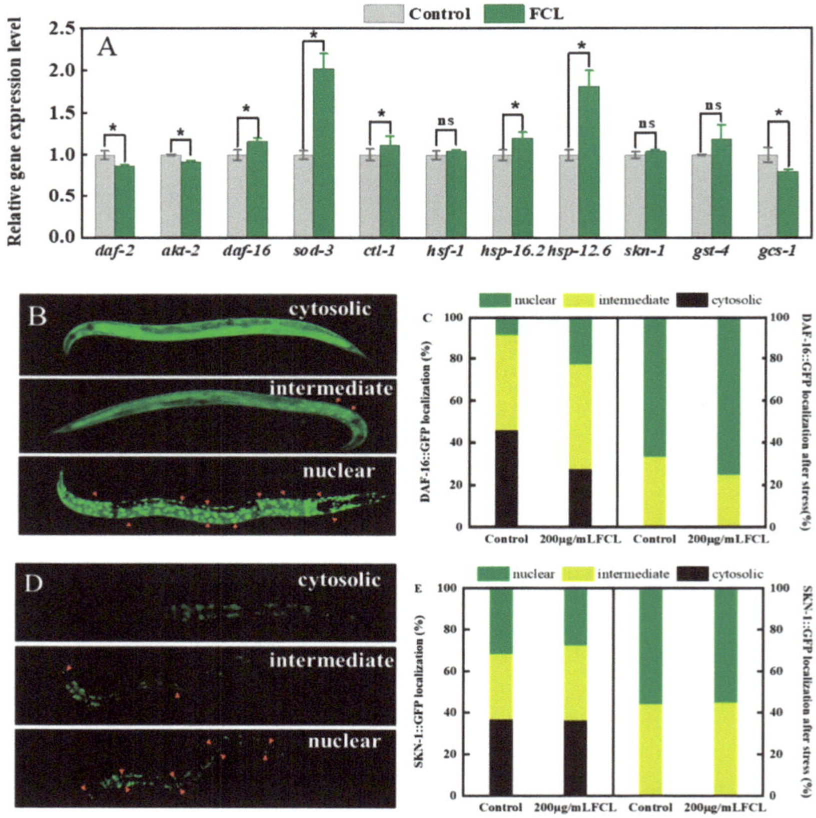

Figure 5. *Cont.*

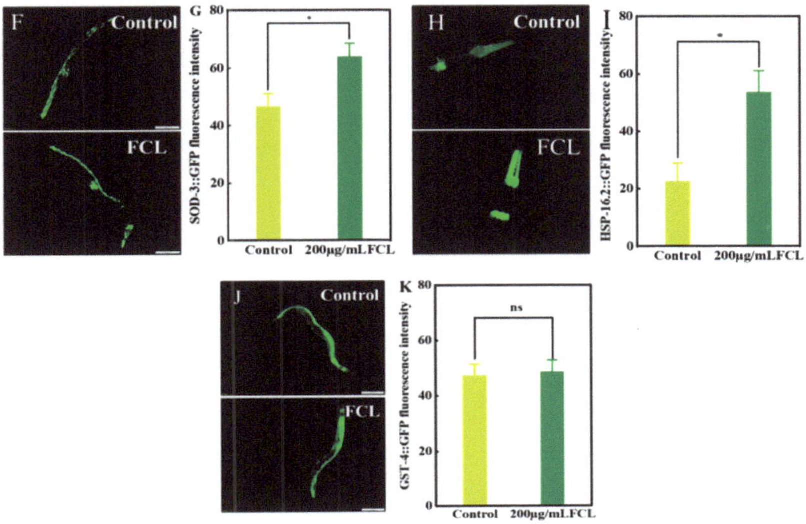

Figure 5. Effect of FCL on stress resistance-related genes and protein expression in *C. elegans*. (**A**) The expression level of stress resistance related genes. (**B**) Three kinds of DAF-16::GFP localization. (**C**) Histogram of the distribution ratio of DAF-16::GFP in *C. elegans*. (**D**) Three types of SKN-1::GFP localization. (**E**) Histogram of the distribution ratio of SKN-1::GFP in *C. elegans* (Red arrows indicate nuclear translocation). (**F**,**G**) SOD-3: GFP fluorescence expression of CF1553 strain. (**H**,**I**) HSP-16.2: GFP fluorescence expression of TJ375 strain. (**J**,**K**) GST-4::GFP fluorescence expression of CL2166 strain (scale 200 μm); * was significantly different ($p < 0.05$), and ns was not significantly different.

FCL treatment increased gene expression of *hsf-1* compared to control, and its downstream genes *hsp-16.2* (small heat shock protein gene) and *hsp-12.6* increased by 20% and 81%, respectively (Figure 5A). Expression of HSP-16.2::GFP protein was significantly up-regulated by FCL treatment in the TJ375 transgenic strain (HSP-16.2::GFP) (Figure 5H,I). HSF-1 is a heat shock transcription factor that maintains protein homeostasis by regulating the expression of chaperone heat shock proteins, which can protect proteins in nematodes from damage caused by external stresses environmental stress [34]. The high expression of HSF-1-regulated HSP-16.2 improves cellular heat resistance and protein homeostasis in unknown organisms by preventing misfolding and translation [35]. The results indicated that HSF-1 and its downstream genes were significantly activated and had an important role in the enhancement of heat stress resistance activity of *C. elegans* by FCL.

No significant difference was observed in the expression level of *skn-1* compared to the control group. *Gst-4* expression was up-regulated but not statistically significant ($p > 0.05$); *gcs-1* expression was down-regulated (Figure 5A). The same results were found in the SKN-1 nuclear translocation assay, where the FCL-treated group had no significant effect on the level of nuclear translocation of SKN-1::GFP compared with the control group (Figure 5D,E). In addition, the expression of GST-4::GFP protein in the CL2166 (GST-4::GFP) transgenic strain was not significantly altered by FCL treatment (Figure 5J,K). *Skn-1*, a gene homologous to mammalian Nrf2, can augment the resistance to oxidative stress in *C. elegans* by modulating phase II detoxification genes [36]. However, the results showed that FCL treatment did not significantly affect *skn-1* gene and protein expression, which suggested that *skn-1* may not be related to the enhancement of *C. elegans* antioxidant capacity by FCL.

4. Discussion

Oxidative stress is caused by an imbalance between the generation of oxidants and the elimination of free radicals by antioxidants, and this imbalance leads to the destruction of biomolecules and cells, which is potentially destructive to the whole organism [12,28].

Therefore, enhancing the antioxidant capacity of the organism is an effective approach to defend against oxidative stress. The antioxidant defense system of organisms mainly includes endogenous antioxidant enzymes, endogenous non-enzymatic antioxidants, and exogenous antioxidants [29]. Some natural and harmless compounds, such as polyphenolics and flavonoids, are good exogenous antioxidants, which could effectively scavenge free radicals and reduce the damage caused by oxidative stress [14].

In this study, we used UPLC-ESI-QTOF-MS/MS and UPLC-PAD techniques to separate and characterize flavonoids with different polarities using binary gradient elution consisting of water–formic acid and acetonitrile [9], and found that FCL was rich in flavonoids. In terms of antioxidant activity, FCL could effectively scavenge DPPH radicals and possessed ferric reducing antioxidant power and oxygen radical absorption capacities comparable to those of Trolox, in which the ORAC value was 3.74 $\pm$ 0.22 mmol TE$\cdot$g^{-1}, higher than that of the forsythia flavonoids extract (0.928 mmol TE$\cdot$g^{-1}) and hawthorn extracts (1.17 mmol TE$\cdot$g^{-1}) [37,38], demonstrating strong in vitro antioxidant activity. Using *C. elegans* as a model, we induced oxidative stress with H_2O_2 and high temperatures (35 °C), which significantly increased ROS levels in nematodes, while FCL-fed nematodes had lower levels of ROS and resulted in a significant increase in their lifespan. This indicated that FCL could enhance the ability of stress resistance in nematodes through directly scavenging ROS. This effect was similar to that of some reported flavonoid compounds, which could also improve the oxidative stress resistance of nematodes, such as flavonol glycoside complanatoside A, Rhodiola rosea extract, quercetin, rutin, etc. [27,29,34].

In addition to exogenous antioxidants, endogenous antioxidant enzymes also play important roles in the cellular antioxidant defense system, such as SOD and CAT. SOD could disproportionate $O_2^{-}\cdot$ to H_2O_2 and O_2, and CAT could catalyze H_2O_2 to produce water and O_2 [39,40]. To verify if the stress resistance ability of FCL is associated with the antioxidant enzyme system, we evaluated antioxidant enzyme activities in nematodes and found that FCL treatment resulted in a significant increase in enzyme activities in nematodes. A number of studies have shown that the regulation of antioxidant enzyme gene expression greatly influences antioxidant enzyme activity, implying that FCL may enhance the defense mechanism of nematodes by modulating endogenous metabolic pathways.

The IIS pathway is a key regulatory mechanism for the growth, development, immune defense, and stress resistance of organisms [36,41]. The initiation of its kinase cascade is dependent on the phosphorylation of DAF-2 (insulin receptor), which then regulates the downstream signaling molecule AGE-1/PI3K (phosphatidylinositol kinase), followed by the further activation of serine/threonine protein kinases, including AKT, which in turn regulates downstream transcription factors [42]. DAF-16, SKN-1, and HSF-1 are three important transcription factors in the IIS pathway. The IIS kinase cascade regulates the nuclear translocation of DAF-16 and SKN-1, which can activate the downstream antioxidant enzyme (SOD, CAT, GST-4, GCS-1) genes and enhance the antioxidant capacity of the organism [33,34]. HSF-1 plays an important role in maintaining protein homeostasis by regulating the expression of heat shock proteins (HSP12.6, HSP16.2) [34]. The research results indicate that FCL treatment decreased *daf-2* and *akt-2* gene relative expression, which negatively regulates lifespan in the IIS pathway of nematodes, and increased the relative expression levels of *daf-16* and its downstream target genes, *sod-3*, *ctl-1*, and *hsp-16.2* in this pathway. At the same time, it promoted DAF-16 entry into the nucleus and enhanced the expression of DAF-16 and HSF-1-regulated downstream proteins SOD-3 and HSP-16.2, indicating that FCL could mediate stress resistance in nematodes through the IIS signaling pathway, which required the involvement of transcription factors DAF-16 and HSF-1.

Taken together, FCL, as an antioxidant, could directly scavenge free radicals to reduce ROS levels and achieve stress resistance and also modulate the IIS signaling pathway, activate the downstream transcription factors DAF-16 and HSF-1, and regulate the expression of antioxidant enzymes and heat shock proteins, to achieve the virtuous circle of alleviating oxidative stress and prolonging lifespan (Figure 6). To further explore the potential contribution of FCL's antioxidant effects and its underlying mechanisms, we will investigate the signature flavonoids and conduct extra experiments, such as protein-specific analyses and more relevant mutational analyses.

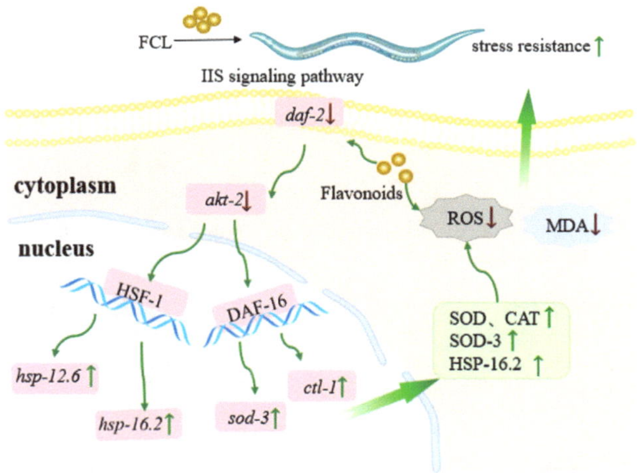

Figure 6. Pattern diagram of FCL regulation of stress resistance in *C. elegans*. The symbol ↓ indicates down-regulation of gene expression, the symbol ↑ indicates up-regulation.

5. Conclusions

In FCL, 57 flavonoids were identified, including 53 flavonols, 3 flavanols, and 1 flavan, with 6 new flavonoid compounds being detected in sea buckthorn leaves. In particular, flavonols, especially isorhamnetin-3-*O*-glucoside-7-*O*-rhamnoside, were predominant. FCL exhibited Trolox-like antioxidant activity and showed potential to enhance nematode stress resistance by modulating ROS and MDA levels, enhancing SOD and CAT activities, and activating antioxidant enzymes and heat shock protein genes in the IIS pathway. These findings highlight the antioxidant and stress resistance properties of Chinese sea buckthorn leaf flavonoids and provide insights for their application in functional foods.

Supplementary Materials: The following supporting information can be downloaded at: https://www.mdpi.com/article/10.3390/antiox13070763/s1, Figure S1: Total ion current chromatography of FCL in negative ion mode; Figure S2: Total ion current chromatography of mixed standards in negative ion mode; Figure S3: Possible fragmentation pathway of isorhamnetin-3-*O*-rutinoside-7-*O*-glucoside in MS/MS; Figure S4: Possible fragmentation pathways of 6 new compounds; Figure S5: UPLC-DAD chromatogram of flavonoids in FCL; Table S1: Content of 16 flavonoids in Chinese sea buckthorn leaves/(mg·g^{-1} DW).

Author Contributions: Writing—Original Draft Preparation, Writing—Review and Editing, Data Curation, J.Z.; Investigation, Formal analysis, Y.J.; Funding Acquisition, Project Administration, Writing—Review and Editing, Y.B.; Methodology, Supervision, Writing—Review and Editing, J.W. All authors have read and agreed to the published version of the manuscript.

Funding: This work was financially supported by the quality improvement of large fruit sea buckthorn and high value utilization of by-products (Burqin Huiyuan Biotechnology Co., Ltd.) and the China–Russian Sea Buckthorn Joint Research Center (Ministry of Education, People's Republic of China).

Institutional Review Board Statement: Not applicable.

Informed Consent Statement: Not applicable.

Data Availability Statement: Data is contained within the article and Supplementary Materials.

Acknowledgments: We thank Youwei Chen (Burqin Huiyuan Biotechnology Co., Ltd., Burqin, Altay, Xinjiang, China) for providing the sea buckthorn leaf samples, and Bin Zhang (International Seabuckthorn Association, Beijing, China) and Siyuan Yang (College of Food Science and Engineering, Gansu Agricultural University, Lanzhou, China) for their guidance on methodology and software use during the experiments.

Conflicts of Interest: The authors declare that this study received funding from the quality improvement of large fruit sea buckthorn and high value utilization of by-products (Burqin Huiyuan Biotechnology Co., Ltd). The funder had the following involvement with the study: Sample collection and methodology.

References

1. Su, T.T.; Wei, J.; Zhao, J.M.; Jiang, Y.M.; Bi, Y.; George, G. Comparative assessment of functional components and antioxidant activities between *Hippophae rhamnoides* ssp. sinensis and *H. tibetana* Berries in Qinghai-Tibet Plateau. *Foods* **2023**, *12*, 341. [CrossRef] [PubMed]
2. Qin, Z.X.; Zhang, Y.; Qi, M.D.; Zang, Q.; Liu, S.; Li, M.H.; Liu, Y.G.; Liu, Y. Rapid analysis of compounds in leaves of Chinese seabuckthorn and Tibetan seabuckthorn by UPLC/Q-TOF-MS. *Chin. J. Chin. Mater. Med.* **2016**, *41*, 1461–1467. [CrossRef]
3. Zheng, J.; Kallio, H.; Yang, B. Sea buckthorn (*Hippophaë rhamnoides* ssp. *rhamnoides*) berries in Nordic environment: Compositional response to latitude and weather conditions. *J. Agric. Food Chem.* **2016**, *64*, 5031–5044. [CrossRef] [PubMed]
4. Liu, S.; Xiao, P.; Kuang, Y.; Hao, J.; Huang, T.; Liu, E. Flavonoids from sea buckthorn: A review on phytochemistry, pharmacokinetics and role in metabolic diseases. *J. Food Biochem.* **2021**, *45*, e13724. [CrossRef] [PubMed]
5. Wei, J.; Su, T.T.; Su, H.L.; Jiang, Y.M.; Li, J.X.; Bi, Y. Comparative assessment of phenolics, antioxidant and antiproliferative activities between *Hippophae rhamnoides* ssp. *sinensis* and *H. tibetana* leaf in Qinghai-Tibet Plateau. *Food Biosci.* **2022**, *46*, 101507. [CrossRef]
6. Kang, Y.; Mao, Y.N.; Wang, F.F.; Wu, W.Q.; Liu, Y. Analysis of chemical components in leaves of *Hippophae rhamnoides* by UPLC-LTQ Orbitrap MS. *Mod. Chin. Med.* **2018**, *20*, 1340–1346, 1366. [CrossRef]
7. Zheng, W.; Bai, H.; Han, S.; Bao, F.; Zhang, K.; Sun, L.; Du, H.; Yang, Z. Analysis on the constituents of branches, berries, and leaves of *Hippophae rhamnoides* L. by UHPLC-ESI-QTOF-MS and their anti-inflammatory activities. *Nat. Prod. Commun.* **2019**, *14*, 1934578X19871404. [CrossRef]
8. Ma, X.; Laaksonen, O.; Zheng, J.; Yang, W.; Trépanier, M.; Kallio, H.; Yang, B. Flavonol glycosides in berries of two major subspecies of sea buckthorn (*Hippophaë rhamnoides* L.) and influence of growth sites. *Food Chem.* **2016**, *200*, 189–198. [CrossRef] [PubMed]
9. Dong, R.F.; Su, J.; Nian, H.; Shen, H.; Zhai, X.; Xin, H.L.; Qin, L.P.; Han, T. Chemical fingerprint and quantitative analysis of flavonoids for quality control of Sea buckthorn leaves by HPLC and UHPLC-ESI-QTOF-MS. *J. Funct. Foods* **2017**, *37*, 513–522. [CrossRef]
10. Yogendra Kumar, M.S.; Tirpude, R.J.; Maheshwari, D.T.; Bansal, A.; Misra, K. Antioxidant and antimicrobial properties of phenolic rich fraction of Seabuckthorn (*Hippophae rhamnoides* L.) leaves *in vitro*. *Food Chem.* **2013**, *141*, 3443–3450. [CrossRef]
11. Sytařová, I.; Orsavová, J.; Snopek, L.; Mlček, J.; Byczyński, Ł.; Mišurcová, L. Impact of phenolic compounds and vitamins C and E on antioxidant activity of sea buckthorn (*Hippophaë rhamnoides* L.) berries and leaves of diverse ripening times. *Food Chem.* **2020**, *310*, 125784. [CrossRef] [PubMed]
12. Gu, M.J.; Lee, H.; Yoo, G.; Kim, D.; Kim, Y.; Choi, I.; Cha, Y.; Ha, S.K. *Hippophae rhamnoides* L. leaf extracts alleviate diabetic nephropathy via attenuation of advanced glycation end product-induced oxidative stress in db/db mice. *Food Funct.* **2023**, *14*, 8396–8408. [CrossRef] [PubMed]
13. Yogendra Kumar, M.S.; Dutta, R.; Prasad, D.; Misra, K. Subcritical water extraction of antioxidant compounds from Seabuckthorn (*Hippophae rhamnoides*) leaves for the comparative evaluation of antioxidant activity. *Food Chem.* **2011**, *127*, 1309–1316. [CrossRef]
14. Vayndorf, E.M.; Lee, S.S.; Liu, R.H. Whole apple extracts increase lifespan, healthspan and resistance to stress in *Caenorhabditis elegans*. *J. Funct. Foods* **2013**, *5*, 1235–1243. [CrossRef]
15. Zhang, M.W.; Zhng, R.F.; Zhang, F.X.; Liu, R.H. Phenolic profiles and antioxidant activity of black rice bran of different commercially available varieties. *J. Agric. Food Chem.* **2010**, *58*, 7580–7587. [CrossRef] [PubMed]
16. Adom, K.K.; Liu, R.H. Rapid peroxyl radical scavenging capacity (PSC) assay for assessing both hydrophilic and lipophilic antioxidants. *J. Agric. Food Chem.* **2005**, *53*, 6572–6580. [CrossRef]
17. Hansen, M.; Hsu, A.L.; Dillin, A.; Kenyon, C. New genes tied to endocrine, metabolic, and dietary regulation of lifespan from a *Caenorhabditis elegans* genomic RNAi screen. *PLOS Genet.* **2005**, *1*, 119–128. [CrossRef]
18. Chen, Y.; Onken, B.; Chen, H.; Xiao, S.; Liu, X.; Driscoll, M.; Cao, Y.; Huang, Q. Mechanism of longevity extension of *Caenorhabditis elegans* induced by pentagalloyl glucose isolated from eucalyptus leaves. *J. Agric. Food Chem.* **2014**, *62*, 3422–3431. [CrossRef]
19. Liu, M.; Li, N.; Lu, X.; Shan, S.; Gao, X.; Cao, Y.; Lu, W. Sweet tea (*Rubus suavissmus* S. Lee) polysaccharides promote the longevity of *Caenorhabditis elegans* through autophagy-dependent insulin and mitochondrial pathways. *J. Food Biochem.* **2022**, *207*, 883–892. [CrossRef]

20. Lu, Y.; Wang, X.; Wu, Y.; Wang, Z.; Zhou, N.; Li, J.; Shang, X.; Lin, P. Chemical characterization of the antioxidant and α-glucosidase inhibitory active fraction of *Malus transitoria* leaves. *Food Chem.* **2022**, *386*, 132863. [CrossRef]
21. Zhao, M.; Linghu, K.; Xiao, L.; Hua, Y.; Zhao, G.; Chen, Q.; Xiong, S.; Shen, L.; Yu, J.; Hou, X.; et al. Anti-inflammatory/anti-oxidant properties and the UPLC-QTOF/MS-based metabolomics discrimination of three yellow camellia species. *Food Res. Int.* **2022**, *160*, 111628. [CrossRef] [PubMed]
22. Zhang, J.; Gao, W.; Cao, M.; Kong, D. Three new flavonoids from the seeds of *Hippophae rhamnoides* subsp. *Sinensis. J. Asian Nat. Prod. Res.* **2012**, *12*, 1122–1129. [CrossRef] [PubMed]
23. Huang, D.S.; Li, C.; Chen, Q.; Xie, X.; Fu, X.; Chen, C.; Huang, Q.; Huang, Z.B.; Dong, H. Identification of polyphenols from *Rosa roxburghii* Tratt pomace and evaluation of in vitro and in vivo antioxidant activity. *Food Chem.* **2022**, *377*, 131922. [CrossRef]
24. Zhang, C.; Xin, X.; Zhang, J.; Zhu, S.; Niu, E.; Zhou, Z.; Liu, D. Comparative evaluation of the phytochemical profiles and antioxidant potentials of olive leaves from 32 cultivars grown in China. *Molecules* **2022**, *27*, 1292. [CrossRef] [PubMed]
25. Zheng, Q.; Tan, W.; Feng, X.; Feng, K.; Zhong, W.; Liao, C.; Liu, Y.; Li, S.; Hu, W. Protective effect of flavonoids from mulberry Leaf on AAPH-Induced oxidative damage in sheep erythrocytes. *Molecules* **2022**, *27*, 7625. [CrossRef]
26. Shumoy, H.; Gabaza, M.; Vandevelde, J.; Raes, K. Soluble and bound phenolic contents and antioxidant capacity of tef injera as affected by traditional fermentation. *J. Food Compos. Anal.* **2017**, *58*, 52–59. [CrossRef]
27. Jiang, S.; Deng, N.; Zheng, B.; Li, T.; Liu, R.H. Rhodiola extract promotes longevity and stress resistance of *Caenorhabditis elegans* via DAF-16 and SKN-1. *Food Funct.* **2021**, *12*, 4471–4483. [CrossRef] [PubMed]
28. Li, H.; Liu, Y.; Gu, Z.; Li, L.; Liu, Y.; Wang, L.; Su, L. p38 MAPK-MK2 pathway regulates the heat-stress-induced accumulation of reactive oxygen species that mediates apoptotic cell death in glial cells. *Oncol. Lett.* **2018**, *15*, 775–782. [CrossRef]
29. Kampkötter, A.; Nkwonkam, C.G.; Zurawski, R.F.; Timpel, C.; Chovolou, Y.; Wätjen, W.; Kahl, R. Investigations of protective effects of the flavonoids quercetin and rutin on stress resistance in the model organism *Caenorhabditis elegans*. *Toxicology* **2007**, *234*, 113–123. [CrossRef]
30. Wang, J.; Gong, H.; Zou, H.; Liang, L.; Wu, X. Isorhamnetin prevents H_2O_2-induced oxidative stress in human retinal pigment epithelial cells. *Mol. Med. Rep.* **2017**, *17*, 648–652. [CrossRef]
31. Maheshwari, D.T.; Yogendra Kumar, M.S.; Verma, S.K.; Singh, V.K.; Singh, S.N. Antioxidant and hepatoprotective activities of phenolic rich fraction of Seabuckthorn (*Hippophae rhamnoides* L.) leaves. *Food Chem. Toxicol.* **2011**, *49*, 2422–2428. [CrossRef]
32. Jin, S.; Li, D.; Shan, L.; Han, L.; Liu, D.; Huang, Z.; Huang, B.; Yan, C. Ethanol extracts of *Panax notoginseng* increase lifespan and protect against oxidative stress in *Caenorhabditis elegans* via the insulin/IGF-1 signaling pathway. *J. Funct. Foods* **2019**, *58*, 218–226. [CrossRef]
33. Tullet, J.M.; Hertweck, M.; An, J.H.; Baker, J.; Hwang, J.Y.; Liu, S.; Oliveira, R.P.; Baumeister, R.; Blackwell, T. Direct inhibition of the longevity-promoting factor SKN-1 by insulin-like signaling in *C. elegans*. *Cell* **2008**, *132*, 1025–1038. [CrossRef] [PubMed]
34. Tan, L.; Zheng, Z.; Huang, L.; Jin, Z.; Li, S.; Wu, G.; Luo, H. Flavonol glycoside complanatoside A requires FOXO/DAF-16, NRF2/SKN-1, and HSF-1 to improve stress resistances and extend the life span of *Caenorhabditis elegans*. *Front. Pharmacol.* **2022**, *13*, 931886. [CrossRef]
35. Dhondt, I.; Petyuk, V.A.; Cai, H.; Vandemeulebroucke, L.; Vierstraete, A.; Smith, R.D.; Depuydt, G.; Braeckman, B.P. FOXO/DAF-16 activation slows down turnover of the majority of proteins in *C. elegans*. *Cell Rep.* **2016**, *16*, 3028–3040. [CrossRef]
36. Wan, Q.; Fu, X.; Meng, X.; Luo, Z.; Dai, W.; Yang, J.; Wang, C.; Wang, H.; Zhou, Q. Hypotaurine promotes longevity and stress tolerance via the stress response factors DAF-16/FOXO and SKN-1/NRF2 in *Caenorhabditis elegans*. *Food Funct.* **2020**, *11*, 347–357. [CrossRef]
37. Luo, X.; Wang, J.; Chen, H.; Zhou, A.; Song, M.; Zhong, Q.; Chen, H.; Cao, Y. Identification of flavoanoids from finger citron and evaluation on their antioxidative and antiaging activities. *Front. Nutr.* **2020**, *7*, 584900. [CrossRef] [PubMed]
38. Li, M.; Chen, X.; Deng, J.; Ouyang, D.; Wang, D.; Liang, Y.; Chen, Y.; Sun, Y. Effect of thermal processing on free and bound phenolic compounds and antioxidant activities of hawthorn. *Food Chem.* **2020**, *332*, 127429. [CrossRef] [PubMed]
39. Das, K.; Muniyappa, H. Age-dependent mitochondrial energy dynamics in the mice heart: Role of superoxide dismutase-2. *Exp. Gerontol.* **2013**, *48*, 947–959. [CrossRef]
40. Rizvi, S.I.; Maurya, P.K. Alterations in antioxidant enzymes during aging in humans. *Mol. Biotechnol.* **2007**, *37*, 58–61. [CrossRef]
41. Zhang, S.Y.; Qin, Z.C.; Sun, Y.Y.; Chen, Y.S.; Chen, W.B.; Wang, H.G.; An, D.; Sun, D.; Liu, Y.Q. Genistein promotes anti-heat stress and antioxidant effects via the coordinated regulation of IIS, HSP, MAPK, DR, and mitochondrial pathways in *Caenorhabditis elegans*. *Antioxidants* **2023**, *12*, 125. [CrossRef]
42. Xu, Q.; Zheng, B.; Li, T.; Liu, R.H. *Hypsizygus marmoreus* extract exhibited antioxidant effects to promote longevity and stress resistance in *Caenorhabditis elegans*. *Food Funct.* **2023**, *14*, 9743–9754. [CrossRef]

antioxidants

Article

Inhibition of Lipid Accumulation and Oxidation in Hepatocytes by Bioactive Bean Extracts

Dya Fita Dibwe [1], Emi Kitayama [2], Saki Oba [2], Nire Takeishi [2], Hitoshi Chiba [3] and Shu-Ping Hui [1,*]

[1] Faculty of Health Sciences, Hokkaido University, Kita-12, Nishi-5, Kita-Ku, Sapporo 060-0812, Japan; dibwedf@hs.hokudai.ac.jp
[2] Graduate School of Health Sciences, Hokkaido University, Kita-12, Nishi-5, Kita-Ku, Sapporo 060-0812, Japan; emi7-04@hotmail.co.jp (E.K.); saki.oba.h@gmail.com (S.O.); n10461991724nasukakstg@docomo.ne.jp (N.T.)
[3] Department of Nutrition, Sapporo University of Health Sciences, Nakanuma Nishi-4-3-1-15, Higashi-Ku, Sapporo 007-0894, Japan; chiba-h@sapporo-hokeniryou-u.ac.jp
* Correspondence: keino@hs.hokudai.ac.jp; Tel./Fax: +81-11-706-3693

Abstract: During our search for natural resources that can inhibit lipid droplet accumulation (LDA) and potentially prevent metabolic dysfunction-associated fatty liver disease (MAFLD) and its progressive stages, such as metabolic dysfunction-associated steatohepatitis (MASH), eight bean extracts (BE1–BE8) were tested for their ability to inhibit lipid accumulation and oxidation in hepatocytes. Substantial inhibitory effects on LDA with bean extracts (BEs) BE2, BE4, BE5, and BE8 were demonstrated. An advanced lipidomic approach was used to quantify the accumulation and inhibition of intracellular triacylglycerol (TAG) and its oxidized species, TAG hydroperoxide (TGOOH), in hepatocytes under fatty acid-loading conditions. The results show that the antioxidants BE2 and BE8 are potential candidates for regulating TAG and TGOOH accumulation in fatty acid-induced lipid droplets (LDs). This study suggests that bean-based foods inhibit LDs formation by decreasing intracellular lipids and lipid hydroperoxides in the hepatocytes. The metabolic profiling of BEs revealed that BE2 and BE8 contained polyphenolic compounds. These may be potential resources for the development of functional foods and drug discovery targeting MAFLD/MASH.

Keywords: functional foods; bean extracts; polyphenols; hepatosteatosis; neutral lipids; oxidative lipidomics; triacylglycerols; hydroperoxides

Citation: Dibwe, D.F.; Kitayama, E.; Oba, S.; Takeishi, N.; Chiba, H.; Hui, S.-P. Inhibition of Lipid Accumulation and Oxidation in Hepatocytes by Bioactive Bean Extracts. *Antioxidants* **2024**, *13*, 513. https://doi.org/10.3390/antiox13050513

Academic Editors: Baojun Xu and Bin Du

Received: 7 March 2024
Revised: 13 April 2024
Accepted: 14 April 2024
Published: 25 April 2024

1. Introduction

Chronic diseases are the primary causes of death and disability worldwide. Metabolic dysfunction-associated fatty liver disease (MAFLD) is the most prevalent chronic liver disease [1]. MAFLD is a chronic liver disorder prevalent worldwide, and numerous scientific studies have suggested that oxidative stress can accelerate its development. This occurs through the activation of Kupffer cells, hepatic stellate cells, and hepatocytes, which can trigger the progression of MAFLD to Metabolic dysfunction-associated steatohepatitis (MASH) [2,3]. Therefore, researchers are investigating the potential benefits of antioxidants in MAFLD treatment. Previous studies have shown that reactive oxygen species (ROS) play a role in MAFLD, supporting the use of antioxidants as potential therapies. Oxidative stress is a key factor in the development of MAFLD, its progression to MASH, and the development of hepatic fibrosis, making antioxidants a potential treatment option for MAFLD [4]. The incidence of MASH-related hepatocellular carcinoma (MASH-HCC) and MAFLD-related hepatocellular carcinoma (MAFLD-HCC) has increased significantly worldwide and is becoming the most common cause of liver cancer [5]. At present, there is a lack of effective therapeutic agents for MASH, as well as tools and indicators to prevent and monitor the progression of MASH to HCC. Therefore, it is of great practical importance to strengthen research on the pathogenesis of MASH-HCC and to find effective drugs to prevent and treat NASH-HCC [6,7].

Lipotoxicity plays an important role in MAFLD pathogenesis. Lipotoxicity manifests as excessive accumulation of free fatty acids in the liver. Hepatocytes are subjected to increased uptake of free fatty acids from the bloodstream. Local and circulating free fatty acids are critical causative agents of insulin resistance, hyperlipidemia, inflammation, and hepatic lipidosis [8–11]. In addition, the major fatty acids present in the human body are palmitic and oleic acids, and these free fatty acids are easily converted to neutral lipid triacylglycerol (TAG) and their oxidized species, namely, triacylglycerol hydroperoxide (TGOOH). Therefore, understanding their roles in intracellular uptake and cytotoxicity is crucial to understanding the mechanisms of MASH [12]. An increasing number of studies have reported an association between excessive intracellular lipid droplet accumulation (LDA) and obesity, diabetes, and other metabolic disorders [13,14]. Hepatic LDA is thought to be involved in the early stages of MAFLD [15]. Inflammation-induced oxidative stress is a common feature of many chronic diseases. Excessive reactive oxygen species cause oxidative damage to important cellular components such as DNA, proteins, and lipid membranes. The hydroxyl radical is highly reactive and rapidly extracts hydrogen atoms from biomolecules in an immediate environment [16]. Phospholipids and membrane triglycerides are the main targets of hydroxyl radical attack and the formation of lipid radicals [17,18]. These lipid radicals are rapidly oxidized, leading to the peroxidation of lipid acyl chains. The excessive intracellular accumulation of neutral lipids and their oxidized species has been associated with various human metabolic diseases such as MAFLD/MASH [19,20]. Agents capable of reducing the accumulation of LDs and oxLDs in hepatocytes/liver via the regulation of neutral lipids TAG and TGOOH represent promising candidates for the prevention and management of MAFLD/MASH and associated diseases. Bioactive compounds of natural origin with LDA inhibitory activity (LDAI) in hepatocytes are potential candidates [12,21–27].

The inhibition of LDA and lipid oxidation is the main approach for preventing hepatic steatosis. In our search for natural crude extracts with LDAI activity, flazin, an alkaloid β-carboline with a C-3 carboxyl group, a C-1 furfuryl alcohol moiety, and its derivative with a piperidine C-ring from oyster extracts were identified as LDAI molecules. They showed significant LDAI activity through the regulation of LDs accumulation. They significantly inhibited the accumulation of TAG species in cells, activated lipolysis, and suppressed lipogenesis. These results suggest that β-carboline alkaloids may be potentially useful in preventing MAFLD. Crude extracts, including medicinal plants and functional foods, are potential sources of LDAI candidates [12,28]. Therefore, functional foods and nutraceuticals may be beneficial in the management of chronic diseases, including MAFLD/MASH, and offer a potential target for drug discovery and developments. Following this approach, we collected familiar beans from the Hokkaido region for use as edible foodstuffs. We examined their effects on hepatocytes as a source of potential natural bioactive substances regulating LDAI, TAG and TGOOH [12,28,29]. Beans are essential crops used as the main vegetable in many countries. Beans, including soybean plants, are rich in protein, ω-3 fatty acids, carbohydrates, dietary fiber, and micronutrients such as folate, manganese, vitamin K, potassium, and zinc. Polyphenols and alkaloids have also been identified in these foods and exhibit numerous bioactivities [30]. Thus, the LDAI capacity of beans was assessed under two fatty acid loading conditions, oleic acid (OA) and linoleic acid (LA) in hepatocytes, and the main secondary metabolites in the bioactive samples were analyzed.

2. Materials and Methods

2.1. Materials

The chemicals and equipment used in this study are described in the supporting information as previously reported [12] (Supplemental Information: Materials and methods). The eight bean samples used in this study were collected by DDF, SO and NT from the Sapporo market in April 2020 and deposited at the Health Innovation Center of the university's health science faculty (Table S1). They were listed as (Code, Scientific name (abbreviation)): BE1, *Glycine max* (*GM1*); BE2, *Glycine max* (*GM2*); BE3, *Vigna angularis*

(*VA*); BE4, *Phaseolus vulgaris* (*PV*); BE5, *Phaseolus coccineus* (*PC1*); BE6, *Phaseolus coccineus* (*PC2*); BE7, *Vicia faba* (*VF*); BE8, *Vanilla planiflolia* (*VP*). To identify the bioactive extracts and metabolites derived from food sources, we prepared eight different food samples, including soybeans, black soybeans, adzuki beans, taisho kintoki, purple kidney beans, white kidney beans, lentils, and vanilla beans (BE1–BE8). Each of these food items was purchased from a market or supermarket in Sapporo in the spring of 2020 (Table S1). First, 10 g of each sample was ground using a mortar and pestle, and the crushed material was transferred to a beaker. Next, 100 mL of methanol was added to the mixture for one hour. The mixture was sonicated for 30 min, which was repeated twice to obtain the extract. Methanol was removed by decompression and dried to obtain the methanol extract. The biological activities of the extracts were evaluated using a lipid accumulation inhibition test.

2.2. Evaluation of Cell Viability and Lipid Droplet Accumulation Inhibition Assay

Cytotoxicity and lipocytotoxicity assays were performed according to the manufacturer's protocol using the Cell Counting Kit-8 (CCK-8; Dojindo Molecular Technologies, Rockville, MD, USA) as described previously [12] (Supplemental Information: Materials and Methods, Figure S1). HepG2 cells were purchased from RIKEN BRC cell bank (Ibaraki, Japan).

LDAI activity was determined using an Oil Red O assay with 24-well plates (n = 4 per treatment) as previously reported [12]. LD staining was performed with modifications based on a previous report [12] (Supplemental Information: Materials and Methods). HepG2 cells were seeded in 24-well plates at a concentration of 5.0×10^4/well and incubated for 24 h. Subsequently, the cells were treated with OA (0.25 mM) and BEs samples in serial dilution concentration from 500 μg/mL and incubated overnight. On the following day, oil red O staining was performed. Specifically, the medium was removed and washed twice with 200 μL of PBS. Next, 200 μL of 10% formalin was added and allowed to stand for 10 min. After removing the formalin and washing twice with 200 μL of PBS, 200 μL of 60% IPA was added and allowed to stand for 10 min. The mixture was then removed, and Oil Red O Working Solution (prepared by mixing Oil Red O Stock Solution and DDW in a 6:1 ratio and filtering) was added and allowed to stand for 15 min. The upper layer was removed, washed with 200 μL of 60% IPA, and then washed twice with 300 μL of PBS. Next, 200 μL of 100% isopropyl alcohol (IPA) was added and de-stained. The upper layer (150 μL) was transferred to a 96-well plate, and the absorbance (at 560 nm) was measured using an AR-VO-MX plate reader. LDA and LDAI quantification were performed for all BEs samples. The quantification of accumulation and inhibition under OA was carried out by comparing the LDA under OA treatment with the untreated control (−OA) group and the LDAI under OA treatment with BE2 and BE8 when compared with the untreated control (+OA) group. All LDA and LDAI data were normalized at the end of the process, and the quantification was expressed as a percentage (with six replicates).

2.3. Analysis of TAG and TGOOH Species

Lipid extraction. HepG2 cells (2.0×10^5/well) were cultured in a dish for 24 h, and after adding fatty acids and extracts, the cells were cultured for another 24 h. On the third day, the media was removed and the cells were washed twice with 300 μL of PBS. Then, 300 μL of trypsin was added and the cells were incubated for 10 min. Then 300 μL of DMEM was added and the cells were collected in 1.5 mL microtubes. After centrifugation at 13,000 rpm for 15 min at 4 °C (High-Speed Micro Centrifuge himac CF 15R, Tokyo, Japan), the supernatant was separated into new 1.5 mL microtubes. The residue was combined with 150 μL MeOH, 400 μL $CHCl_3$ and 10 μL internal standard and vortexed (MSV-3500, Biosan, Riga, Latvia) at a speed of 3500 rpm for 10 min. Then centrifugation at 15,000 rpm for 15 min at 4 °C was performed, and the supernatant was carefully separated into new 1.5 mL microtubes. To the residue, 150 μL MeOH and 400 μL $CHCl_3$ were added, and extraction was carried out. The resulting extracts were combined, and the solvent was removed under vacuum (CC-105, Tommy Seiko Co., Ltd., Tokyo, Japan) for about 1

h. To the dried lipids, 100 μL MeOH was added and vortexed at 3500 rpm for 1 min, then centrifuged at 15,000 rpm at 4 °C for 10 min, and the supernatant was transferred to LC-MS vials and stored at −80 °C until the measurement date.

LC-MS measurements were performed on a Shimadzu Prominence HPLC system (Shimadzu Corp., Kyoto, Japan). It was coupled to an LTQ Orbitrap mass spectrometer (Thermo-Fisher Scientific Inc., Waltham, MA, USA). Lipids extracted from HepG2 cells were purified using an Atlantis T3 C18 column (2.1 × 150 mm, 3 μm, Waters, Milford, MA, USA) at a flow rate of 200 μL/min. LC gradient elution was performed using a mobile phase consisting of 10 mM ammonium acetate solution, isopropanol, and methanol. The measurements were carried out in positive mode: the volTGe of the MS capillary was 4.04 kV, the flow rate of the sheath gas (nitrogen) was 50 psi, and the auxiliary gas (nitrogen) was 20 psi. The high-resolution MS data were obtained in a scan range of m/z 150–1200.

The obtained data were subjected to peak alignment, peak area integration, and quantitative calculations using Thermo Fisher Scientific Xcalibur 2.2. Software with a mass tolerance of 5.0 ppm. The exported quantitative data for TAG and TGOOH lipid species were subjected to further calculation using Excel, then exported to GraphPad Prism V7.0/10.1.2 software for further graph processes and statistical analysis (Supplemental Information: Material and Methods). The data are portrayed in heat maps depicting the accumulated and inhibited TAG and TGOOH species in the cells. The fluctuation of the accumulated TAG and TGOOH species under OA and LA treatment with BE2 and BE8 was quantified and normalized at the end of the process. The accumulation and inhibition of TAG and TGOOH species in the cells were expressed as percentages, and six replicates were used. The graph illustrates the mean values of the accumulation of TAG and TGOOH species in the control (+OA and +LA) group compared with the untreated control (−OA and −LA) groups, and the inhibition of accumulated TAG and TGOOH species in the BE2- and BE8-treated group under +OA and +LA conditions compared with the untreated control (+OA and +LA) groups. The results were statistically significant with p-values of **** $p < 0.0001$, *** $p < 0.001$, ** $p < 0.01$ and * $p < 0.05$ (Supplemental Information: Materials and Methods).

2.4. Metabolites Profiles

NMR profiling of selected bean samples: Methanol extracts were weighed to 0.6 mg each and dissolved in 600 μL of DMSO-d6 to make 10 mg/mL samples for NMR. ^{1}H-NMR proton data were obtained using a 400 MHz JNM-ECX400P (JEOL, Tokyo, Japan). The spectra were processed using JOEL v6.3.0 software, and chemical shift values are expressed in ppm.

HPLC profiling of selected bean samples: Methanol extracts were weighed to 10 mg each and dissolved in 1 mL of methanol to be used for HPLC measurement. The HPLC Solvent System, the mobile phase, consisted of solvents A, Mill-Q, and B, HPLC-grade methanol, at a flow rate of 5 mL/min. For the HPLC method, the gradient method was 0–1 min. The gradient method was 0–1 min, 20% B; 1–20 min, 20–100% B; 20–23 min, 100% B; and a post-run time of 2 min including 25 min, 100–20% B; and stop the measurement. Chromatograms were recorded at 200 nm, 275 nm, 343 nm.

LC-MS profiling of selected bean samples: LC-MS measurements were performed on a Shimadzu Prominence HPLC system (Shimadzu Corp., Kyoto, Japan) coupled to an LTQ Orbitrap system. LC-MS measurements were performed using a Shimadzu Prominence HPLC system (Shimadzu Corp., Kyoto, Japan) coupled to an LTQ Orbitrap mass spectrometer (Thermo Fisher Scientific Inc., San Jose, CA, USA) with an electrospray ionization (ESI) source. The conditions of the LC-MS instrument were as follows: samples of methanol extracts of beans were separated using an Atlantis T3 C18 column (2.1 × 150 mm, 3 μm, 155 Waters, Milford, MA, USA) at a flow rate of 200 μL/min. LC gradient elution was performed using a mobile phase consisting of a 10 mM ammonium acetate solution, isopropanol, and methanol. The measurements were carried out in positive mode: the volTGe of the MS capillary was 4.04 kV, the flow rate of the sheath gas (nitrogen) was 50 psi, and the auxiliary gas (nitrogen) was 20 psi. The high-resolution MS data were obtained in a scan range of m/z 50–500.

3. Results and Discussion

3.1. Evaluation of Cell Viability under Co-Treatment with Fatty Acids and Bean Extracts

In our previous study, the viability of HepG2 cells treated with oleic acid (OA) was measured 24 h after exposure to OA. Briefly, we induced lipid droplets by supplementing DMEM with 0.1, 0.25, and 0.1 mM oleic acid in HepG2 cells after incubating for 24 h to determine the appropriate concentration for the study. Lipid droplets were significantly increased, by 2.6- and 2.8-fold, for 0.25 and 0.5 mM. Thus, 0.25 mM of oleic acid was chosen to induce lipid droplets in HepG2 cells. Moreover, HepG2 cells were tested for viability after treatment with 0.25 mM of oleic acid at different concentrations. No cytotoxicity was observed after 24 h of treatment (Figure S1, Scheme S1, capture image of real-time lipid droplet accumulation). Therefore, 0.5 mM was determined to be a non-toxic concentration of OA and the optimal concentration for inducing LDs [12]. Thus, in the present study, cell viability was assessed 24 h after treatment with the eight different bean extracts at 125–500 μg/mL concentrations. The results show that the IC_{50} values were above 500 μg/mL for all samples (Figure S1). Based on these results, 500 μg/mL was selected as the final concentration of bean extract for the LDAI assay.

3.2. Lipid Droplet Accumulation Inhibition Activity of Selected Bean Extracts

Our previous report revealed that the accumulation of intracellular LDs increases in a concentration-dependent manner under OA conditions. Lipid droplets were significantly induced at a factor of 2.6 and 2.8 for 0.25 and 0.5 mM, respectively. Based on these results, we decided to use 0.25 mM of oleic acid to induce lipid droplet formation in HepG2 cells. HepG2 cells were incubated for 24 h under OA loading (0.1~0.5 mM in the non-toxic range) [12]. In particular, LD accumulation predominantly increased at 0.25 mM and above. The real-time observation of this phenomenon showed that intracellular LD levels increased while the cells were still alive [12]. Based on these results, in the present study, the conditions to induce steatosis in HepG2 cells were set at 0.25 mM OA and 24 h of incubation. Eight selected bean extracts were simultaneously co-treated with 0.25 mM OA and incubated for 24 h. The results are shown in Figure 1A (Scheme S2). All bean samples showed a decreasing trend in LD. In particular, samples BE2, BE4, BE5, and BE8 exhibited a pronounced inhibition of lipid droplet accumulation with LDAI values of 56% at 125 μg/mL and 100% at 250 μg/mL for BE2; 33% at 125 μg/mL and 76% at 250 μg/mL for BE4; 60% at 250 μg/mL and 9.1% at 500 μg/mL for BE5; and 63% at 125 μg/mL and 100% at 500 μg/mL for BE8 (Figure 1). To visualize the inhibition of neutral lipid accumulation in lipid droplets (LDs), we stained LD formation in control +OA and LDAI experiments treated with BE2 and BE8 samples in HepG2 cells loaded with oleic acid. The formation and inhibition of LDs, as well as cell morphological changes, were observed at 24 h by staining with Oil Red O. The significant accumulation of LDs was observed at 24 h after incubation, as demonstrated by imaging with LDA and Oil Red O staining (Figure 1B(B-1,B-2)). LDA was observed in control cells and inhibited in treated samples after 24 h. Both the control and treated groups were used to determine the LDAI activity of BEs samples in HepG2 cells using Oil Red O staining of LDs. In the treated BE8 sample, the number of accumulated LDs was significantly reduced at 24 h, with a clear change in the morphology and size of the LDs compared to those in the control at the corresponding time points, as shown in Figure 1(B-2). The quantification of TAG and TGOOH accumulation in control OA and LA and inhibition by BE2 and BE8 in LDs was performed using Orbitrap-LC-MS/MS analysis under OA and LA conditions.

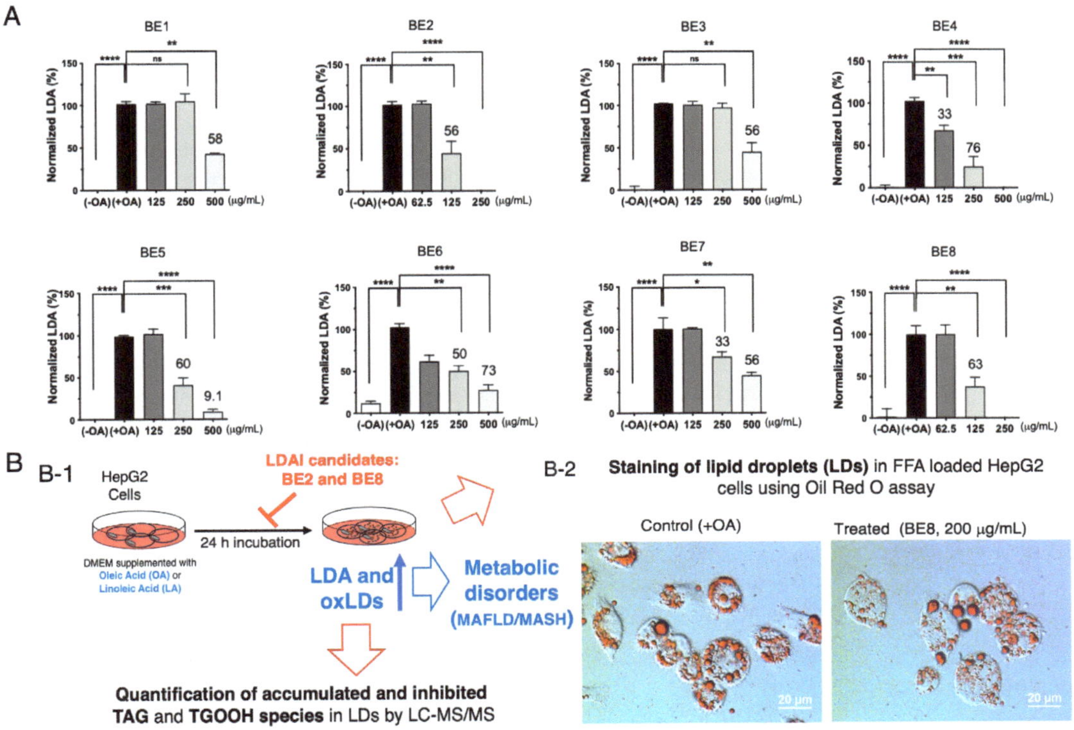

Figure 1. (**A**) LDAI activity of selected bean extracts on OA-loaded HepG2 cells. Graph showing the mean values of the LDAI (four replications). **** $p < 0.0001$, *** $p < 0.001$, ** $p < 0.01$, * $p < 0.05$ and ns not significance, one-way analysis of variance (ANOVA) with Tukey's multiple comparisons test when compared with the untreated control. (**B**) (**B-1**) Discovery of the LDAI candidate strategy: Quantification of TAG and TGOOH accumulation and inhibition in LDs using Orbitrap-LC-MS/MS analysis under OA and LA conditions. (**B-2**) LDAI activity of bioactive BE8 samples in HepG2 cells using an Oil Red O staining technique: phase contrast images showing LDs (red) in control (OA) and treated BE8 cells. The blue colors indicate free fatty acids (OA and LA) used in this study; the small blue arrow indicates the augmentation of LDA and oxLDs in HepG2 cells, and the big blue arrow indicates the metabolic disorders induced in LDA and oxLD conditions. The red colors indicate LDAI candidates that inhibit LDA and oxLDs, as illustrated by the red inhibition mark; the large red arrows indicate quantification of fluctuation of TAG and TGOOH under OA and LA conditions using LC-MS, and Oil Red O staining of LDs.

3.3. Quantification of the Effect on Inhibiting the Accumulation of TAG and TGOOH Species

The accumulation of hepatic lipids, primarily in the form of triacylglycerols (TAGs), is a defining characteristic of MAFLD. Researchers have implicated oxidative stress, as evidenced by the presence of lipid oxidation products, in the progression of more severe forms of MAFLD, such as MASH, and associated diseases [31]. To obtain more accurate information on how changes in molecular species occur, researchers have recently reported the limitations of traditional methods that rely solely on total oxidation determination, and the importance of adopting advanced approaches that overcome these limitations. These investigations may lead to a better understanding of the diagnosis, prevention, and treatment of MAFLD, including MASH and associated diseases. In our ongoing studies, we measured and analyzed fluctuations in TAG and TGOOH species in FFA-loaded HepG2 cells induced by food-derived bioactive compounds. Starting with the abundant individual FFA under physiological conditions, researchers used PA-, OA-, and

LA-loaded HepG2 cells to compare traditional, unique imaging, and lipidomics approaches TAG levels for PA, OA, and LA were observed at 6 h A, 12 h B, 24 h C, and 48 h D, while TGOOH levels for PA, OA, and LA were observed at the same time points. This opens up opportunities to investigate fluctuations in food-derived metabolites that target the regulation of pronounced accumulated TAG and TGOOH neutral lipid species in lipid droplets. Among the BEs samples, BE2 and BE8 significantly inhibited excessive lipid droplet accumulation. Thus, both bioactive LDA extracts were subjected to a detailed investigation of their effects on TAG and TGOOH neutral lipid species in the LDs. In this study, the fluctuation in TAG and TGOOH neutral lipid species induced in lipid droplets of OA- and LA-loaded HepG2 cells after treatment with BE2 and BE8 bioactive samples was quantified in detail.

In our previous study, we assessed the antioxidant capacity of eight selected bean extracts (BE1–BE8) using a DPPH radical scavenging assay. Among these extracts, BE3, BE-7, and BE-8 demonstrated the highest activities, with values of 5.96 nmolTE/mg, 7.38 nmolTE/mg, and 7.60 nmolTE/mg, respectively. Additionally, we evaluated the total phenolic content (TPC) of the eight selected bean samples (BE1–BE2). The results reveal that both BE5 and BE8 showed the highest phenolic contents at 52 and 60 μg/mg, respectively. This finding suggests that the observed antioxidant activity of BE8 may be attributed to the high concentration of phenolic compound derivatives. Previous research has also suggested that phenolic metabolites play a role in preventing metabolic diseases such as MAFDL/MASH. In this study, we tested the ability of the bioactive LDAI extracts to inhibit both neutral lipids TAG and oxidized species of TGOOH, which are implicated in metabolic disorders through oxidative stress [12,28,32].

3.3.1. Analysis of Accumulated TAG and TGOOH Species Induced by Oleic Acid in Hepatocytes

OA-treated HepG2 cells showed a significant increase in LDA, and a significant LDAI was observed for the BE2 and BE8 samples, as shown in Figure 1A. Bioactive BE2 and BE8 samples inhibited LDA in a concentration-dependent manner in HepG2 cells treated with oleic acid. We analyzed the effects of LDAI bioactive bean extracts on TAG species accumulation in hepatocytes using LC-MS. In this study, BE8 (0.25 mM and 0.125 mM) significantly inhibited LDA in HepG2 cells loaded with 0.25 mM OA at 100% and 56%, respectively. The profiles of neutral lipid TAG and TGOOH species in cells incubated for 24 h were analyzed using Orbitrap LC-MS. All the accumulated TAG and TGOOH species induced by OA were detected (Supplementary Materials, Tables S2 and S3). Seventy-five triacylglycerol molecular species were detected using an in-house lipidomic system. More than 30 species were accumulated. Eighteen of the accumulated species were inhibited by BE8 treatment (Figure 2(B-1)). BE8 inhibited the accumulation of seventeen and eighteen TAG molecular species in cells after 24 h of treatment at 125 and 250 μg/mL, respectively. Most of the accumulated TAG molecular species were inhibited by BE8 in the range of 25–49%, except for TAGs 52:4, 52:5, 52:8, 54:7, and 56:10. Eight accumulated TAG molecular species (TAGs:44:0, 46:1, 46:2, 48:1, 48:2, 50:6, 52:1, and 54:1) were inhibited by BE2 by 50–74%. The accumulated triacylglycerol molecular species (TAG 50:1) were particularly inhibited by BE2 in the range of 25% to 49%.

Fourteen accumulated TAG molecular species were inhibited by BE8 at concentrations ranging from 25 to 49% (TAGs 46:1, 46:2, 48:1, 48:2, 48:3, 50:2, 50:4, 50:5, 50:6, 58:12, 56:10, 52:7, 54:0, and 58:12). Between 50% and 74%, significant inhibition was observed, especially by BE2, for eight accumulated TAG molecular species (TAGs:44:0, 46:1, 46:2, 48:1, 48:2, 50:6, 52:1, and 54:1) and only for five accumulated TAGs molecular species (TAGs 52:4, 52:5, 52:8, 54:7, and 56:10) using the BE8.

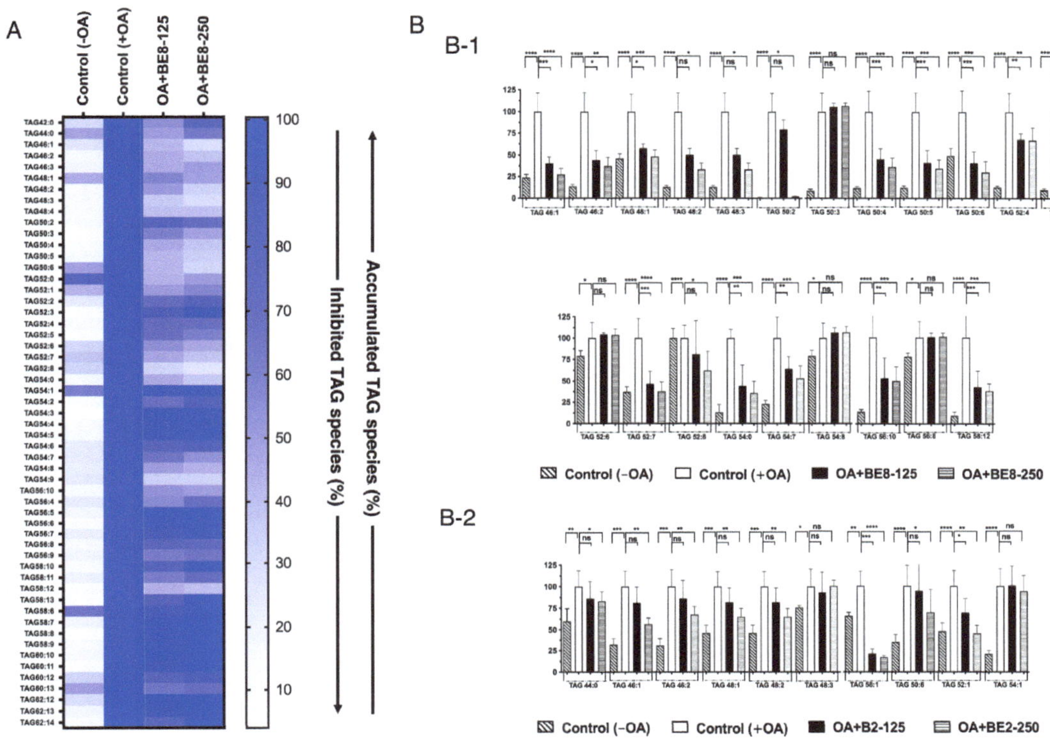

Figure 2. (**A**) Heat map of accumulated and inhibited TAG species in cells. (**B**) Quantification of fluctuation of accumulated TAG species under OA treatment. Analysis of triacylglycerol molecular species under OA treatment with: (**B-1**) BE8 sample and (**B-2**) BE2 sample. Graph showing the mean values of the LDA and LDAI (six replications) **** $p < 0.0001$, *** $p < 0.001$, ** $p < 0.01$, * $p < 0.05$ and ns not significance, one-way analysis of variance (ANOVA) with Tukey's multiple comparisons test when compared with the untreated control (+OA) group. ns, not significant; TAG, triacylglycerol; OA, oleic acid.

Among all accumulated triacylglycerol molecular species, total inhibition was observed only for TAGs 50:2 during BE8 treatment (Figure 2(B-1)). BE2 and BE8 ameliorated LDA by inhibiting the accumulation of triacylglycerol molecular species in human HepG2 cells (Figure 2(B-1,B-2)).

In addition, the fluctuation of the TGOOH species was analyzed under OA conditions. As a result of this analysis, eight TGOOH molecular species accumulated under the OA treatment (Figure 3A and Table S3). Furthermore, among the eight species, three accumulated TGOOH molecular species (TGOOHs 48:1, 48:5 and 54:10) decreased significantly after BE8 sample treatment. The BE2 sample treated under OA conditions showed only a tendency to decrease two TGOOH species (TGOOHs 58:10 and 58:11) (Figure 3B). Among the accumulated TGOOH molecular species TGOOHs 48:1 and 54:10 were significantly inhibited by both BE2 and BE8 under OA conditions.

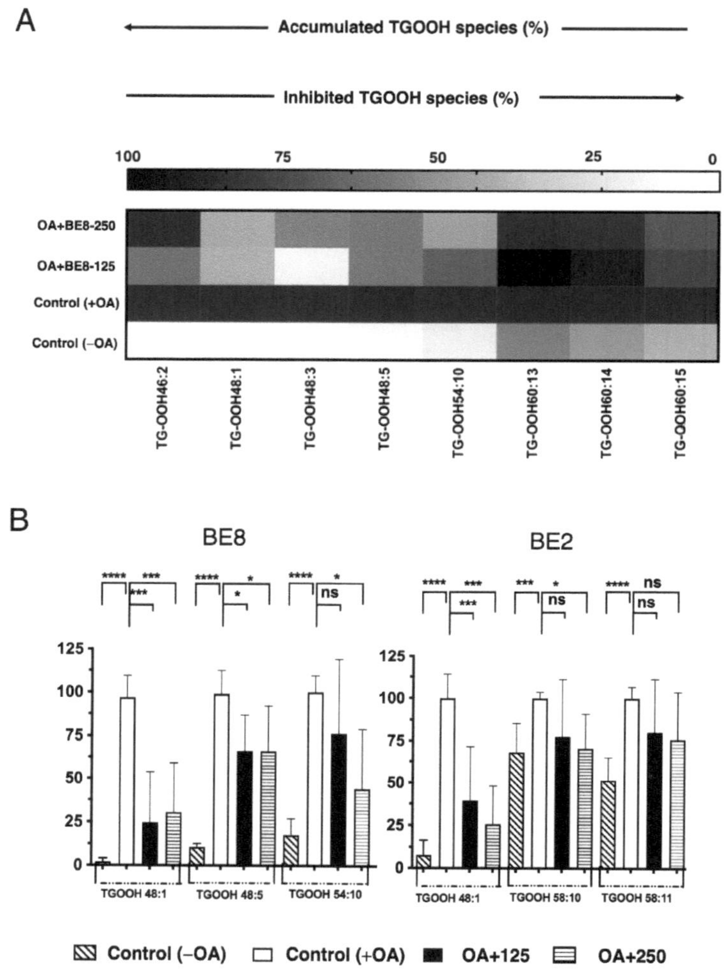

Figure 3. (**A**) Heat map of accumulated and inhibited TGOOH species in cells. (**B**) Quantification of fluctuation of accumulated TGOOH species under OA treatment. Analysis of TAG molecular species under OA treatment with BE2 and BE8. Graph showing the mean values of the LDA and LDAI (six replications) **** $p < 0.0001$, *** $p < 0.001$, * $p < 0.05$ and ns not significance, one-way analysis of variance (ANOVA) with Tukey's multiple comparisons test, when compared with the untreated control (+OA) group. ns, not significant; TGOOH, triacylglycerol hydroperoxide; OA, oleic acid.

3.3.2. Quantification of TAG and TGOOH Species Induced by Linoleic Acid From Traditional and Imaging Approaches to Advanced MS Approaches

A fundamental first step in lipid oxidation is the formation of lipid hydroperoxides, which possess an added hydroperoxyl group and undergo C=C double bond rearrangement. Depending on the source of the oxidation, such as radical oxidation (including thermal oxidation and auto-oxidation), enzymatic oxidation (including lipoxygenase), and singlet oxygen oxidation (including photo-oxidation and inflammation), the generated hydroperoxides (-OOH) exhibit distinct and specific features. These reactions subsequently lead to a series of continuous chain reactions [32,33].

We focused on advanced methods and strategies for oxidative lipidomics to fill the gap in the traditional approaches to oxidized lipid analysis in LDs lipidome. Typically, the analysis of oxidized lipids is based on the indirect investigation of lipid oxidation end products, which can be measured using colorimetric assays, immunoassays, or electron spin resonance. These methods include thiobarbituric acid-reactive substances (TBARS) and malondialdehyde (MDA) tests, ferrous oxidation in xylenol orange (FOX), and gated diene tests. However, the results obtained from these methods, including recent fluorescence approaches, are usually considered an indicator of the degree of lipid oxidation rather than as detailed analyses of changes in lipid species. To address this limitation, the MS/MS-based detection of oxidized molecular lipid species has emerged as a promising approach [32–35].

MS/MS-Based Detection of Oxidized Molecular Lipid Species

MS technology addresses the issues of sensitivity, specificity, accuracy, and dynamic range to a large extent, and enables high-throughput analysis, whereas other methods may require selection and adaptation to measure lipid oxidation products. This is particularly beneficial for the detection of oxidized lipids, generated in complex mixtures with varying levels and structural diversities. The MS-based detection of oxidized lipids is emerging as a promising approach [33].

Currently, LC-MS techniques, particularly LC coupled with tandem MS (LC-MS/MS), are the dominant approaches for oxidative lipidomic applications. Oxidized lipids are structurally characterized by the presence of extra oxygen atoms attached to lipid molecules, such as the insertion of peroxyl/hydroperoxyl/epoxyl groups or deletion of C=C double bonds. By using MS, particularly high-resolution MS, lipid molecules with oxidized functional groups can be captured. MS/MS can then be used for structural confirmation, including the confirmation of the presence of oxygen, the determination of the oxidizing position, and the discovery of specific fragmentation from precursor ions to product ions [33]. Our previous report indicated that linoleic acid (LA) induced LDA and oxLDs in hepatocytes [32]. A unique imaging approach under FA-loaded conditions was reported in relation to conventional approaches for determining the degree of oxidation, where a fluorescence microscopic method for characterizing the size, quantity, and oxidation of lipid droplets (LDs) in HepG2 cells was developed. LDs were induced by palmitic acid (PA), oleic acid (OA), or linoleic acid (LA), and stained with two fluorescent probes for neutral lipids and lipid peroxide; each fatty acid increased the number of LDs and oxidized LDs (oxLDs) and the degree of LD oxidation in a time-dependent manner, as well as increasing intracellular triglyceride hydroperoxide levels. LDs induced by LA without 2,2′-azobis(2-amidinopropane)dihydrochloride (AAPH) showed the most significant degree of oxidation over PA and OA, especially in large LDs (area $\geq 3\ \mu m^2$, oxLD/LD = 52.3 $\pm$ 21.7%). Under these conditions, two food-derived antioxidants (chlorogenic acid and DHMBA) were evaluated, and both significantly improved LD characteristics. Moreover, chlorogenic acid reduced the quantity of large LDs by 74.0–87.6% in a dose-dependent manner, providing a new approach for evaluating the effects of dietary antioxidants on LD characteristics. This novel fluorescence-based method was developed to characterize the number, size distribution, and degree of oxidation of LDs in human hepatocytes. Two fluorescent probes were used to individually analyze intact and oxLDs. TG hydroperoxides (TG-OOH) were determined in hepatocytes using liquid chromatography–mass spectrometry (LC–MS)/MS for comparison with the proposed method, and to determine the changes in molecular species to complete the information on the degree of oxidation. The above information shows the differences between the effects of two antioxidants, chlorogenic acid and DHMBA, on the changes in intracellular lipid species of LDs [32]. Thus, following this new finding, an experiment similar to that on the OA-loaded HepG2 condition was performed by replacing fatty acids with LA to analyze in detail the changes in the fluctuation of accumulated TAG and TGOOH induced by the BE2 and BE8-LDAI extracts. All accumulated TAG and TGOOH species induced by LA were detected by Orbitrap LC-MS (Supplementary Materials, Tables S4 and S5). Fifty-seven TAG species were significantly increased in the positive control (+LA) compared with the negative control (−LA) (Figure 4). During treatment with BE8 under LA conditions, TAGs of the five molecular species decreased in a concentration-dependent manner. The most commonly reduced species was TAG 54:9. During treatment with BE2 under LA conditions, TAG 54:10 levels decreased in a concentration-dependent manner, as shown in Figure 4.

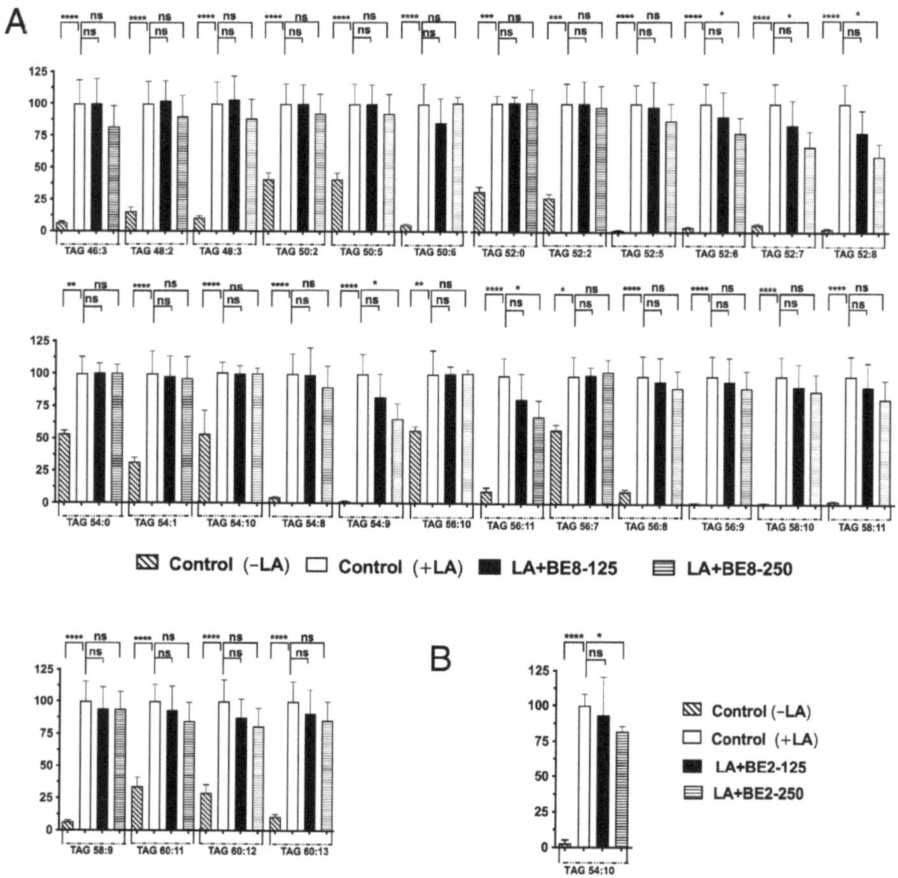

Figure 4. Quantification of the fluctuation of accumulated TAG species under LA treatment. Analysis of triacylglycerol molecular species under OA treatment with (**A**) BE8 and (**B**) BE2. Graph showing the mean values of the LDA and LDAI (six replications) **** $p < 0.0001$, *** $p < 0.001$, ** $p < 0.01$, * $p < 0.05$ and ns not significance, one-way analysis of variance (ANOVA) with Tukey's multiple comparisons test. when compared with the untreated control (+LA) group. ns, not significant; TAG, triacylglycerol; LA, linoleic acid.

In addition, the fluctuation of the TGOOH species was analyzed under LA conditions. The results reveal that approximately fifteen molecular TGOOH species were detected under LA conditions (Supplementary Materials, Figure S4). In addition, among the fifteen species, three accumulated molecular species (TGOOHs 56:4, 56:5, and 60:11) were reduced during the treatment with BE8. The greatest reduction was observed for the TGOOH60:11. However, during treatment with BE2, only the TGOOH52:2 level was significantly reduced. Compared to the results of BE2 and BE8, no common TGOOH species were reduced in either the BE2 or BE8 samples (Figure 5). Under both OA and LA conditions, the BE2 experiment demonstrated inhibition ranging from 0–49% of one TAG in OA and no inhibition in LA, and in the range of 50–85%, inhibition of eight TAGs was observed in OA, but only one TAG in LA. The BE8 experiment showed inhibition results ranging from 0–49% of 14 TAGs in

OA and 1 TAG in LA, and in the range of 50–85%, inhibition of 4 TAGs in OA and 14 TAGs in LA, as shown in Table 1. The evaluation of TGOOH species' inhibition by BE2 under OA and LA conditions showed a range of 0–49% inhibition for one in OA and no inhibition of TGOOH in LA, and a range of 50–85% inhibition for one TGOOH species under OA and no significant inhibition in LA. Meanwhile, BE8 under the same conditions showed a range of 0–49% inhibition for two TGOOH species in OA and three in LA, and a range of 50–85% inhibition for one TGOOH under OA and three in LA as shown in Table 2. The TGOOH comparison results for BE2 and BE8 under OA and LA conditions showed that both BE2 and BE8 significantly reduced TGOOH 48:1 and 58:10 under OA conditions. However, in LA, BE8 was more effective in reducing TGOOH, whereas no significant inhibition was observed for BE2. Therefore, it was suggested that the BE8 may reduce lipid droplets by inhibiting lipid accumulation and their oxidized species in hepatocytes. BE2 was studied in the same manner to compare the components, and it was found that both BE2 and BE8 reduced several different neutral lipids and their oxidized species. These results suggest that different secondary metabolites may be involved in the regulation of LDA by BE2 and BE8 samples.

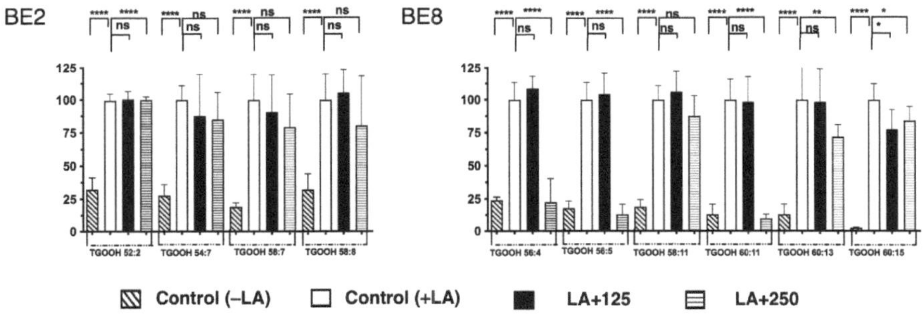

Figure 5. Quantification of fluctuation of accumulated TGOOH species under LA treatment. Analysis of triacylglycerol molecular species under LA treatment with BE2 and BE8. Graph showing the mean values of the LDA and LDAI (six replications) **** $p < 0.0001$, ** $p < 0.01$, * $p < 0.05$ and ns not significance, one-way analysis of variance (ANOVA) with Tukey's multiple comparisons test when compared with the untreated control (+LA) group. ns: not significant; TGOOH, triacylglycerol hydroperoxide; LA, linoleic acid.

Table 1. Comparison of inhibited TAG species by BE2 and BE8 under both OA and LA conditions in the range of inhibition.

Range of Inhibition (%)	Inhibited TAG Species			
	OA		LA	
	BE2	BE8	BE2	BE8
0–49	1	14	0	1
50–85	8	4	1	14

Table 2. Comparison of inhibited TGOOH species by BE2 and BE8 under both OA and LA conditions in the range of inhibition.

Range of Inhibition (%)	Inhibited TGOOH Species			
	OA		LA	
	BE2	BE8	BE2	BE8
0–49	1	2	0	3
50–85	1	1	0	2

The MS/MS-based detection of oxidized molecular lipid species enabled us to clearly observe fluctuations induced by BE2 and BE8 LDAI-bioactive food extracts in individual minor and major accumulated lipid molecular species of TAG and TGOOH. This method allowed for sensitive and accurate quantification with clear discrimination between the two OA- and LA-loading conditions. By extending these results to various FFAs and their combinations, we may gain more insights into the mechanism and diagnosis of oxidative stress via LOOH molecular species. This could potentially expand our understanding of the prevention and discovery of MAFLD/MASH and related metabolic disorders.

3.4. Metabolite Profiling of Bioactive Extracts of Beans

Research on the bioactive compounds present in medicinal plants and edible resources, such as crude natural and food extracts, may lead to the discovery of known or new molecules with potential biological activity. Therefore, this method may be beneficial for drug discovery and development. To achieve this, researchers are utilizing metabolomics to identify and profile the chemical components of these resources with the aim of identifying lead molecules for future drug development, as well as the development of functional foods and nutraceuticals. We screened 50 varieties of vegetables, alkaloids, and phenolic compounds for LDAI activity in FFA-loaded cells. Of these, four BEs samples showed significant LDA activity. Thus, this strategy has been applied to bioactive bean extracts to prevent excessive LD formation in cells, which may serve as a lead molecule for drug discovery and development. The significance of combining bioactive metabolites and their key metabolic signatures in functional dietary foods and supplements using NMR and LC-MS metabolomics and complementary techniques, such as diagnostic fragmentation, has been emphasized. BEs samples were compared using NMR and LC-MS. Furthermore, the results of this approach allow for the targeted isolation of key metabolites, and provide a range of possibilities for synthetic routes for LDAI drug discovery and development, as there are currently no chemotherapeutic compounds available.

^{1}H-NMR analyses of the BE1–BE8 bean samples indicated the presence of polyphenols as glycosides and aglycone metabolites as one of the chemical constituent classes. Additionally, HPLC analysis of the selected samples revealed prominent peaks in BE8 that were not present in the other samples, including BE2, which was consistent with the NMR and LC-MS results (Supplementary Materials, Figures S2–S11). Previous research has suggested that vanilla beans (*V. planifolia*) contain various components, including phenolic acid derivatives [30]. In this study, four LDAI-bioactive bean extracts were analyzed using LC-MS for metabolite profiling. Chemical metabolite screening from the LC-MS/MS datasets of BE2, BE4, BE5, and BE8 was previously investigated and indicated unknown compounds in molecular networking. In this study, a complementary analysis using diagnostic fragmentation filtering was performed. The results for all four bioactive extracts are presented as 3D images (Figure S9). Several minor phenolic compounds and polyphenolic metabolites (**1–11**) were identified from the LC-MS data of the bioactive BEs (Table 3, Figures S9). The results show that seven phenolic compounds were detected in BE8. These compounds were classified into acid and aldehyde groups. Vanillin, one of the major components of the BE8 sample, was an aldehyde, and its presence was indicated by both the LC-MS/MS and NMR data (Figures S5 and S11). In contrast, four polyphenolic metabolites (**7–11**) were identified in the BE8 as its minor constituents. Total polyphenol measurements showed that the BE8 sample contained 60 μg of polyphenols in 1 mg. Five polyphenolic metabolites were detected in BE2, four in BE4, and three in BE5. The chemical metabolite analysis of extracts BE2 and BE8 revealed that both extracts are rich sources of minor polyphenolic compounds that could be promising targets for the discovery of candidate LDA and oxLD inhibitors. Our previous report revealed that both BE5 and BE8 samples have the highest phenolic contents, with total phenolic content (TPC) values of 52 and 60 mg/mg, respectively, among the eight selected bean samples (BE1–BE8). This finding suggests that the observed antioxidant activity in BE8 may be attributed to the high concentration of phenolic compound derivatives. Previous research has also suggested that phenolic metabolites play a role in preventing metabolic diseases such as MAFDL/MASH. These results suggest that bioactive bean extracts could be a potential resource for the development of nutraceuticals and furthering natural drug discovery for the prevention and management of MAFLD/MASH and oxidative stress-related diseases [28].

Table 3. Detection of phenolic and polyphenolic compounds from bioactive BEs.

Compounds	RT	Ion	Calc. *m/z*	Exptl. *m/z*	ppm
1. Gallic acid hydrate	1.83	$[M+H]^+$	171.0288	171.0289	0.58
2. Syringic acid	2.16	$[M+H]^+$	198.0528	198.0524	−2.02
3. 4-Hydroxybenzoic acid	2.1	$[M+H]^+$	139.0390	139.0382	−5.75
4. Vanillic acid	2.02	$[M+H]^+$	169.0495	169.0491	−2.37
5. 3,4-Dihydroxybenzoic aldehyde	2.1	$[M+H]^+$	139.0390	139.0382	−5.75
6. 4-Hydroxybenzoic aldehyde	3.13	$[M+H]^+$	123.0441	123.0439	−1.63
7. Vanillin	3.13	$[M+H]^+$	153.0546	153.0545	−0.65
8. Daidzein	2.75	$[M+H]^+$	255.0642	255.0650	3.14
9. Genistein	2.96	$[M+H]^+$	271.0601	271.0599	−0.74
10. Anthocyanin derivative	2.24	$[M+H]^+$	291.0863	291.0851	−4.12
11. Genistein glucoside	2.69	$[M+H]^+$	431.0984	431.0991	1.62

3.5. Strengths, Limitations and Future Prospects of the Oxidative Lipidomic Approach Based on the LDAI Study

Additional research is required to assess the bioavailability of bioactive metabolites from various natural and food extracts. Examining the relationship between the structures of these natural compounds and their biological activities may offer valuable insights into their potential for use as effective interventions for preventing and treating the accumulation of LDs and oxLDs via TAG-neutral lipid accumulation and TGOOH oxidized species in hepatocytes.

Advanced lipidomics approaches are powerful and more appropriate for the detection of oxidized intracellular neutral lipid species of lipid droplets, mainly TAGs and their hydroperoxide species, because they can accurately characterize the neutral lipids of individual species and their lipid hydroperoxide (LOOHs) species using mass fragmentation MS2; in contrast, traditional approaches mainly indicate the degree of oxidation instead of the changes occurring in molecular lipid species. Molecular species of LOOHs are prominent non-radical intermediates of lipid peroxidation in both model systems and cells. Advanced lipidomic approaches can fill the gap in information regarding the value of total lipid accumulation and oxidation, and provide valuable potential diagnostic biomarker candidates for more detailed mechanistic investigations. Recently, cholesterol-derived hydroperoxides (ChOOHs) and intracellular neutral lipids such as TGOOHs and CEOOHs have been targeted in this regard for both model systems and cells to clarify their potential effects on prevention, or their use in drug discovery and development investigations connected to metabolic deregulation, by also revealing the capacity of this approach to address the limitation of the degree of oxidation achievable by conventional methods, and providing more sensitive and accurate comprehensive changes as well as the quantification of molecular lipid species under oxidative stress, as detailed in the following four selected reports among our ongoing studies on oxidative lipidomics [32–35].

Oxidative stress and other disorders, such as cardiovascular diseases, neurodegenerative diseases, and cancers, can be detected using oxidized lipids as potential markers. Lipidomic analysis, which is an advanced approach for studying lipid molecules, is beneficial for biomarker discovery, clinical diagnosis, and pathway exploration. Recent advances in mass spectrometry technology have made it possible to extend the investigation of oxidized lipids, known as oxidative lipidomics. This technique combines the complexity of lipidomics with a focus on oxidation, making it more challenging than regular lipidomics, particularly for sample preparation, standards, and spectral analysis. Despite these challenges, oxidative lipidomics has been applied to clinical and animal experiments to study various diseases. Hui et al. recently described the methods and strategies used in oxidative lipidomics and summarized their current applications [33]. Thus, the development of chromatographic–MS spectrometric technology is expected to further our understanding of the relationships among lipids, oxidation, and human health. A recent study by our research group examined the lipidomic profile of a MASH mouse model and demonstrated that oxidative stress significantly influences lipid regulation and peroxidation at the molecular level. Lipid hydroperoxides, which are involved in the abnormal cycle of lipid metabolism and the disruption of energy production, have emerged as potential diagnostic markers and evaluation indices of the effects of antioxidants. As crucial metabolic organs, the liver and kidneys exhibit oxidative stress-associated dysregulation of the lipidome, and display organ-specific characteristics. Our findings contribute to a better understanding of the role of oxidative stress in NASH development, by highlighting the effects of oxidative damage-induced lipid hydroperoxidation. Chronic kidney disease (CKD), characterized by the progressive loss of kidney function, is associated with renal lipid droplet (LD) accumulation and oxidative damage. LD quality may contribute to CKD development. This study aimed to examine the chemical makeup of LDs formed in human kidney cells exposed to free fatty acids as sources of LDs and oxidized lipoproteins under oxidative stress. LDs were extracted directly from cells using nanotips, followed by in-tip microextraction, and the LD lipidomic profile was determined using nanoelectrospray mass spectrometry. As a result, free fatty acids elevated the LD lipid content and altered their composition considerably. Oxidized lipoproteins distort the proportion of intact lipids, such as triacylglycerols (TAG), phosphatidylcholines (PC), phosphatidylethanolamines (PE), and cholesteryl esters (CE). Notably, the levels of oxidized lipids, including TAG, PC, and PE hydroperoxides, were significantly increased in a dose-dependent manner. Furthermore, the dysregulation of intact lipids is accompanied by the accumulation of lipid hydroperoxides. This study revealed that lipid oxidation and lipid metabolism dysregulation coexist in LDs in kidney cells, presenting a potential new target facilitating diagnosis and offering fresh insights into CKD [34,35].

The fluorescence imaging approach proposed by Tsuikui et al. has the potential to provide precise physicochemical details of LDs in hepatocytes, which could be beneficial to exploring antioxidants in foods and drugs to prevent and alleviate health issues associated with LD accumulation and lipid oxidation. In that study, chlorogenic acid was found to reduce the numbers of large LDs and oxLDs, possibly through TG hydrolysis rather than antioxidative effects. Therefore, the proposed method could be a valuable tool enabling future research on interactions between antioxidants and LDs. In conclusion, the proposed imaging method has the potential to offer detailed physicochemical information regarding LDs in hepatocytes, which may be useful for exploring antioxidants in foods and drugs to prevent and alleviate health issues involving LD accumulation and lipid oxidation. However, one major limitation of the current study is the technical aspect, and it should be noted that LD metabolism involves several cross-talks between ROS, ER stress, mitochondrial function, and lipogenic and lipolytic enzymes, which may also be involved in LD oxidation. Understanding these interactions may help us gain a better understanding of LDs. Additionally, it is important to note that data from primary cell cultures may differ from those retrieved from cultured cancer cell lines, which may not perfectly represent the physiological conditions. Further studies are required to verify these differences. However, it is noteworthy that HepG2 cells are more suitable for use in the stable screening of antioxidants than primary cells because of the availability and easy growth of HepG2 cells. Thus, the proposed method using HepG2 cells is a promising tool that can be used worldwide [12,28,32]. Finally, an oxidative lipidomics study of food extracts and derived food compounds under individual FFAs and the combination of FFAs to clarify the effect of LDAI on prevention and drug discovery related to MAFLD/MASH and related metabolic disorders was linked to the excessive production of LDs and oxLDs.

4. Conclusions

In this study, the inhibitory effect of bean extract on lipid droplet accumulation in FFA-loaded HepG2 cells was investigated. The results show that BE2, BE4, BE5, and BE8 significantly inhibited lipid droplet accumulation in hepatocytes. In addition, the inhibition of accumulation of TAG neutral lipids and their oxidized species in hepatocytes by most bioactive bean extracts, BE2 and BE8, was quantified using a lipidomic approach. BE2 and BE8 significantly inhibited TAG and TGOOH accumulation under OA and LA conditions. Analysis of the chemical constituents of BE2 and BE8 revealed that both extracts were rich sources of polyphenolic metabolites that could be promising targets for the discovery of LDA and oxLD regulator candidates. Beans may be effective foods owing to the limitation of excessive accumulation of lipid droplets via their intracellular neutral lipids, which may potentially impact the prevention of early MAFLD symptoms and serve as potential sources for drug discovery and development.

Supplementary Materials: The following supporting information can be downloaded at https://www.mdpi.com/article/10.3390/antiox13050513/s1. Supplementary Materials and Methods: 1. Chemicals and instruments; 2. Liquid chromatography/mass spectrometry profiling of vanilla bean extract; 3. Lipid droplet accumulation inhibition assay; 4. Metabolite identification using DFF and ^{1}H-NMR analyses of BEs; Table S1: List of selected beans used in this study; Figure S1: Cytotoxicity of selected bean samples in HepG2 cell lines; Table S2: Accumulation of TAG species induced by OA; Table S3: Accumulation of TGOOH species induced by OA; Table S4: Accumulation of TAG species induced by LA; Table S5: Accumulation of TGOOH species induced by LA; Figure S2: ^{1}H NMR spectrum of BE2 in DMSO-d6; Figure S3: ^{1}H NMR spectrum of BE4 in DMSO-d6; Figure S4: ^{1}H NMR spectrum of BE5 in DMSO-d6; Figure S5: ^{1}H NMR spectrum of BE8 in DMSO-d6; Figure S6: ^{1}H NMR spectra of bioactive extract BEs (BE2 and BE8) in DMSO-d6; Figure S7: ^{1}H NMR spectra of bioactive extract BEs (BE2, BE4, BE5 and BE8) in DMSO-d6; Figure S8: A schema of the general data processing workflow of LC-MS data; Figure S9: LC-MS profiling of bioactive bean extracts. (A) Diagnostic fragmentation filtering (DFF) plot for metabolites analysis. (B) 3D visualization of MS data; Figure S10: Identification of vanillin in BE8. (A) Three-dimensional (3D) liquid chromatography/mass spectrometry (LC-MS) of BE8 and vanillin. (B) Plot of the vanillin in BE8 using diagnostic fragmentation filtering (DFF); Figure S11. HPLC profile of selected bean samples BE2 and BE8 at 200 nm. Scheme S1: (A) Capture of the LDA real-time images on HepG2 cells with the interval of 6 h. (B) Vial ability staining with AO and EB. (C) Evaluation of LDA and oxLDs inhibition strategy' Scheme S2: (A) Comparison of the capture of the LDA under −OA and +OA in real-time images on HepG2 cells with the interval of 12 h. (B) LDAI activity of BE1-BE8. (C) Comparison of the vial ability staining with AO and EB under −OA and +OA.

Author Contributions: Conceptualization and design of the study, D.F.D.; data curation, D.F.D., E.K., S.O. and N.T.; formal analysis, D.F.D., E.K., S.O. and N.T.; funding acquisition, D.F.D.; investigation and performance of the experiments, D.F.D., E.K., S.O. and N.T.; methodology, D.F.D.; project administration, D.F.D.; resources, H.C. and S.-P.H.; software, D.F.D.; supervision, D.F.D.; H.C. and S.-P.H.; validation, D.F.D., H.C. and S.-P.H.; visualization, D.F.D. and S.-P.H.; writing—original draft, D.F.D.; writing—review and editing, D.F.D. All authors have read and agreed to the published version of the manuscript.

Funding: This research was supported by the Startup Research grant, Young Teacher Encouragement Grant (YTEG), and a research grant from the Faculty of Health Science at Hokkaido University (870411H) to D.F.D.

Institutional Review Board Statement: Not applicable.

Informed Consent Statement: Not applicable.

Data Availability Statement: Data are contained within the article and Supplementary Materials.

Acknowledgments: The authors thank the Central Research Laboratory, Faculty of Health Sciences, Hokkaido University, for kindly providing the laboratory equipment.

Conflicts of Interest: The authors declare no conflicts of interest.

Abbreviations

LDA, lipid droplet accumulation; LDAI, lipid droplet accumulation inhibition; MAFLD, metabolic dysfunction-associated fatty liver disease; MASH, metabolic dysfunction-associated steatohepatitis, HCC, hepatocellular carcinoma; oxLD, oxidized lipid droplet; OA, oleic acid; LA, linoleic acid; EB, ethidium bromide; AO, acridine orange; LD, lipid droplet; TAGs, triacylglycerol; TGOOH, triacylglycerol hydroperoxide; FFA, free fatty acid; LC/MS, liquid chromatography/mass spectrometry; NMR, nuclear magnetic resonance; HPLC, high-performance liquid chromatography.

References

1. Müller, F.A.; Sturla, S.J. Human in vitro models of nonalcoholic fatty liver disease. *Curr. Opin. Toxicol.* **2019**, *16*, 9–16. [CrossRef]
2. Shao, G.; Lui, Y.; Lu, L.; Zhang, G.; Zhou, W.; Wu, T.; Wang, L.; Xu, H.; Ji, G. The Pathogenesis of HCC Driven by NASH and the Preventive and Therapeutic Effects of Natural Products. *Front. Pharmacol.* **2022**, *13*, 9440088. [CrossRef] [PubMed]
3. Ziamajidi, N.; Khaghani, S.; Hassanzadeh, G.; Vardasbi, S.; Ahmadian, S.; Nowrouzi, A.; Ghaffari, S.M.; Abdirad, A. Amelioration by chicory seed extract of diabetes- and oleic acid-induced non-alcoholic fatty liver disease (NAFLD)/non-alcoholic steatohepatitis (NASH) via modulation of PPARα and SREBP-1. *Food Chem. Toxicol.* **2013**, *58*, 198–209. [CrossRef] [PubMed]
4. Younossi, Z.M. Non-Alcoholic Fatty Liver Disease—A Global Public Health Perspective. *J. Hepatol.* **2019**, *70*, 531–544. [CrossRef] [PubMed]
5. Ganguli, S.; DeLeeuw, P.; Satapathy, S.K. A Review of Current and Upcoming Treatment Modalities in Non-Alcoholic Fatty Liver Disease and Non-Alcoholic Steatohepatitis. *Hepatic Med. Evid. Res.* **2019**, *11*, 159–178. [CrossRef] [PubMed]
6. Basaranoglu, M.; Neuschwander-Tetri, B.A. Nonalcoholic Fatty Liver Disease: Clinical Features and Pathogenesis. *Gastroenterol. Hepatol.* **2006**, *2*, 282–291.
7. Wang, T.; Zhao, J.; Li, X.; Xu, Q.; Liu, Y.; Khan, I.A.; Yang, S. New Alkaloids from Green Vegetable Soybeans and Their Inhibitory Activities on the Proliferation of Concanavalin A-Activated Lymphocytes. *J. Agric. Food Chem.* **2016**, *64*, 1649–1656. [CrossRef] [PubMed]
8. Dolgova, N.; Uhlemann, E.-M.E.; Boniecki, M.T.; Vizeacoumar, F.S.; Ralle, M.; Tonelli, M.; Abbas, S.A.; Patry, J.; Elhasasna, H.; Freywald, A.; et al. MEMO1 is a Metal Containing Regulator of Iron Homeostasis in Cancer Cells. *bioRxiv* **2023**. [CrossRef]
9. Walther, T.C.; Chung, J.; Farese, R.V., Jr. Lipid Droplet Biogenesis. *Annu. Rev. Cell Dev. Biol.* **2017**, *33*, 491–510. [CrossRef] [PubMed]
10. Cruz, A.L.S.; Barreto, E.A.; Fazolini, N.P.B.; Viola, J.P.B.; Bozza, P.T. Lipid droplets: Platforms with multiple functions in cancer hallmarks. *Cell Death Dis.* **2020**, *11*, 105. [CrossRef]
11. Greenberg, A.S.; Coleman, R.A.; Kraemer, F.B.; McManaman, J.L.; Obin, M.S.; Puri, V.; Yan, Q.W.; Miyoshi, H.; Mashek, D.G. The role of lipid droplets in metabolic disease in rodents and humans. *J. Clin. Investig.* **2011**, *121*, 2102–2110. [CrossRef]
12. Dibwe, D.F.; Oba, S.; Takeishi, N.; Sakurai, T.; Tsukui, T.; Chiba, H.; Hui, S.P. Food-Derived β-Carboline Alkaloids Ameliorate Lipid Droplet Accumulation in Human Hepatocytes. *Pharmaceuticals* **2022**, *15*, 578. [CrossRef] [PubMed]
13. Gluchowski, N.L.; Becuwe, M.; Walther, T.C.; Farese, R.V. Lipid Droplets and Liver Disease: From Basic Biology to Clinical Implications. *Nat. Rev. Gastroenterol. Hepatol.* **2017**, *14*, 343–355. [CrossRef]
14. Krahmer, N.; Farese, R.V.; Walther, T.C. Balancing the Fat: Lipid Droplets and Human Disease. *EMBO Mol. Med.* **2013**, *5*, 973–983. [CrossRef] [PubMed]
15. Hauck, A.K.; Bernlohr, D.A. Oxidative stress and lipotoxicity. *J. Lipid Res.* **2016**, *57*, 1976–1986. [CrossRef] [PubMed]
16. Ayala, A.; Muñoz, M.F.; Argüelles, S. Lipid peroxidation: Production, metabolism, and signaling mechanisms of malondialdehyde and 4-hydroxy-2-nonenal. *Oxid. Med. Cell. Longev.* **2014**, *2014*, 360438. [CrossRef] [PubMed]
17. Valgimigli, L. Lipid Peroxidation and Antioxidant Protection. *Biomolecules* **2023**, *13*, 1291. [CrossRef] [PubMed]
18. Mashek, D.G. Hepatic Lipid Droplets: A Balancing Act between Energy Storage and Metabolic Dysfunction in NAFLD. *Mol. Metab.* **2021**, *50*, 101115. [CrossRef] [PubMed]
19. Souza, M.R.d.A.; Diniz, M.F.F.M.; de Medeiros-Filho, J.E.M.; de Araújo, M.S.T. Metabolic Syndrome and Risk Factors for Non-Alcoholic Fatty Liver Disease. *Arq. Gastroenterol.* **2012**, *49*, 89–96. [CrossRef]
20. López Panqueva, R.d.P. Pathological Aspects of Fatty Liver Disease. *Rev. Colomb. Gastroenterol.* **2014**, *29*, 82–88.
21. Hernández-Pérez, E.; León García, P.E.; López-Díazguerrero Henne, W.M.; Reese, M.L.; Goodman, J.M. The Assembly of Lipid Droplets and Their Roles in Challenged Cells. *EMBO J.* **2018**, *37*, e98947. [CrossRef] [PubMed]
22. Hernández-Pérez, E.; León García, P.E.; López-Díazguerrero, N.E.; Rivera-Cabrera, F.; Del Ángel Benítez, E. Liver steatosis and nonalcoholic steatohepatitis: From pathogenesis to therapy. *Medwave* **2016**, *16*, e6535. [CrossRef] [PubMed]
23. Sumida, Y.; Okanoue, T.; Nakajima, A. Phase 3 Drug Pipelines in the Treatment of Non-Alcoholic Steatohepatitis. *Hepatol. Res.* **2019**, *49*, 1256–1262. [CrossRef] [PubMed]
24. Pei, K.; Gui, T.; Kan, D.; Feng, H.; Jin, Y.; Yang, Y.; Zhang, Q.; Du, Z.; Gai, Z.; Wu, J.; et al. An Overview of Lipid Metabolism and Nonalcoholic Fatty Liver Disease. *Biomed Res. Int.* **2020**, *2020*, 4020249. [CrossRef] [PubMed]
25. Alkhatib, A.; Tsang, C.; Tiss, A.; Bahorun, T.; Arefanian, H.; Barake, R.; Khadir, A.; Tuomilehto, J. Functional Foods and Lifestyle Approaches for Diabetes Prevention and Management. *Nutrients* **2017**, *9*, 1310. [CrossRef] [PubMed]

26. Rana, J.C. *Molecular Breeding and Nutritional Aspects of Buckwheat*; Elsevier: Amsterdam, The Netherlands, 2016; pp. 109–129.
27. Cheng, C.; Li, Z.; Zhao, X.; Liao, C.; Quan, J.; Bode, A.M.; Cao, Y.; Luo, X. Natural alkaloid and polyphenol compounds targeting lipid metabolism: Treatment implications in metabolic diseases. *Eur. J. Pharmacol.* **2020**, *870*, 172922. [CrossRef]
28. Dibwe, D.F.; Takeishi, N.; Oba, S.; Sakurai, A.; Sakurai, T.; Tsukui, T.; Chiba, H.; Hui, S.P. Identification of a b-carboline alkaloid from chemoselectively derived vanilla bean extract and its prevention of lipid droplet accumulation in human hepatocytes (HepG2). *Molecules* **2023**, *15*, 8024. [CrossRef] [PubMed]
29. Cheng, C.; Zhuo, S.; Zhang, B.; Zhao, X.; Liu, Y.; Liao, C.; Quan, J.; Li, Z.; Bode, A.M.; Cao, Y.; et al. Treatment implications of natural compounds targeting lipid metabolism in nonalcoholic fatty liver disease, obesity and cancer. *Int. J. Biol. Sci.* **2019**, *15*, 1654–1663. [CrossRef] [PubMed]
30. Busconi, M.; Lucini, L.; Soffritti, G.; Bernardi, J.; Bernardo, L.; Brunschwig, C.; Lepers-Andrzejewski, S.; Raharivelomanana, P.; Fernandez, J.A. Phenolic Profiling for Traceability of *Vanilla* × *tahitensis*. *Front. Plant Sci.* **2017**, *8*, 1746. [CrossRef] [PubMed]
31. Ipsen, D.H.; Lykkesfeldt, J.; Tveden-Nyborg, P. Molecular mechanisms of hepatic lipid accumulation in non-alcoholic fatty liver disease. *Cell. Mol. Life Sci.* **2018**, *75*, 3313–3327. [CrossRef] [PubMed]
32. Tsukui, T.; Chen, Z.; Fuda, H.; Furukawa, T.; Oura, K.; Sakurai, T.; Hui, S.P.; Chiba, H. Novel Fluorescence-Based Method to Characterize the Antioxidative Effects of Food Metabolites on Lipid Droplets in Cultured Hepatocytes. *J. Agric. Food Chem.* **2019**, *67*, 9934–9941. [CrossRef] [PubMed]
33. Chen, Z.; Wu, X.; Chiba, H.; Hui, S.-P. Investigating oxidized lipids in an omics way: Oxidative lipidomics in biological applications using mass spectrometry. *Med. Mass Spectrom.* **2022**, *6*, 72–84. [CrossRef]
34. Sazaki, I.; Sakurai, T.; Yamahata, A.; Mogi, S.; Inoue, N.; Ishida, K.; Kikkai, A.; Takeshita, H.; Sakurai, A.; Takahashi, Y.; et al. Oxidized Low-Density Lipoproteins Trigger Hepatocellular Oxidative Stress with the Formation of Cholesteryl Ester Hydroperoxide-Enriched Lipid Droplets. *Int. J. Mol. Sci.* **2023**, *24*, 4281. [CrossRef] [PubMed]
35. Wu, Y.; Chen, Z.; Fuda, H.; Tsukui, T.; Wu, X.; Shen, N.; Saito, N.; Chiba, H.; Hui, S.-P. Oxidative Stress Linked Organ Lipid Hydroperoxidation and Dysregulation in Mouse Model of Nonalcoholic Steatohepatitis: Revealed by Lipidomic Profiling of Liver and Kidney. *Antioxidants* **2021**, *10*, 1602. [CrossRef] [PubMed]

antioxidants

MDPI

Article

Taurine Chloramine-Mediated Nrf2 Activation and HO-1 Induction Confer Protective Effects in Astrocytes

Song-I Seol [1], In Soon Kang [2], Ji Seok Lee [2,3], Ja-Kyeong Lee [1,*] and Chaekyun Kim [2,3,*]

1 Department of Anatomy, Inha University School of Medicine, Incheon 22212, Republic of Korea; 22192191@inha.edu
2 Laboratory of Leukocyte Signaling Research, Department of Pharmacology, Inha University School of Medicine, Incheon 22212, Republic of Korea; round001@inha.ac.kr (I.S.K.); jiseok@inha.edu (J.S.L.)
3 BK21, Program in Biomedical Science & Engineering, Inha University, Incheon 22212, Republic of Korea
* Correspondence: jklee@inha.ac.kr (J.-K.L.); chaekyun@inha.ac.kr (C.K.); Tel.: +82-32-860-9893 (J.-K.L.); +82-32-860-9874 (C.K.); Fax: +82-32-885-8302 (J.-K.L. & C.K.)

Abstract: Taurine is ubiquitously distributed in mammalian tissues, with the highest levels in the brain, heart, and leukocytes. Taurine reacts with hypochlorous acid (HOCl) to produce taurine chloramine (Tau-Cl) via the myeloperoxidase (MPO) system. In this study, we elucidated the antioxidative and protective effects of Tau-Cl in astrocytes. Tau-Cl increased the expression and nuclear translocation of nuclear factor E2-related factor (Nrf2) and the expression of Nrf2-regulated antioxidant genes, including heme oxygenase 1 (HO-1). Nrf2 activity is negatively regulated by Kelch-like ECH-associated protein 1 (Keap1). Tau-Cl decreased the level of the reduced thiol groups of Keap1, resulting in the disruption of the Keap1-Nrf2 complex. Consequently, Tau-Cl rescued the H_2O_2-induced cell death by enhancing HO-1 expression and suppressing reactive oxygen species. In conclusion, Tau-Cl confers protective effects in astrocytes by disrupting the Keap1-Nrf2 complex, thereby promoting Nrf2 translocation to the nucleus, wherein it binds to the antioxidant response element (ARE) and accelerates the transcription of antioxidant genes. Therefore, in astrocytes, the activation of the Keap1-Nrf2-ARE pathway by Tau-Cl may increase antioxidants and anti-inflammatory mediators as well as other cytoprotective proteins, conferring protection against brain infection and injury.

Keywords: astrocytes; myeloperoxidase (MPO); nuclear factor E2-related factor (Nrf2); heme oxygenase 1 (HO-1); Kelch-like ECH-associated protein 1 (Keap1)

check for updates

Citation: Seol, S.-I.; Kang, I.S.; Lee, J.S.; Lee, J.-K.; Kim, C. Taurine Chloramine-Mediated Nrf2 Activation and HO-1 Induction Confer Protective Effects in Astrocytes. *Antioxidants* **2024**, *13*, 169. https://doi.org/10.3390/antiox13020169

Academic Editors: Baojun Xu and Bin Du

Received: 16 December 2023
Revised: 23 January 2024
Accepted: 25 January 2024
Published: 29 January 2024

1. Introduction

Taurine (2-aminoethansulfolic acid) is one of the most abundant nonproteinogenic amino acids; it is ubiquitously distributed in mammalian tissues, with the highest levels found in the heart, brain, and leukocytes [1]. Taurine concentrations range from 1 to 20 μmol/g in the brain and from 20 to 50 mM in human neutrophils [2–4]. It plays a vital role in many biological processes, including central nervous system (CNS) and retinal development, membrane stabilization, calcium mobilization, neurotransmission, reproduction, and detoxification [1,5,6]. Furthermore, taurine confers protective effects on infections and inflammation [6,7]. It reacts with the highly toxic hypochlorous acid (HOCl), which is released by the myeloperoxidase (MPO) system of leukocytes, to generate the less toxic taurine chloramine (Tau-Cl) [8]. Tau-Cl inhibits the production of proinflammatory mediators and increases the expression of several antioxidant enzymes. Therefore, taurine ameliorates inflammation by eliminating the highly toxic HOCl as well as modulating inflammatory mediators via Tau-Cl action [5,6].

The brain contains high concentrations of taurine, which functions as an osmoregulator and a neuromodulator [9,10]. Taurine exhibits neuroprotective effects in neurons and astrocytes and animal models of neurological disorders such as ischemic stroke and inflammation [11–16]. Furthermore, it decreases the expression of apoptotic protein during ischemic injury [12,17] and maintains intracellular Ca^{2+} homeostasis [18], thereby attenuating apoptotic neuronal death [12,19]. Under pathological conditions, neutrophils infiltrate the regions of inflammatory areas or infected tissues. Neutrophils are the most abundant leukocytes and contain high concentrations of MPO and taurine; therefore, activated neutrophils produce considerable amounts of Tau-Cl. In our previous study, we demonstrated that neutrophils infiltrate ischemic injury areas and that Tau-Cl exerts robust neuroprotective effects using a rat middle cerebral artery occlusion (MCAO) model [20]. Tau-Cl significantly decreased the infarct volume and neurological deficits and promoted motor function. Furthermore, it significantly increased the levels of antioxidant enzymes such as heme oxygenase-1 (HO-1), NAD(P)H:quinone oxidoreductase 1 (NQO1), glutamate–cysteine ligase catalytic (GCLC), and peroxiredoxin-1 (Prx-1) in the post-ischemic brain and in BV2 cells, a microglial cell line [20].

Astrocytes are the most abundant glial cell type in the CNS that outnumbers neurons [21]. Astrocytes perform various functions in the normal brain, including maintaining a stable extracellular environment by regulating the balance of ions and fluid, which are essential for neurons [22]. Furthermore, they control the energy supply to the neurons [23,24] and regulate synapse formation and neurotransmitter turnover [25,26]. Astrocytes are the primary protective cells in the brain. Under pathological conditions, astrocytes confer neuroprotection by regulating ionic balance and energy metabolism and by inducing scar formation, which protects adjacent neural tissues via separating brain injury-induced lesions [21].

Nuclear factor E2-related factor (Nrf2) is a basic-region leucine zipper transcription factor that plays a vital role in oxidative stress response. Nrf2 binds to the antioxidant response element (ARE) located in the promoter regions of several antioxidant and detoxifying genes, including HO-1, NQO1, glutathione S-transferases (GST), and GCL modifier (GCLM) [27–31]. Under basal conditions, Nrf2 activity is tightly regulated by Kelch-like ECH-associated protein 1 (Keap1), an E3 ubiquitin ligase substrate adaptor [32,33]. Keap1 targets Nrf2 for proteasomal degradation. However, during oxidative or electrophilic stress, Keap1 is inactivated by the modification of its highly reactive cysteine residues; as a result, Nrf2 escapes ubiquitination, accumulates within the cell, and translocates to the nucleus, where it promotes its antioxidant transcription program [34–36]. Tau-Cl increases the cytosolic accumulation and nuclear translocation of Nrf2, resulting in the high expression of several Nrf2-regulated antioxidant enzymes [20,37–40]. However, to the best of our knowledge, the mechanism by which Tau-Cl regulates Nrf2 activation remains unclear.

In this study, we investigated the antioxidant effects of Tau-Cl in astrocytes C6 cells. We determined that Tau-Cl regulates the expression and nuclear translocation of Nrf2 and the subsequent expression of Nrf2-induced genes in C6 cells. Furthermore, we examined the molecular mechanism by which Tau-Cl triggers Nrf2 activation. Ultimately, we explored the protective effects exerted by Tau-Cl on astrocytes.

2. Materials and Methods

2.1. C6 Cell Culture and Tau-Cl Treatment

C6 astroglioma cells (American Type Culture Collection, Manassas, VA, USA) were grown in Dulbecco's modified Eagle's medium (DMEM; Sigma, St. Louis, MO, USA) supplemented with 5% fetal bovine serum (FBS; Hyclone, Logan, UT, USA), 1% penicillin, and streptomycin at 37 °C in a 5% CO_2 incubator. Cells were prepared a day before Tau-Cl treatment. Tau-Cl was synthesized by adding equimolar amounts of NaOCl (Aldrich Chemical, Milwaukee, MI, USA) to taurine (Sigma), and the formation of Tau-Cl was monitored by measuring UV absorption (200~400 nm) [41]. Tau-Cl was diluted to 200 μM in DMEM containing 10% FBS. C6 cells were treated with Tau-Cl for the indicated time points.

2.2. Nuclear and Cytoplasmic Extract Preparation

C6 cells were lysed with solution A (0.5% Triton X-100, 0.5% NP-40, 10 mM HEPES [pH 7.9], 10 mM KCl, 0.1 mM EDTA, and 1 mM DTT) containing complete Mini Protease

Inhibitor Cocktail tablet (Roche diagnostics, Basel, Switzerland). Lysates were centrifuged at 17,500× *g* for 5 min at 4 °C. The supernatant containing cytoplasmic protein was collected and stored at −80 °C. Pellets were lysed with solution B (10% glycerol, 20 mM HEPES [pH 7.9], 0.4 M NaCl, 1 mM EDTA, and 1 mM DTT) containing complete Mini Protease Inhibitor Cocktail tablet. Lysates were centrifuged at 17,500× *g* for 10 min at 4 °C. The supernatant containing the nuclear protein was collected and stored at −80 °C.

2.3. Immunoblotting

Cells were lysed with RIPA buffer (0.25% sodium-deoxycholate, 150 mM NaCl, 50 mM Tris-HCl [pH 7.4], and 1% NP-40) containing complete Mini Protease Inhibitor Cocktail tablet. Lysates were centrifuged at 12,000× *g* for 10 min at 4 °C and the supernatants were loaded into 12% SDS-PAGE gels. The primary antibodies used were as follows: anti-Nrf2 (sc-722, 1:1000; Santa Cruz Biotechnology, Santa Cruz, CA, USA), anti-α-tubulin (GTX112141, 1:10,000; GeneTex, Irvine, CA, USA), anti-lamin B1 (12987-1-AP, 1:2000; Proteintech, Rosemont, IL, USA), anti-HO-1 (43966S, 1:2000; Cell Signaling, Danvers, MA, USA), anti-NQO1 (ab34173, 1:5000; Abcam, Cambridge, UK), anti-GCLC (ab207777, 1:2000; Abcam), and anti-GCLM (ab153967, 1:1000; Abcam). Horseradish peroxidase-conjugated secondary antibodies (AP132P, 1:4000; Merck Millipore, Darmstadt, Germany) were used. Immunoblots were detected using a chemiluminescence kit (Merck Millipore).

2.4. Immunocytochemistry

C6 cells (4×10^4) were treated with Tau-Cl (200 μM) for 3 or 9 h or with taurine (200 μM) for 9 h, and then fixed with 4% paraformaldehyde for 20 min. The anti-Nrf2 antibody (16396-1-AP, 1:200; Proteintech) was incubated overnight, followed by incubation with FITC-labeled anti-IgG antibody (AP132F, 1:200; Merck Millipore). Cells were counterstained with 4′,-6-diamidino-2-phenylindole (DAPI; H-1200; Vector Laboratories, Newark, CA, USA) to visualize the nuclei. Images were observed under a Zeiss LSM 510 META microscope (Carl Zeiss Meditec AG, Jena, Germany).

2.5. Modification of the Thiol Groups of Keap1 by Tau-Cl

C6 cells were incubated with 200 or 500 μM Tau-Cl for 5, 10, and 20 min, and then washed three times with PBS. Thereafter, the cells were lysed with RIPA buffer containing 2 mM PMSF and 20 μg/mL chymostatin. The lysate was incubated with 1 μg/mL biotin polyethyleneoxide iodoacetamide (Biotin-IAA) (Sigma) for 18 h. The mixtures were incubated with 20 μL of streptavidin–agarose beads (50% slurry) (Pierce, Rockland, IL, USA) for 1 h at 4 °C with rotation and centrifuged at 10,000× *g* for 1 min. The pellets containing reduced Keap1 were washed three times with PBS and subjected to 10% SDS-PAGE, followed by immunoblotting with an anti-Keap1 antibody (SC-15246, 1:500; Santa Cruz Biotechnology). The supernatants containing oxidized Keap1 were incubated with 1 μg/mL biotin maleimide (Sigma) or Biotin-IAA for 18 h. Then, the mixtures were incubated with streptavidin–agarose beads, and the pellets were analyzed by immunoblotting as with reduced Keap1 samples.

2.6. Reaction of the Thiol Groups of Keap1 with Tau-Cl

Titrants of thiol groups, including 2.2′-dipyridyl disulfide and 4,4′-dipyridyl disulfide, react with the thiols of Keap1 [42]. To determine the potency with which Tau-Cl reacts with thiol groups, 0.5 and 2 μM Tau-Cl or 4,4′-dipyridyl disulfide (Sigma) were added to a solution of 0.5 μM Keap1 (Sino Biological, Beijing, China) solution dissolved in PBS. Immediately after mixing, the changes in the absorbance at 325 nm were monitored for 20 min at intervals of 5 min.

2.7. Cell Viability Assay

The 3-(4,5-dimethylthiazol-2-yl)-2,5-diphenyltetrazolium bromide (MTT) assay was performed to measure cell viability. C6 cells were pre-treated with 50, 100, or 200 μM of Tau-Cl or 200 μM of taurine for 9 h, followed by treatment with 300 μM of H_2O_2 for 1 h in DMEM containing 10% FBS. After 24 h, the cells were incubated with 1 mg/mL MTT (Sigma) for 1 h. The medium was removed and 500 μL of DMSO was added to solubilize the formazan product. The mixture (100 μL) was added to a 96-well plate and the absorbance was measured at 540 nm using a microplate reader.

2.8. siRNA Transfection

C6 cells (4×10^4) were seeded in 24-well plates at 24 h before transfection. Rat HO-1-specific siRNA (siHO-1; 5′-GUC AUG GCC ACU UUG AUA UCA GUG T-3′ and 5′-ACA CUG AUA UCA AAG UGG CCA UGA CGC-3′) and a nonspecific siRNA (siCon; 5′-CGU UAA UCG CGU AUA AUA CGC GUA T-3′ and 5′-AUA CGC GUA UUA UAC GCG AUU AAC GAC-3′) were purchased from Integrated DNA Technologies, Inc. (Coralville, IA, USA). The siHO-1 or siCon (both 40 pM) was mixed with 1 μL/well of Oligofectamine reagent (Invitrogen, Carlsbad, CA, USA) in Opti-MEM (Invitrogen) according to the manufacturer's instructions. The siRNA-Oligofectamine complexes were added to C6 cells. After incubation for 15 h, the cells were treated with 200 μM Tau-Cl.

2.9. Reactive Oxygen Species (ROS) Quantification

C6 cells (4×10^4) were seeded into 24-well plates and cultured for 24 h. C6 cells were pre-treated with 200 μM of Tau-Cl or 200 μM of taurine for 9 h, followed by treatment with 300 μM of H_2O_2 for 1 h. Cells were then incubated in DMEM containing 10 μM 5-(and-6)-chloromethyl-2′,7′-dichlorodihydrofluorescein diacetate (CM-H_2DCFDA; Thermo Fisher Scientific, Waltham, MA, USA) for 30 min. After washing cells with PBS, fluorescence and differential interference contrast images were obtained using a Zeiss microscope. Quantified fluorescence changes were obtained using ImageJ (http://rsbweb.nih.gov/ij/).

2.10. Statistical Analysis

The statistical analyses were performed using analysis of variance (ANOVA) followed by the Newman–Keuls test. Results are presented as the mean ± standard error of the mean (SEM). Statistical significance was accepted for p value < 0.05. The analyses were performed using PRISM software 5.0 (Graph Pad Software, Boston, MA, USA).

3. Results

3.1. Tau-Cl Enhances the Expression and Nuclear Translocation of Nrf2 in C6 Cells

To investigate whether Tau-Cl induces Nrf2 activation in C6 cells, the cells were treated with 200 μM Tau-Cl for 3, 6, 9, 12, or 24 h. The total amount of Nrf2 gradually increased and reached the maximum level after 9 h of treatment (Figure 1A,B). The elevated Nrf2 levels persisted until 24 h (Figure 1A,B). Next, we investigated whether Tau-Cl induces Nrf2 translocation from the cytoplasm to the nucleus. Nuclear Nrf2 levels significantly increased after 9 h of Tau-Cl treatment and the enhanced level was maintained until 12 h (Figure 1C,D); in contrast, cytoplasmic Nrf2 levels gradually decreased during the same period (Figure 1C,D). In the case of taurine (200 μM for 9 h), while total Nrf2 levels moderately increased (Figure 1A,B), the nuclear translocation of Nrf2 remained inconclusive (Figure 1C,D). To confirm the translocation of Nrf2 from the cytoplasm to the nucleus by Tau-Cl, double immunofluorescence staining was performed using an anti-Nrf2 antibody and DAPI. Under normal conditions, Nrf2 was detected in the cytoplasm of C6 cells (Figure 1E). However, it was detected in the nucleus as early as 3 h after Tau-Cl treatment, with a significant increase in nuclear accumulation at 9 h, when co-localization of Nrf2 and DAPI was evident (Figure 1F,G). Consistent with the results obtained from immunoblotting, the nuclear translocation of Nrf2 was not evident after taurine treatment (200 μM for 9 h) (Figure 1H). Together, these results suggest that Tau-Cl increases the total amount of Nrf2 and induces its translocation from the cytoplasm to the nucleus in astrocytes.

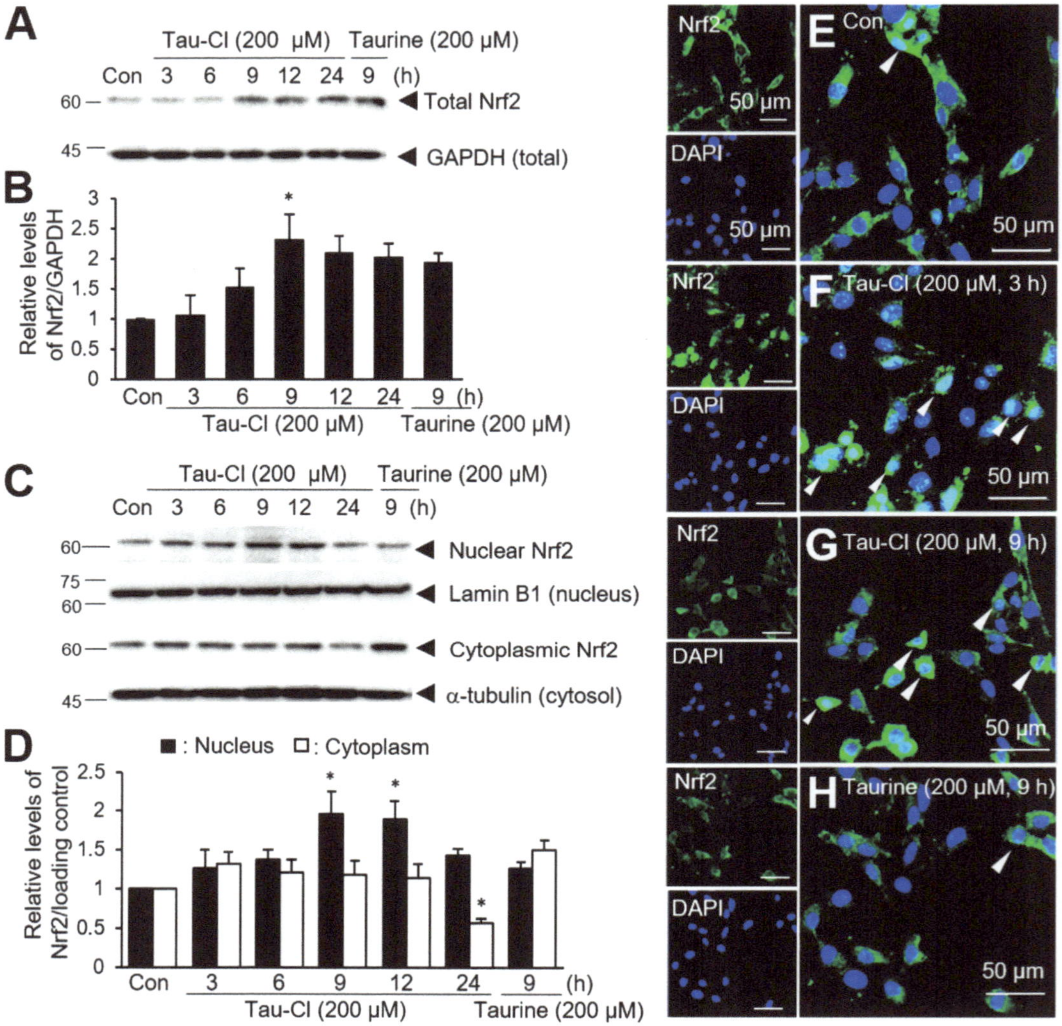

Figure 1. The upregulation and cytoplasmic-to-nuclear translocation of Nrf2 by Tau-Cl in C6 cells. C6 cells were treated with 200 μM of Tau-Cl for 3, 6, 9, 12, or 24 h or with 200 μM of taurine for 9 h. Nrf2 levels in total cell lysates (**A**,**B**) or in nuclear and cytosolic fractions (**C**,**D**) were determined by immunoblotting. Representative images are presented in A and C, and quantified results are presented as mean ± SEM (n = 3) in (**B**,**D**). (**E**–**H**) C6 cells were treated with 200 μM of Tau-Cl for 3 or 9 h or with 200 μM of taurine for 9 h. Double immunofluorescence staining was performed using anti-Nrf2 antibody and DAPI. Arrowheads indicate Nrf2 in the nucleus. The scale bar represents 50 μm. * $p < 0.05$ versus untreated controls.

3.2. *Tau-Cl Induces the Upregulation of Various Antioxidant Enzymes Downstream of Nrf2*

Because Nrf2 induces the expression of various antioxidant genes, we investigated whether Tau-Cl upregulates the expression of antioxidant genes in C6 cells. C6 cells were treated with 200 μM Tau-Cl for 3, 6, 9, 12, or 24 h. HO-1 expression significantly increased after 6 h of 200 μM Tau-Cl treatment, reached the maximum level at 12 h, and subsequently decreased (Figure 2A,B). Similarly, Tau-Cl significantly induced NQO1, GCLC, and GCLM in C6 cells; however, the timing of induction differed for each gene (Figure 2). Collectively, these results suggest that Tau-Cl activates Nrf2 and induces the upregulation of various antioxidant genes in astrocytes.

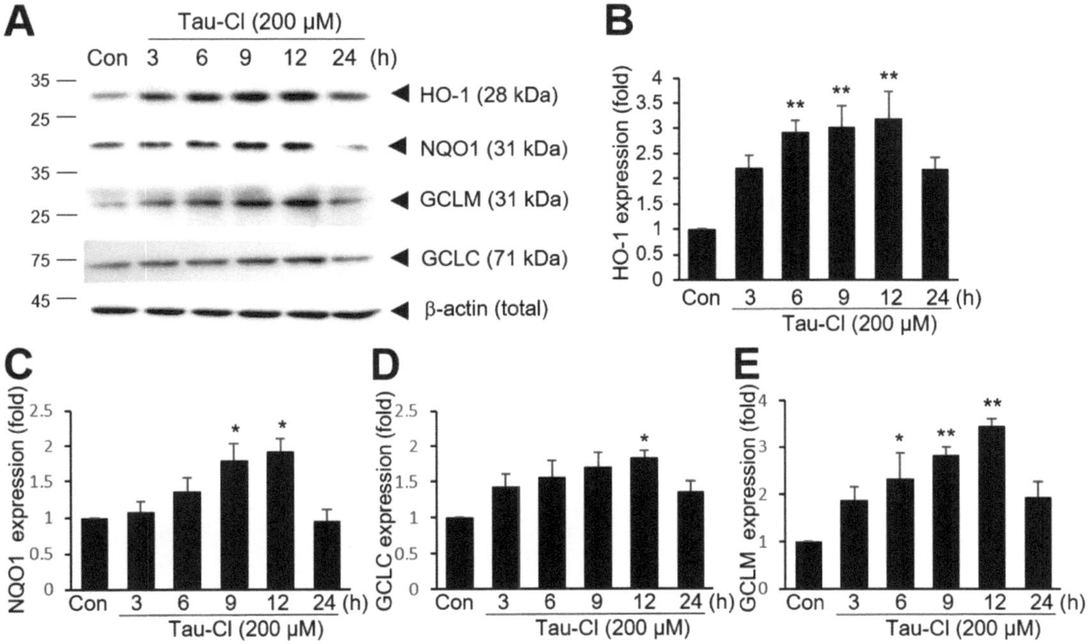

Figure 2. Induction of various antioxidant genes by Tau-Cl in C6 cells. C6 cells were treated with Tau-Cl (200 μM) for 3, 6, 9, 12, or 24 h and protein levels of HO-1, NQO1, GCLM, GCLC, and β-actin were determined by immunoblotting. Representative images of the protein levels are in (**A**) and quantified results are presented in (**B–E**) as mean ± SEM (n = 3). * $p < 0.05$ and ** <0.01 versus untreated controls.

3.3. Tau-Cl Induces the Disulfide Bond Formation of Keap1

Keap1 binding leads to the anchoring of Nrf2 in the cytoplasm, which is attached to the actin cytoskeleton. Inducers and electrophiles disrupt the Keap1-Nrf2 complex, and Nrf2 translocates to the nucleus, where it binds to the ARE region and initiates transcription [42]. Because Tau-Cl activates Nrf2 and increases the expression of downstream genes including HO-1, NQO-1, and GCLM (Figures 1 and 2), we elucidated whether Tau-Cl disrupts the binding between Keap1 and Nrf2. Maleimide and iodoacetamide (IAA) are known to bind to free thiol (-SH) groups [43]. Tau-Cl treatment decreased IAA-bound Keap1 at 5 and 10 min (Figure 3A,B), suggesting that Tau-Cl increases the oxidation of the thiol groups of Keap1 to generate disulfide bonds. Next, we attempted to identify oxidized Keap1; however, it remained undetectable within our experimental system. In addition, Tau-Cl decreased IAA-bound beta-actin (Figure 3B,C; β-actin (R)), suggesting its ability to oxidize thiol groups.

The thiol groups of Keap1 react with 4,4′-dipyridyl disulfide and reveal distinct UV absorptions; therefore, 4,4′-dipyridyl disulfide is used as a spectroscopic titration reagent for the thiol groups [42]. The addition of Tau-Cl to Keap1 solution decreased its UV absorption at 325 nm in a manner similar to 4,4′-dipyridyl disulfide (Figure 3D,E), suggesting that Tau-Cl modifies the thiol groups of Keap1. Nevertheless, this method has experimental limitations owing to the difficulties associated with obtaining a sufficiently high concentration of Keap1 to reveal significant absorbance changes. Moreover, Keap1 undergoes rapid oxidation under normal laboratory conditions.

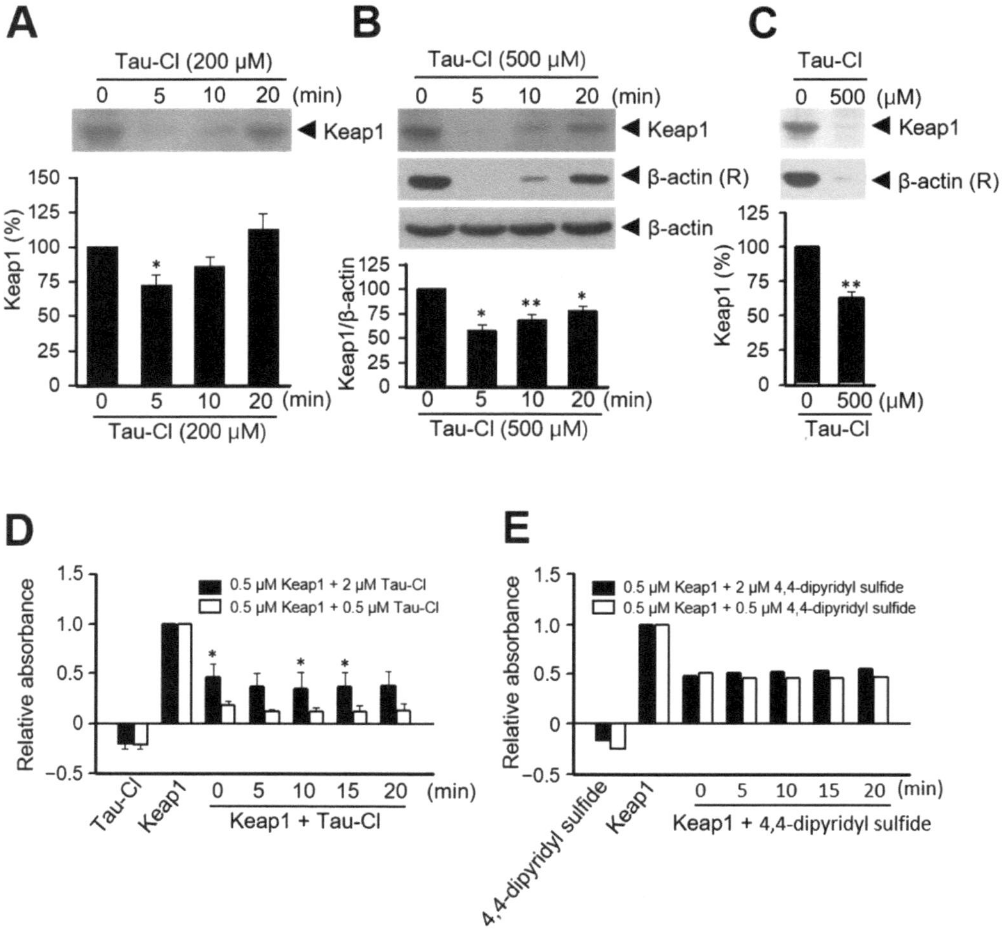

Figure 3. Modification of the thiol groups of Keap1 by Tau-Cl. (**A–C**) C6 cells were incubated with 200 or 500 μM Tau-Cl and the cell lysates were reacted with Biotin-iodoacetamide (IAA). IAA-bound Keap1 was precipitated, subjected to SDS-PAGE, and determined by immunoblotting with anti-Keap1 antibody. The quantified results are presented as mean ± SEM (n = 5 for (**A**,**B**), n = 4 for (**B**)). * $p < 0.05$ and ** <0.01 versus untreated controls. (**D**,**E**) 0.5 and 2 μM Tau-Cl (**D**) or 4,4′-dipyridyl disulfide (**E**) were added to 0.5 μM Keap1 solution, and the changes in the absorbance at 325 nm were monitored for 20 min at intervals of 5 min. The quantified results are presented as mean ± SEM (n = 3 for (**D**) and n = 1 for (**E**)). * $p < 0.05$ versus 0.5 μM Keap1.

3.4. Tau-Cl Suppresses H_2O_2-Induced Cell Death in C6 Cells

To determine whether Tau-Cl confers protective effects in C6 cells, we examined the cell viability of C6 cells after H_2O_2 treatment in the presence or absence of Tau-Cl (Figure 4A). The viability of C6 cells was decreased to 49.0 ± 2.2% after treatment with 300 μM H_2O_2 for 1 h (Figure 4B). However, the pretreatment of C6 cells with 100 and 200 μM Tau-Cl for 9 h significantly improved cell survival to 59.2 ± 2.0% and 64.2 ± 3.0%, respectively (Figure 4B). In contrast, pretreatment with 200 μM taurine for 9 h did not improve the cell viability (Figure 4B). Moreover, cell survival was increased after treatment of 200 μM Tau-Cl for 6, 9, and 12 h (Figure 4C). These findings suggest that Tau-Cl inhibits astrocyte cell death induced by H_2O_2.

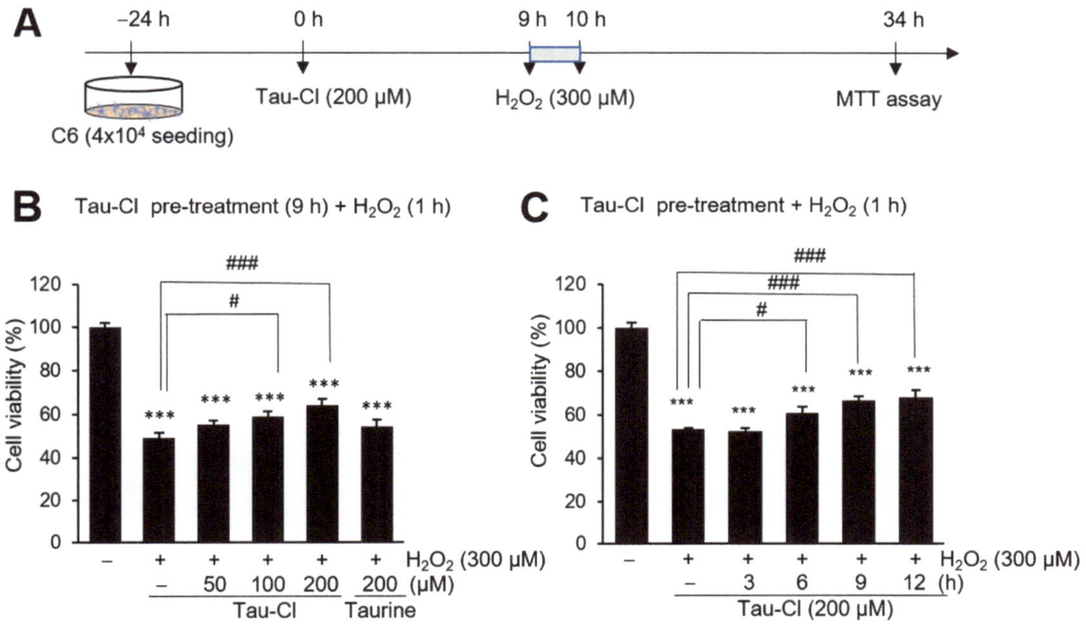

Figure 4. Protection of H_2O_2-treated C6 cells by Tau-Cl. (**A**) Schematic diagram of time points for treatment of Tau-Cl and H_2O_2 and cell viability assay. (**B**) C6 cells were pre-treated with Tau-Cl (50, 100, or 200 μM) or taurine (200 μM) for 9 h and then treated with H_2O_2 (300 μM) for 1 h. (**C**) C6 cells were pre-treated with 200 μM of Tau-Cl for 3, 6, 9, or 12 h and then treated with H_2O_2 (300 μM) for 1 h. For all experiments, MTT assays were performed 24 h after H_2O_2 treatment and cell viabilities are presented as mean ± SEM (n = 8). *** $p < 0.001$ versus untreated controls, # $p < 0.05$ and ### $p < 0.001$ between indicated groups.

3.5. Tau-Cl-Mediated HO-1 Induction Is Responsible for the Protective Effects in C6 Cells

We examined whether the Tau-Cl-mediated protective effect is related to the upregulation of antioxidant genes in C6 cells, particularly HO-1, using siRNA-mediated HO-1 knockdown (Figure 5A). HO-1 was significantly induced in C6 cells after H_2O_2 treatment (100 μM, 1 h) (2.2-fold) (Figure 5B,C). Furthermore, preincubation with 200 μM Tau-Cl for 9 h further enhanced H_2O_2-induced HO-1 upregulation by 6.8-fold (Figure 5B,C), indicating that Tau-Cl augmented H_2O_2-induced HO-1 upregulation. Moreover, Tau-Cl preincubation enhanced H_2O_2-induced Nrf2 expression and other Nrf2-regulated antioxidant enzymes such as GCLM, GCLC, and NQO1 (Figure S1). When C6 cells were transfected with HO-1 siRNA, Tau-Cl-induced HO-1 expression decreased to 43.5 ± 0.3% compared with siRNA-non-transfected control cells (Figure 5D). However, this reduction was not observed in cells transfected with nonspecific siRNA (siCon) (Figure 5D). Importantly, the increased viability observed for Tau-Cl-pretreated/H_2O_2-treated cells was significantly reduced in HO-1 siRNA-transfected cells, that is, it decreased from 75.1 ± 4.2% to 54.7 ± 2.4% (Figure 5E). However, a similar level of protective effect was observed in siCon-transfected cells (Figure 5E), indicating that HO-1 is responsible for the protective effect of Tau-Cl. Taken together, these results suggest that Tau-Cl-mediated HO-1 upregulation is responsible for the protective effects of Tau-Cl in H_2O_2-treated astrocytes.

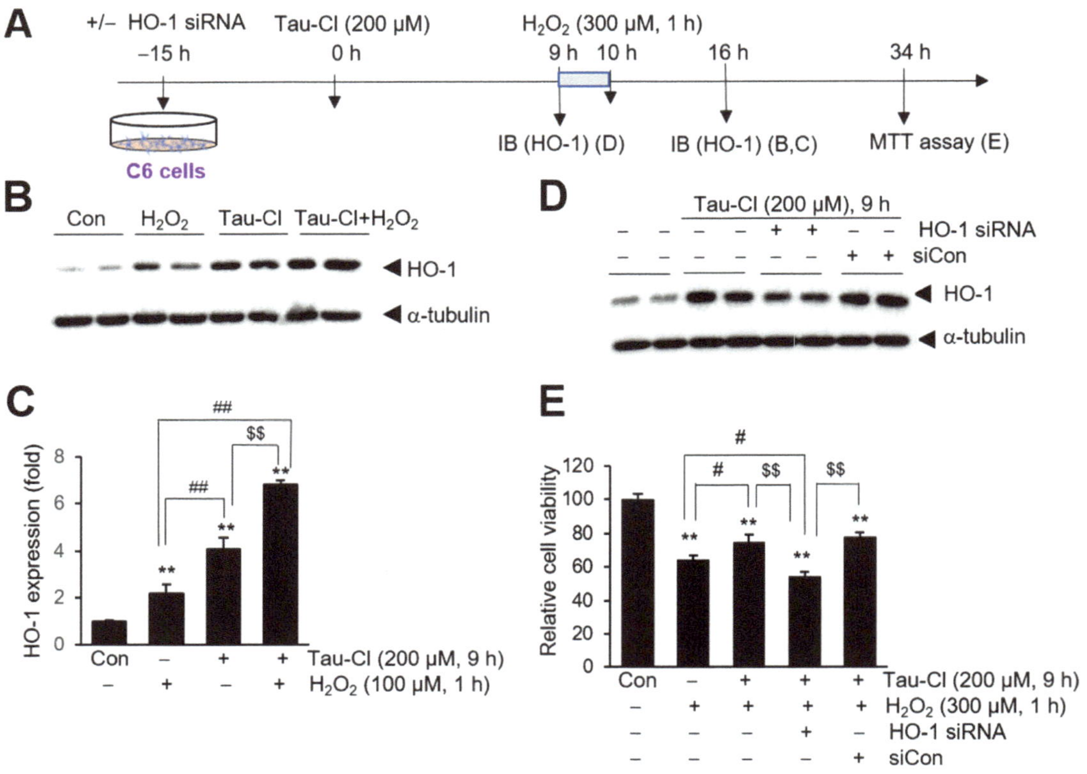

Figure 5. Suppression of Tau-Cl-mediated protective effects in H_2O_2-treated C6 cells by HO-1 knockdown. (**A**) Schematic diagram of time points for HO-1 knockdown, treatment of Tau-Cl and H_2O_2, and cell viability assay. (**B**,**C**) C6 cells were pre-treated with Tau-Cl (200 μM) for 9 h and then treated with H_2O_2 (100 μM) for 1 h. Protein levels of HO-1 were determined by immunoblotting at 6 h after H_2O_2 treatment. (**D**) C6 cells were transfected with HO-1 siRNA or non-specific siRNA (siCon). After 15 h, cells were treated with Tau-Cl (200 μM) for 9 h. HO-1 levels were determined by immunoblotting. Representative images of the protein levels are in B and D and quantified results are presented D as mean ± SEM (n = 3). (**E**) Cell viabilities of C6 cells after Tau-Cl pretreatment/H_2O_2 treatment with and without HO-1 siRNA transfection were examined using MTT assays at 24 h after H_2O_2 treatment. Results are presented as mean ± SEM (n = 3). ** $p < 0.01$ versus untreated controls, ## $p < 0.01$, # $p < 0.05$ and $$ $p < 0.01$ between indicated groups.

3.6. Tau-Cl Inhibits ROS Production in H_2O_2-Treated C6 Cells

We then investigated whether Tau-Cl suppresses ROS induction in H_2O_2-treated C6 cells (Figure 6A). When C6 cells were treated with H_2O_2 (300 μM, 1 h) and stained with CM-H_2DCFDA, an intracellular ROS indicator, induction of DCF fluorescence was observed (Figure 6B,C). Interestingly, pretreatment of C6 cells with 200 μM of Tau-Cl for 9 h significantly suppressed the induction of DCF to 69.8 ± 2.9% of that in treatment-naïve H_2O_2 control cells (Figure 6B,C). In contrast, pretreatment with 200 μM of taurine for 9 h failed to suppress DCF induction (Figure 6B,C), demonstrating that Tau-Cl suppressed H_2O_2-induced ROS generation in H_2O_2-treated C6 cells. Importantly, suppression of ROS induction by Tau-Cl was not detected in HO-1 siRNA-transfected C6 cells, however, ROS induction was suppressed in siCon-transfected C6 cells (Figure 6D,E), suggesting that HO-1 is responsible for this suppression. Taken together, these results indicate that Tau-Cl suppresses ROS induction in H_2O_2-treated C6 cells and HO-1 upregulation is responsible for this effect.

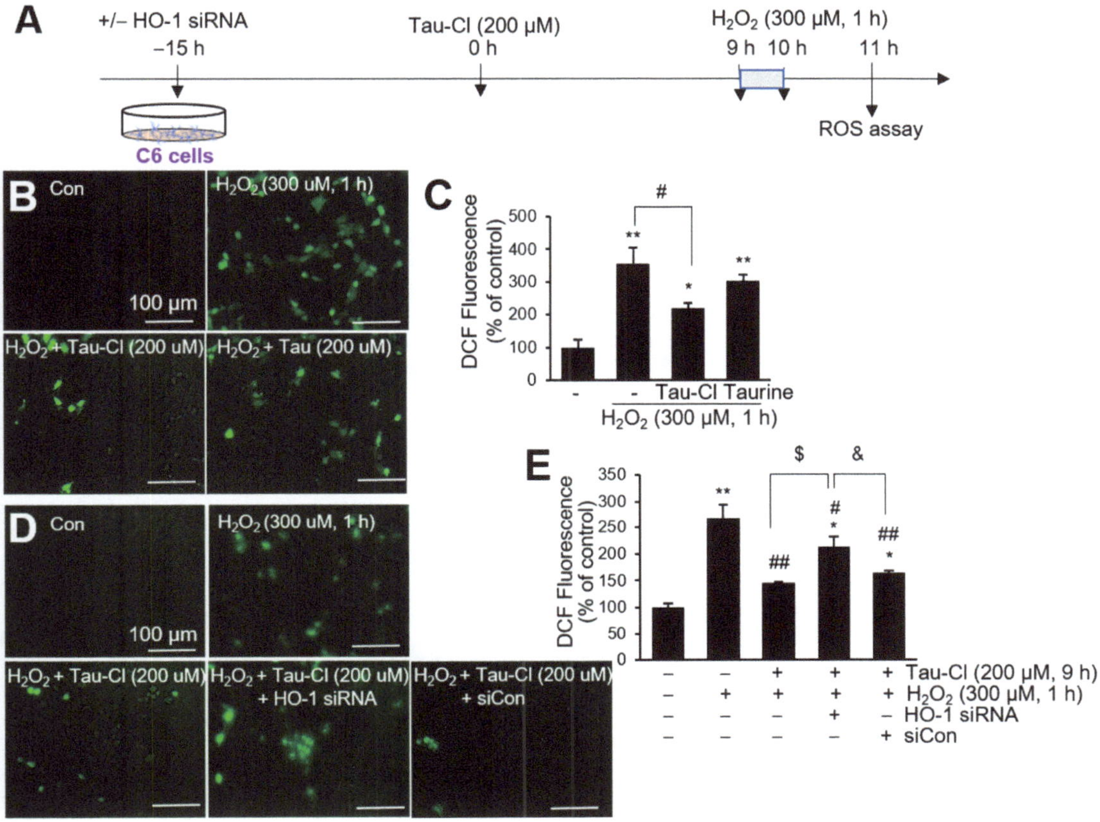

Figure 6. Suppression of H_2O_2-induced ROS production by Tau-Cl in C6 cells. (**A**) Schematic diagram of time points for HO-1 knockdown, treatment of Tau-Cl and H_2O_2, and ROS assay. (**B**,**C**) C6 cells were pre-treated with Tau-Cl (200 μM) for 9 h and then treated with H_2O_2 (300 μM) for 1 h. Intracellular ROS levels were measured using CM-H_2DCFDA at 1 h after H_2O_2 treatment. (**D**,**E**) C6 cells were transfected with HO-1 siRNA or non-specific siRNA (siCon), pre-treated with Tau-Cl (200 μM) for 9 h, and then treated with H_2O_2 (300 μM) for 1 h. ROS levels in C6 cells with and without HO-1 siRNA transfection were examined using CM-H_2DCFDA at 1 h after H_2O_2 treatment. Results are presented as mean ± SEM (n = 3). The scale bar represents 100 μm. ** $p < 0.01$, * $p < 0.05$ versus untreated controls, # $p < 0.05$ and ## $p < 0.01$ versus H_2O_2 controls, \$ $p < 0.05$ and & $p < 0.05$ between indicated groups.

4. Discussion

Astrocytes exert protective effects on neurons via their antioxidant activity [44]. They produce antioxidant molecules in the brain, playing a vital role in preventing ROS elevation and neuronal cell death in various CNS disorders [45]. HO-1 expression is weak in the brain and is limited to small groups of neurons and neuroglia [46]. However, the protective effects of HO-1 in the brain have been reported under several pathological conditions [47–49]. In an animal model of intracerebral hemorrhage, the strong neuroprotective effects of HO-1 overexpression in astrocytes have been reported [50]. In the present study, we demonstrated the autocrine function of Tau-Cl-mediated HO-1 induction, i.e., its ability to suppress H_2O_2-induced astrocyte cell death. However, astrocyte HO-1 may confer paracrine effects on neighboring cells, including neurons and microglia. In previous studies, we demonstrated the neuroprotective effect of HO-1 in astrocytes using conditioned media [47,48] and in animal models of MCAO [49]. Recently, Zhang et al. [51] reported that astrocyte-derived exosomes protect hippocampal neurons after traumatic brain injury by activating Nrf2 signaling in both rat and mouse models. In addition to neurons, studies have reported the beneficial effects of astrocyte HO-1 on microvascular function after various acute injuries, including post-ischemic myocardial injury [52], hemorrhagic shock [53], and seizures [54]. Alfieri et al. [55] reported that preconditioning stimuli increase HO-1 expression, primarily in perivascular astrocytes, which are responsible for preserving the barrier function in a transient rat MCAO model. Nevertheless, additional studies are warranted to investigate the protective effects of enhanced astrocyte HO-1 levels in other brain cell types.

Taurine is one of the most abundant free amino acids in the brain. However, its concentration in the brain decreases with age. Astrocytes are the primary taurine producers in the CNS [56]. Astrocytes release taurine as a gliotransmitter and provide neurons with hypotaurine as a substrate for taurine production [56]. As mentioned above, astrocytes play vital roles in maintaining normal brain function as well as protecting against inflammatory responses in the brain. Reactive astrocytes release inflammatory cytokines, produce various antioxidant molecules, including GSH, and activate ROS-detoxifying enzymes such as GST, GSH peroxidase, thioredoxin reductase, and catalase to improve neuronal survival [57–59]. In our previous study, we observed that neutrophils infiltrate into the ischemic region and that Tau-Cl exerts neuroprotective effects in the post-ischemic brain after MCAO and BV2 cells via increasing the levels of antioxidant enzymes, including HO-1 [20]. In the present study, we elucidated the effects of Tau-Cl in astrocytes. Tau-Cl-induced upregulation and nuclear translocation of Nrf2 as well as upregulated Nrf2-regulated antioxidant genes (Figures 1 and 2), suggesting that Tau-Cl-derived antioxidants protect astrocytes and ameliorate neuronal injury. Tau-Cl recovered cell survival against oxidative stress (H_2O_2)-induced cell death (Figure 4), which was associated with increased HO-1 expression and decreased ROS production (Figures 5 and 6). Therefore, the induction of HO-1 by Tau-Cl mitigates H_2O_2-induced astrocyte cell death through its anti-oxidative properties.

Under homeostatic conditions, the levels of Nrf2 protein are maintained at a relatively low level owing to constitutive ubiquitin-mediated proteasomal degradation of Nrf2 by Keap1 [27,60]. Nrf2 is primarily localized in the cytoplasm with interaction with Keap1. In response to oxidative stress or electrophilic attacks, Keap1 undergoes oxidation at specific cysteine residues, leading to the disruption of its interaction with Nrf2. The disruption results in the stabilization of Nrf2, a critical step in initiating its translocation into the nucleus to regulate transcriptional activities. The translocation of Nrf2 into the nucleus occurs by interacting with the importins, facilitated by the three nuclear localization sequences (NLS) located in Neh1, Neh2, and Neh3 domains [61,62]. Importins recognize the NLS of Nrf2, allowing Nrf2-importin complexes to traverse the nuclear pore complex, comprising various nucleoporins that regulate the transport of molecules between the cytoplasm and the nucleus.

There are two major mechanisms by which Nrf2 dissociates from Keap1: one is a conformational change in Keap1 and the other is the phosphorylation of the serine residues of Nrf2 [63,64]. For the phosphorylation of the serine residues of Nrf2, various protein kinases, particularly mitogen-activated protein kinases (MAPKs), can trigger transcriptional activity via Nrf2 phosphorylation. We observed that Tau-Cl exhibited no stimulatory effects on various protein kinases, including MAPKs. Moreover, Tau-Cl inhibited lipopolysaccharide-induced MAPK activation in RAW 264.7 cells [41]. Therefore, we determined the effect of Tau-Cl on the conformational changes of Keap1, with particular attention on determining whether Tau-Cl modifies the reactive cysteine residues of Keap1, which play a pivotal role in its association with Nrf2. Human Keap1 contains 27 cysteines with thiol groups that form covalent bonds with other cysteine thiols, bind to metals and metalloids, and react with oxidants [65]. The chemical modification of the cysteine residues of Keap1 results in the loss of Nrf2 repressor function and changes the expression of Nrf2 target genes, which collectively restore redox balance and resolve inflammation, thereby ensuring a comprehensive cytoprotection. Tau-Cl decreased IAA-bound Keap1 (Figure 3), suggesting that Tau-Cl converts the thiol groups into disulfide bonds. Most Nrf2 inducers are electrophiles that readily react with the cysteine thiol groups of Keap1 [42]. As a titrant for thiols, 4,4′-dipyridyl disulfide exhibits unique UV spectra; its absorbance changes when it reacts with thiols. Tau-Cl decreased the UV absorption of Keap1 similarly to 4,4′-dipyridyl disulfide, suggesting that Tau-Cl changes the thiol groups of Keap1 (Figure 3). Although we could not identify the cysteine residues targeted by Tau-Cl, we provided tentative evidence that Tau-Cl changes the thiol groups of Keap1.

5. Conclusions

In the present study, we investigated the antioxidative effects of Tau-Cl in astrocytes. Tau-Cl augmented Nrf2 expression and activation and upregulated Nrf2-regulated antioxidants such as HO-1, NQO1, GCLC, and GCLM. The activation of Nrf2 by Tau-Cl occurs by disrupting the Keap1-Nrf2 complex via the oxidation of the thiol groups of Keap1. Consequently, Tau-Cl rescued the cells from H_2O_2-induced death by further enhancing HO-1 expression and suppressing ROS production. Nevertheless, additional studies are required to comprehensively understand the role of Tau-Cl not only in astrocytes but also in other glial cells and neurons. The neuroprotective effects of Tau-Cl in astrocytes can be a promising therapeutic target for various neuroinflammatory and neurodegenerative diseases.

Supplementary Materials: The following supporting information can be downloaded at: https://www.mdpi.com/article/10.3390/antiox13020169/s1, Figure S1: Up-regulations of genes downstream of Nrf2 by Tau-Cl-pre-treatment in H_2O_2-treated C6 cells.

Author Contributions: Conceptualization, J.-K.L. and C.K.; Methodology, S.-I.S., I.S.K., J.S.L., and C.K.; Validation, J.-K.L. and C.K.; Writing—Original Draft Preparation, J.-K.L. and C.K.; Writing—Review and Editing, J.-K.L. and C.K.; Funding Acquisition, J.-K.L. and C.K. All authors have read and agreed to the published version of the manuscript.

Funding: This work was supported by the Mid-carrier Research Grant (2021R1A2C2010920 to J.-K.L.) and Basic Science Research Program (2022R1I1A1A01055015 to I.S.K) funded by the National Research Foundation of Korea.

Informed Consent Statement: Not applicable.

Data Availability Statement: The data presented in this study are included in the article and are also available on request from the corresponding authors.

Conflicts of Interest: The authors declare no conflicts of interest.

References

1. Huxtable, R.J. Physiological actions of taurine. *Physiol. Rev.* **1992**, *72*, 101–163. [CrossRef]
2. Fukuda, K.; Hirai, Y.; Yoshida, H.; Nakajima, T.; Usui, T. Free amino acid content of lymphocytes and granulocytes compared. *Clin. Chem.* **1982**, *28*, 1758–1761. [CrossRef] [PubMed]
3. Green, T.R.; Fellman, J.H.; Eicher, A.L.; Pratt, K.L. Antioxidant role and subcellular location of hypotaurine and taurine in human neutrophils. *Biochim. Biophys. Acta* **1991**, *1073*, 91–97. [CrossRef] [PubMed]
4. Wojcik, O.P.; Koenig, K.L.; Zeleniuch-Jacquotte, A.; Costa, M.; Chen, Y. Temporal reproducibility of taurine measurements in frozen serum of healthy postmenopausal women. *Br. J. Nutr.* **2010**, *104*, 629–632. [CrossRef] [PubMed]
5. Schuller-Levis, G.B.; Park, E. Taurine: New implications for an old amino acid. *FEMS Microbiol. Lett.* **2003**, *226*, 195–202. [CrossRef] [PubMed]
6. Kim, C.; Cha, Y.N. Taurine chloramine produced from taurine under inflammation provides anti-inflammatory and cytoprotective effects. *Amino Acids* **2014**, *46*, 89–100. [CrossRef] [PubMed]
7. Marcinkiewicz, J.; Kontny, E. Taurine and inflammatory diseases. *Amino Acids* **2014**, *46*, 7–20. [CrossRef] [PubMed]
8. Weiss, S.J.; Klein, R.; Slivka, A.; Wei, M. Chlorination of taurine by human neutrophils. Evidence for hypochlorous acid generation. *J. Clin. Investig.* **1982**, *70*, 598–607. [CrossRef]
9. Albrecht, J.; Schousboe, A. Taurine interaction with neurotransmitter receptors in the CNS: An update. *Neurochem. Res.* **2005**, *30*, 1615–1621. [CrossRef]
10. Oja, S.S.; Saransaari, P. Significance of taurine in the brain. *Adv. Exp. Med. Biol.* **2017**, *975 Pt 1*, 89–94. [CrossRef]
11. Wang, G.H.; Jiang, Z.L.; Fan, X.J.; Zhang, L.; Li, X.; Ke, K.F. Neuroprotective effect of taurine against focal cerebral ischemia in rats possibly mediated by activation of both GABAA and glycine receptors. *Neuropharmacology* **2007**, *52*, 1199–1209. [CrossRef]
12. Sun, M.; Zhao, Y.; Gu, Y.; Xu, C. Anti-inflammatory mechanism of taurine against ischemic stroke is related to down-regulation of PARP and NF-kappaB. *Amino Acids* **2012**, *42*, 1735–1747. [CrossRef]

13. Sun, M.; Zhao, Y.M.; Gu, Y.; Xu, C. Therapeutic window of taurine against experimental stroke in rats. *Transl. Res.* **2012**, *160*, 223–229. [CrossRef]
14. Menzie, J.; Prentice, H.; Wu, J.Y. Neuroprotective mechanisms of taurine against ischemic stroke. *Brain Sci.* **2013**, *3*, 877–907. [CrossRef]
15. Su, Y.; Fan, W.; Ma, Z.; Wen, X.; Wang, W.; Wu, Q.; Huang, H. Taurine improves functional and histological outcomes and reduces inflammation in traumatic brain injury. *Neuroscience* **2014**, *266*, 56–65. [CrossRef]
16. Jakaria, M.; Azam, S.; Haque, M.E.; Jo, S.H.; Uddin, M.S.; Kim, I.S.; Choi, D.K. Taurine and its analogs in neurological disorders: Focus on therapeutic potential and molecular mechanisms. *Redox Biol.* **2019**, *24*, 101223. [CrossRef] [PubMed]
17. Taranukhin, A.G.; Taranukhina, E.Y.; Saransaari, P.; Djatchkova, I.M.; Pelto-Huikko, M.; Oja, S.S. Taurine reduces caspase-8 and caspase-9 expression induced by ischemia in the mouse hypothalamic nuclei. *Amino Acids* **2008**, *34*, 169–174. [CrossRef] [PubMed]
18. Gao, X.; Yang, X.; Zhang, B. Neuroprotection of taurine against bilirubin-induced elevation of apoptosis and intracellular free calcium ion in vivo. *Toxicol. Mech. Methods* **2011**, *21*, 383–387. [CrossRef] [PubMed]
19. Wu, J.Y.; Wu, H.; Jin, Y.; Wei, J.; Sha, D.; Prentice, H.; Lee, H.H.; Lin, C.H.; Lee, Y.H.; Yang, L.L. Mechanism of neuroprotective function of taurine. *Adv. Exp. Med. Biol.* **2009**, *643*, 169–179. [CrossRef] [PubMed]
20. Seol, S.I.; Kim, H.J.; Choi, E.B.; Kang, I.S.; Lee, H.K.; Lee, J.K.; Kim, C. Taurine protects against postischemic brain Injury via the antioxidant activity of taurine chloramine. *Antioxidants* **2021**, *10*, 372. [CrossRef] [PubMed]
21. Sofroniew, M.V.; Vinters, H.V. Astrocytes: Biology and pathology. *Acta Neuropathol.* **2010**, *119*, 7–35. [CrossRef]
22. Sofroniew, M.V. Molecular dissection of reactive astrogliosis and glial scar formation. *Trends Neurosci.* **2009**, *32*, 638–647. [CrossRef]
23. Dringen, R.; Gebhardt, R.; Hamprecht, B. Glycogen in astrocytes: Possible function as lactate supply for neighboring cells. *Brain Res.* **1993**, *623*, 208–214. [CrossRef] [PubMed]
24. Pellerin, L.; Bouzier-Sore, A.K.; Aubert, A.; Serres, S.; Merle, M.; Costalat, R.; Magistretti, P.J. Activity-dependent regulation of energy metabolism by astrocytes: An update. *Glia* **2007**, *55*, 1251–1262. [CrossRef] [PubMed]
25. Pfrieger, F.W.; Barres, B.A. Synaptic efficacy enhanced by glial cells in vitro. *Science* **1997**, *277*, 1684–1687. [CrossRef]
26. Eroglu, C.; Barres, B.A. Regulation of synaptic connectivity by glia. *Nature* **2010**, *468*, 223–231. [CrossRef] [PubMed]
27. Itoh, K.; Wakabayashi, N.; Katoh, Y.; Ishii, T.; Igarashi, K.; Engel, J.D.; Yamamoto, M. Keap1 represses nuclear activation of antioxidant responsive elements by Nrf2 through binding to the amino-terminal Neh2 domain. *Genes. Dev.* **1999**, *13*, 76–86. [CrossRef]
28. Wild, A.C.; Moinova, H.R.; Mulcahy, R.T. Regulation of gamma-glutamylcysteine synthetase subunit gene expression by the transcription factor Nrf2. *J. Biol. Chem.* **1999**, *274*, 33627–33636. [CrossRef]
29. Jeyapaul, J.; Jaiswal, A.K. Nrf2 and c-Jun regulation of antioxidant response element (ARE)-mediated expression and induction of gamma-glutamylcysteine synthetase heavy subunit gene. *Biochem. Pharmacol.* **2000**, *59*, 1433–1439. [CrossRef]
30. Jaiswal, A.K. Regulation of genes encoding NAD(P)H:quinone oxidoreductases. *Free Radic. Biol. Med.* **2000**, *29*, 254–262. [CrossRef]
31. Nguyen, T.; Sherratt, P.J.; Huang, H.C.; Yang, C.S.; Pickett, C.B. Increased protein stability as a mechanism that enhances Nrf2-mediated transcriptional activation of the antioxidant response element. Degradation of Nrf2 by the 26 S proteasome. *J. Biol. Chem.* **2003**, *278*, 4536–4541. [CrossRef] [PubMed]
32. Hayes, J.D.; McMahon, M. NRF2 and KEAP1 mutations: Permanent activation of an adaptive response in cancer. *Trends Biochem. Sci.* **2009**, *34*, 176–188. [CrossRef] [PubMed]
33. Nguyen, T.; Nioi, P.; Pickett, C.B. The Nrf2-antioxidant response element signaling pathway and its activation by oxidative stress. *J. Biol. Chem.* **2009**, *284*, 13291–13295. [CrossRef] [PubMed]
34. Eggler, A.L.; Liu, G.; Pezzuto, J.M.; van Breemen, R.B.; Mesecar, A.D. Modifying specific cysteines of the electrophile-sensing human Keap1 protein is insufficient to disrupt binding to the Nrf2 domain Neh2. *Proc. Natl. Acad. Sci. USA* **2005**, *102*, 10070–10075. [CrossRef] [PubMed]
35. Zhang, D.D. Mechanistic studies of the Nrf2-Keap1 signaling pathway. *Drug Metab. Rev.* **2006**, *38*, 769–789. [CrossRef] [PubMed]
36. Kaspar, J.W.; Niture, S.K.; Jaiswal, A.K. Nrf2:INrf2 (Keap1) signaling in oxidative stress. *Free Radic. Biol. Med.* **2009**, *47*, 1304–1309. [CrossRef] [PubMed]
37. Muz, B.; Kontny, E.; Marcinkiewicz, J.; Maslinski, W. Heme oxygenase-1 participates in the anti-inflammatory activity of taurine chloramine. *Amino Acids* **2008**, *35*, 397–402. [CrossRef]
38. Sun Jang, J.; Piao, S.; Cha, Y.N.; Kim, C. Taurine chloramine activates Nrf2, increases HO-1 expression, and protects cells from death caused by hydrogen peroxide. *J. Clin. Biochem. Nutr.* **2009**, *45*, 37–43. [CrossRef]
39. Kim, C.; Jang, J.S.; Cho, M.R.; Agarawal, S.R.; Cha, Y.N. Taurine chloramine induces heme oxygenase-1 expression via Nrf2 activation in murine macrophages. *Int. Immunopharmacol.* **2010**, *10*, 440–446. [CrossRef]
40. Piao, S.; Cha, Y.N.; Kim, C. Taurine chloramine protects RAW 264.7 macrophages against hydrogen peroxide-induced apoptosis by increasing antioxidants. *J. Clin. Biochem. Nutr.* **2011**, *49*, 50–56. [CrossRef]
41. Kim, J.W.; Kim, C. Inhibition of LPS-induced NO production by taurine chloramine in macrophages is mediated though Ras-ERK-NF-kappaB. *Biochem. Pharmacol.* **2005**, *70*, 1352–1360. [CrossRef] [PubMed]

42. Dinkova-Kostova, A.T.; Holtzclaw, W.D.; Cole, R.N.; Itoh, K.; Wakabayashi, N.; Katoh, Y.; Yamamoto, M.; Talalay, P. Direct evidence that sulfhydryl groups of Keap1 are the sensors regulating induction of phase 2 enzymes that protect against carcinogens and oxidants. *Proc. Natl. Acad. Sci. USA* **2002**, *99*, 11908–11913. [CrossRef]
43. Paulsen, C.E.; Carroll, K.S. Cysteine-mediated redox signaling: Chemistry, biology, and tools for discovery. *Chem. Rev.* **2013**, *113*, 4633–4679. [CrossRef]
44. Takahashi, S. Neuroprotective function of high glycolytic activity in astrocytes: Common roles in stroke and neurodegenerative diseases. *Int. J. Mol. Sci.* **2021**, *22*, 6568. [CrossRef]
45. Takahashi, S.; Mashima, K. Neuroprotection and disease modification by astrocytes and microglia in Parkinson disease. *Antioxidants* **2022**, *11*, 170. [CrossRef] [PubMed]
46. Barañano, D.E.; Snyder, S.H. Neural roles for heme oxygenase: Contrasts to nitric oxide synthase. *Proc. Natl. Acad. Sci. USA* **2001**, *98*, 10996–11002. [CrossRef] [PubMed]
47. Shin, J.H.; Kim, S.W.; Jin, Y.; Kim, I.D.; Lee, J.K. Ethyl pyruvate-mediated Nrf2 activation and hemeoxygenase 1 induction in astrocytes confer protective effects via autocrine and paracrine mechanisms. *Neurochem. Int.* **2012**, *61*, 89–99. [CrossRef]
48. Luo, L.; Kim, S.W.; Lee, H.K.; Kim, I.D.; Lee, H.; Lee, J.K. Anti-oxidative effects of 4-hydroxybenzyl alcohol in astrocytes confer protective effects in autocrine and paracrine manners. *PLoS ONE* **2017**, *12*, e0177322. [CrossRef]
49. Luo, L.; Kim, S.W.; Lee, H.K.; Kim, I.D.; Lee, H.; Lee, J.K. Anti-Zn(2+)-Toxicity of 4-hydroxybenzyl alcohol in astrocytes and neurons contribute to a robust neuroprotective effects in the postischemic brain. *Cell Mol. Neurobiol.* **2018**, *38*, 615–626. [CrossRef]
50. Chen-Roetling, J.; Song, W.; Schipper, H.M.; Regan, C.S.; Regan, R.F. Astrocyte overexpression of heme oxygenase-1 improves outcome after intracerebral hemorrhage. *Stroke* **2015**, *46*, 1093–1098. [CrossRef]
51. Zhang, W.; Hong, J.; Zhang, H.; Zheng, W.; Yang, Y. Astrocyte-derived exosomes protect hippocampal neurons after traumatic brain injury by suppressing mitochondrial oxidative stress and apoptosis. *Aging* **2021**, *13*, 21642–21658. [CrossRef]
52. Ockaili, R.; Natarajan, R.; Salloum, F.; Fisher, B.J.; Jones, D.; Fowler, A.A., 3rd; Kukreja, R.C. HIF-1 activation attenuates postischemic myocardial injury: Role for heme oxygenase-1 in modulating microvascular chemokine generation. *Am. J. Physiol. Heart Circ. Physiol.* **2005**, *289*, H542–H548. [CrossRef] [PubMed]
53. Kubulus, D.; Mathes, A.; Pradarutti, S.; Raddatz, A.; Heiser, J.; Pavlidis, D.; Wolf, B.; Bauer, I.; Rensing, H. Hemin arginate-induced heme oxygenase 1 expression improves liver microcirculation and mediates an anti-inflammatory cytokine response after hemorrhagic shock. *Shock* **2008**, *29*, 583–590. [CrossRef] [PubMed]
54. Carratu, P.; Pourcyrous, M.; Fedinec, A.; Leffler, C.W.; Parfenova, H. Endogenous heme oxygenase prevents impairment of cerebral vascular functions caused by seizures. *Am. J. Physiol. Heart Circ. Physiol.* **2003**, *285*, H1148–H1157. [CrossRef] [PubMed]
55. Alfieri, A.; Srivastava, S.; Siow, R.C.M.; Cash, D.; Modo, M.; Duchen, M.R.; Fraser, P.A.; Williams, S.C.R.; Mann, G.E. Sulforaphane preconditioning of the Nrf2/HO-1 defense pathway protects the cerebral vasculature against blood-brain barrier disruption and neurological deficits in stroke. *Free Radic. Biol. Med.* **2013**, *65*, 1012–1022. [CrossRef] [PubMed]
56. Vitvitsky, V.; Garg, S.K.; Banerjee, R. Taurine biosynthesis by neurons and astrocytes. *J. Biol. Chem.* **2011**, *286*, 32002–32010. [CrossRef]
57. Wilson, J.X. Antioxidant defense of the brain: A role for astrocytes. *Can. J. Physiol. Pharmacol.* **1997**, *75*, 1149–1163. [CrossRef]
58. Dringen, R.; Kussmaul, L.; Gutterer, J.M.; Hirrlinger, J.; Hamprecht, B. The glutathione system of peroxide detoxification is less efficient in neurons than in astroglial cells. *J. Neurochem.* **1999**, *72*, 2523–2530. [CrossRef]
59. Belanger, M.; Magistretti, P.J. The role of astroglia in neuroprotection. *Dialogues Clin. Neurosci.* **2009**, *11*, 281–295. [CrossRef]
60. Kobayashi, A.; Kang, M.I.; Okawa, H.; Ohtsuji, M.; Zenke, Y.; Chiba, T.; Igarashi, K.; Yamamoto, M. Oxidative stress sensor Keap1 functions as an adaptor for Cul3-based E3 ligase to regulate proteasomal degradation of Nrf2. *Mol. Cell Biol.* **2004**, *24*, 7130–7139. [CrossRef] [PubMed]
61. Theodore, M.; Kawai, Y.; Yang, J.; Kleshchenko, Y.; Reddy, S.P.; Villalta, F.; Arinze, I.J. Multiple nuclear localization signals function in the nuclear import of the transcription factor Nrf2. *J. Biol. Chem.* **2008**, *283*, 8984–8994. [CrossRef]
62. Ishii, T.; Warabi, E.; Mann, G.E. Stress activated MAP kinases and cyclin-dependent kinase 5 mediate nuclear translocation of Nrf2 via Hsp90alpha-Pin1-Dynein motor transport machinery. *Antioxidants* **2023**, *12*, 274. [CrossRef] [PubMed]
63. Baird, L.; Yamamoto, M. The molecular mechanisms regulating the KEAP1-NRF2 pathway. *Mol. Cell Biol.* **2020**, *40*, e00099-20. [CrossRef] [PubMed]
64. Liu, T.; Lv, Y.F.; Zhao, J.L.; You, Q.D.; Jiang, Z.Y. Regulation of Nrf2 by phosphorylation: Consequences for biological function and therapeutic implications. *Free Radic. Biol. Med.* **2021**, *168*, 129–141. [CrossRef] [PubMed]
65. Marino, S.M.; Gladyshev, V.N. Analysis and functional prediction of reactive cysteine residues. *J. Biol. Chem.* **2012**, *287*, 4419–4425. [CrossRef]

antioxidants

MDPI

Article

Changes in Phenolic Compounds and Antioxidant Capacity of *Artocarpus heterophyllus* Lam. (Jackfruit) Pulp during In Vitro Gastrointestinal Digestion

Ming Cheng [1,2], Jiali He [1,3], Yu Gu [1,4,5], Gang Wu [1,5], Lehe Tan [1,6], Chuan Li [2,*], Fei Xu [1,4,5] and Kexue Zhu [1,4,5,6,*]

1 Spice and Beverage Research Institute, Chinese Academy of Tropical Agricultural Sciences, Wanning 571533, China
2 School of Food Science and Engineering, Hainan University, Haikou 570228, China
3 College of Tropical Crop Science, Yunnan Agricultural University, Pu'er 665099, China
4 Key Laboratory of Processing Suitability and Quality Control of the Special Tropical Crops of Hainan Province, Wanning 571533, China
5 National Center of Important Tropical Crops Engineering and Technology Research, Wanning 571533, China
6 Key Laboratory of Nutritional Quality and Health Benefits of Tropical Agricultural Products of Haikou City, Haikou 571100, China
* Correspondence: lichuan@hainanu.edu.cn (C.L.); zhukexue@catas.cn (K.Z.)

Abstract: An in vitro gastrointestinal digestion model was applied to investigate the effect of digestion on the phenolic compounds and antioxidant capacity of *Artocarpus heterophyllus* Lam. (jackfruit) pulp. The total phenol content (TPC) was determined using Folin–Ciocalteu method, and the antioxidant activities were evaluated by DPPH and ABTS assays. Phenolic compounds were analyzed using ultra-performance liquid chromatography coupled with electrospray ionization, followed by quadrupole time-of-flight mass spectrometry (UPLC-ESI-Q-TOF-MS/MS). The results showed that TPC was significantly higher after gastric digestion. Thirty phenolic compounds (hydroxybenzoic acids and derivatives, hydroxycinnamic acids and derivatives, and flavonoids) were identified. The antioxidant activities of the digested samples varied with the TPC, and there was a correlation between antioxidant activity and TPC. The present study implies that gastrointestinal digestion may improve TPC and increase the amount of free phenolic compounds, mainly related to changes in pH value and digestive enzymes.

Keywords: *Artocarpus heterophyllus* Lam.; phenolic compounds; in vitro digestion; antioxidant capacity

Citation: Cheng, M.; He, J.; Gu, Y.; Wu, G.; Tan, L.; Li, C.; Xu, F.; Zhu, K. Changes in Phenolic Compounds and Antioxidant Capacity of *Artocarpus heterophyllus* Lam. (Jackfruit) Pulp during In Vitro Gastrointestinal Digestion. *Antioxidants* **2024**, *13*, 37. https://doi.org/10.3390/antiox13010037

Academic Editors: Baojun Xu and Bin Du

Received: 8 November 2023
Revised: 21 December 2023
Accepted: 21 December 2023
Published: 23 December 2023

1. Introduction

Artocarpus heterophyllus Lam. (Jackfruit) is a species of the Moraceae (Mulberry) family and is renowned as the "queen of tropical fruits" [1]. The jackfruit plant can reach a height of 20 m, and the fruits are large with different shapes. The fruit weights normally range from 5 to 20 kg, and the largest fruit can weigh up to 50 kg. Jackfruit contains many bioactive compounds, such as dietary fiber, volatile sterols, pectin, carotene, etc. [2]. Nowadays, the polyphenols of jackfruit have received extensive attention [3]. The polyphenols of jackfruit peel include organic acids, phenolic acids, and flavonoids [4].

Fruits and vegetables are the primary sources of dietary phenolics. Phenolic compounds are important secondary metabolites widely used as natural antioxidants, as well as producers of sensory properties such as color and flavor. Moreover, phenolic compounds have many benefits for the human body, such as inhibiting reactive oxygen and nitrogen species, transferring electrons to free radicals, activating antioxidant enzymes, and alleviating oxidative stress and inflammation. Studies have shown that phenolic compounds positively impact various diseases, including diabetes, obesity, cancer, cardiovascular disease, osteoporosis, and neurodegenerative diseases [5–9]. Thus, phenolic compounds may not only have great application potential in the field of food but also in medicine and healthcare.

Due to the drawbacks, including the time-consuming and high cost of animal and human studies, in vitro digestion models are developed to assess the bioaccessibility or absorption of phytochemicals during digestion [10]. Current studies on in vitro simulated digestion have focused on phenolic compounds in food matrices and extracts, such as coffee (*Coffea arabica* L.) pulp, oranges (*Citrus sinensis*), raspberry, and flours from persimmon fruit (*Diospyros kaki*) co-products, among others [11–14]. Konsue et al. [15] determined the bioaccessibility of phytochemical compounds in jackfruits using simulated in vitro gastrointestinal digestion. Zhu et al. [16] investigated the effects of in vitro saliva, gastric and intestinal digestion on the chemical properties and antioxidant activity of polysaccharides from *Artocarpus heterophyllus* Lam. (Jackfruit) pulp. After in vitro digestion, the digested jackfruit flake has enhanced protection against acrylamide-induced oxidative damage [17]. The antioxidant activities of papaya, jackfruit, and araticum extracts were evaluated using in vitro gastrointestinal digestion [18]. However, the qualitative analyses and molecular structure of phenolic compounds after in vitro gastrointestinal digestion need further investigation.

Thus, the present study aimed to investigate the influences of in vitro gastrointestinal digestion on TPC values and antioxidant activity of jackfruit pulp. Phenolic compounds were identified using ultra-performance liquid chromatography coupled with electrospray ionization, followed by quadrupole time-of-flight mass spectrometry (UPLC-ESI-Q-TOF-MS/MS) (fig:antioxidants-2734457-f001). The findings may provide new insights into the consumptions of jackfruit, leading to potential health benefits.

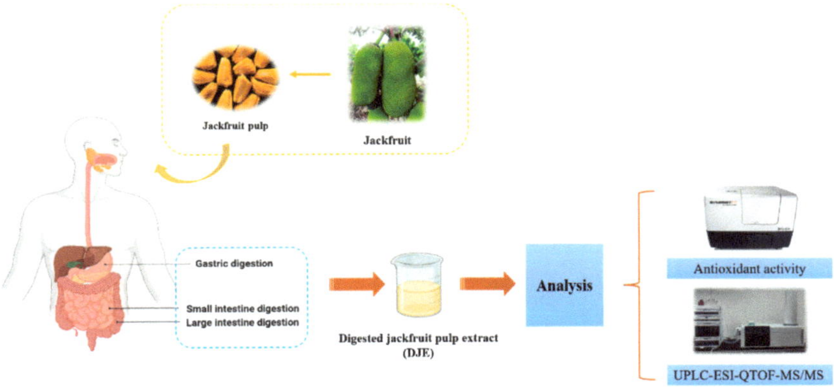

Figure 1. Schematic and design roadmap of in vitro simulated gastrointestinal digestion on phenolic components and the antioxidant activity of jackfruit pulp.

2. Materials and Methods

2.1. Materials and Reagents

Malaysia 1 jackfruit was obtained from the Spice and Beverage Research Institute, Chinese Academy of Tropical Agricultural Sciences (Wanning, Hainan, China). The fruit at the fully ripe stage (14−16 weeks) was selected, and the pulp samples were collected, frozen, and stored at −20 °C for subsequent studies.

Pepsin (from porcine gastric mucosa, ≥500 U/mg), pancreatin (from porcine pancreas), bile salts (porcine bile extract), DPPH (1-diphenyl-2-picrylhydrazyl), and ABTS (2,2′-azinobis (3-ethylbenzothiazoline-6-sulfonic acid) diammonium salt) were from Sigma-Aldrich, Co., Ltd. (St. Louis, MO, USA). Gallic acid was from Shanghai Yuanye Bio-Technology Co., Ltd. (Shanghai, China). Folin-Ciocalteu reagent was from Beijing Solarbio Science & Technology Co.,Ltd. (Beijing, China). All solvents and chemicals utilized were of LC-MS quality or analytical grade (>98%).

2.2. *In Vitro Gastrointestinal Digestion*

Human upper gastrointestinal (GI) digestion was simulated using a two-step in vitro digestion model adapted from the protocol released by INFOGEST [19]. The simulated gastric fluid (SGF) and simulated intestinal fluid (SIF) adhered to the guidelines in the INFOGEST publication (Table S1). To simulate the gastric phase, pepsin solution (0.5 mL, 80,000 U/mL), SGF (8 mL), and $CaCl_2$ solution (5 μL, 0.3 mol/L) were mixed in ultra-pure water, and the pH value was adjusted to 3.0 using 1 mol/L HCl. The homogenized jackfruit pulp samples were blended with an identical volume of simulated gastric juice and incubated at 37 °C for 0 h, 1 h, and 2 h, respectively. The digested samples (2.0 mL) were collected and placed in a boiling water bath for 10 min to neutralize them for subsequent assay.

Simulated intestinal digestion was initiated by adding the gastric chyme with SIF (8.5 mL), pancreatin solution (5.0 mL, 100 U/mL), fresh bile solution (2.5 mL, 160 mmol/L), $CaCl_2$ (0.04 mL, 0.3 mol/L), and ultrapure water. The pH value was adjusted to 7.0 using 1 mol/L NaOH solution. The mixture solutions were placed in a water bath at 37 °C, and the digested samples (2.0 mL) were collected at 1 h, 2 h, and 4 h, respectively, and then immediately immersed in a boiling water bath for 10 min to stop the digestion.

2.3. *Extraction of Digested Jackfruit Pulp Samples*

Digested jackfruit pulp samples were extracted using an Ultrasonic-microwave Co-operative Extractor/Reactor (Model CW-2000, XTrust Analytical Instrument Technology Co., Ltd., Shanghai, China) according to our previous method [20]. Briefly, digested jackfruit pulp samples and 60% ethanol were mixed and vortexed, at a solid-to-liquid ratio of 1:30 (g: mL). The mixture was then subjected to microwave extraction for 165 s at a power level of 550 W. The extracted blend was centrifugated at 10,000 rpm, 4 °C for 10 min, and the residue was re-extracted twice. The combined supernatants were evaporated using a rotary evaporator R-215 (BUCHI Labortechnik AG, Flawil, Switzerland) to remove ethanol. Subsequently, the concentrate was recovered with ultrapure water until a final volume of 25 mL, and a working solution of digested jackfruit pulp extract (DJE) was obtained.

2.4. *Determination of Total Phenol Content (TPC)*

The Folin–Ciocalteu method was used to measure the phenolic contents. Briefly, DJE (0.5 mL) were mixed with Folin–Ciocalteu reagents (2 mL) for 5 min, and then 7.5% Na_2CO_3 (2 mL) was added. A blank was setup by following the same procedure wherein the DJE was replaced with an equal volume of methanol. The absorbance was read at 760 nm after incubation for 40 min in the dark for color development. The results were estimated as gallic acid equivalent (mg GAE/g).

2.5. *Analysis of Phenolic Compounds*

According to our previous method [20], DJE samples were analyzed by Agilent 1290 UPLC and 6530B hybrid Q-TOF-MS (Agilent Technologies, Santa Clara, CA, USA). Mass spectral signals were acquired in positive and negative electrospray ionization (ESI) scanning mode, respectively (tabref:antioxidants-2734457-t001).

2.6. *Antioxidant Activity Assay*

DPPH radical scavenging ability assay. The DPPH method was applied in accordance with a prior report with slight modifications [21]. Briefly, DPPH was dissolved in 80% methanol. A mixture of DPPH working solution (150 μL) and the diluted sample extract (150 μL) or 80% methanol (control) was prepared, shaken, and allowed to stand in the dark at room temperature for 30 min. The absorbance was then read at 517 nm (SynergyH1, BioTek, Santa Clara, CA, USA). Trolox was used as a standard, and the results were presented as μg Trolox equivalent /100 g fresh mass (the calibration curve was $Y = -0.0058X + 0.3554$, $R^2 = 0.9909$).

Table 1. The analysis conditions of UPLC-ESI-QTOF-MS/MS.

Parameters	Agilent 1290 Infinity II UPLC, 6530B Hybrid Q-TOF-MS
Column	Agilent Zorbax Eclipse Plus C_{18} column (3.0 mm × 150 mm, 1.8 μm)
Mobile phase	0.1%formic acid in water (A) and acetonitrile (B)
Phase gradient	0–1.5 min, 5% B; 1.5–15 min, 5–60% B; 15–25 min, 60–100% B; 30–30.10 min, 100–5% B; 30.10–35 min, 5% B
Injection volume	3 μL
Column temperature	35 °C
Flow rate	0.4 mL/min
Sheath gas	Helium
Sheath gas temperature	325 °C
Sheath gas flow rate	11 L/min
Crash voltage	140 V

ABTS radical scavenging ability assay. The method of Cheng et al. (2020) was used to determine the ABTS radical scavenging ability [22]. In summary, ABTS solution (7 mmol/L) was combined with potassium persulfate solution (2.45 mmol/L) in equal volumes and reacted in the dark for 12–16 h to produce ABTS+ solution. Prior to usage, the stock solution was appropriately diluted with ultrapure water to obtain an absorbance of 0.70 ± 0.02 at 734 nm. For the assay, the diluted sample (10 μL) was mixed with ABTS+ solution (190 μL) and incubated in the dark for 6 min. Subsequently, the absorbance of the resulting mixture was read at 734 nm. Trolox was used as a standard, and the results were presented as μg Trolox equivalent /100 g fresh mass (the calibration curve was Y = −1.65X + 0.3682, R^2 = 0.9957).

2.7. Statistical Analysis

Experiments were performed in triplicate, and the results were expressed as mean ± standard error of the mean (SEM). Data were analyzed through one-way analysis of variance (ANOVA) in conjunction with Duncan's multiple range test using SPSS software (version 26.0, SPSS, Inc., Chicago, IL, USA). The *p*-value was set to 0.05 for a significant difference.

The MS spectra were analyzed and converted to compound Exchange Format (.CEF) files with the help of the "find compounds by molecular feature" tool using Agilent Mass Hunter Qualitative Analysis Software (version B.07.00). The exported files were then imported into Mass Profiler Professional (MPP) software (version 14.0, Agilent Technologies, Santa Clara, CA, USA) for further statistical analysis. Alignment parameters were: RT window = 0.5% + 0.1 min, mass window = 10 ppm + 1 mDa. A principal component analysis (PCA) was performed with MPP to visualize the sample groupings at the different stages of in vitro simulated digestion. The resulting entity list was then processed in the ID browser, which allowed for chemical formulas to be generated and searched against a proprietary database.

3. Results and Discussion

3.1. TPC

As shown in fig:antioxidants-2734457-f002, compared with 0 h of gastric digestion, TPC in jackfruit pulp increased significantly (from 1.99 ± 0.07 to 2.85 ± 0.51 mg GAE/g), with an increase of 43.22% after 2 h of gastric digestion. This result is similar to previous studies, showing that the TPC values exhibited a continuous increase during simulated digestion [15,17]. Konsue et al. [15] reported that TPC values of jackfruit during simulated digestion varied from 4.48 to 124.84 mg GAE/100 g at four ripening stages. TPC also exhibited a significant increase in undigested and digested extracts, measuring 23.3 ± 0.004 and 33.9 ± 0.002 mg GAE/100 g freeze-dried fruit, respectively [18]. Bouayed et al. [23] reported that TPC in fresh apple fruit increased after gastric digestion. Numerous physical factors (temperature, pH value, ion force) and biological factors (bile salts and enzymes) affect phenolic compound stability during in vitro simulated digestion [24,25]. Pepsin can hydrolyze chemical bonds (covalent bonds, hydrogen bonds, etc.) that are formed by polyphenols bound to some macromolecules (such as proteins and carbohydrates) inside and outside the cell, making the polyphenols that were initially bound to these macromolecules free. It also weakens some of the ester bonds where phenolic acids interact with the cell wall, releasing phenolic acids [26]. Meanwhile, in an acidic environment, polyphenols may also undergo hydrolysis, and some glycosides may be converted into aglycones, increasing the phenolic content [27].

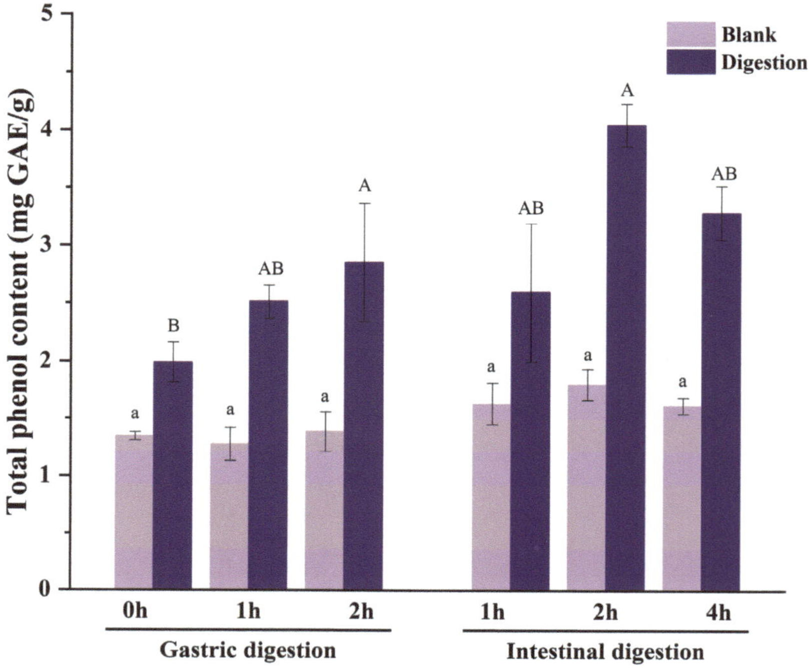

Figure 2. TPC of DJE after in vitro simulated digestion. Different letters point to significant differences in the same digestion solution at different times ($p < 0.05$).

However, TPC was significantly reduced (by 9.11%, $p < 0.05$) after 1 h of intestinal digestion. This result is consistent with previous studies on red- and yellow-colored pea shells [28], and sweet orange (*Citrus sinensis*) [29]. This is due to the dilution of polyphenols caused by the addition of intestinal digestive juices, in addition to an increase in pH value and changes in the acid-base environment, resulting in a decrease in polyphenol content [30]. After 2 h of intestinal digestion, TPC significantly increased. This may be related to the action of intestinal digestive enzymes, which decompose bound phenols in the substrate and release free phenols. Then, TPC started to decrease, which could be caused by decreased pancreatic enzyme activity, slow decomposition of bound phenols, or the conversion of free phenols into other compounds [31]. During gastrointestinal digestion, the TPC of the intestinal digest was higher than that of the gastric digest, indicating that gastrointestinal digestion enhanced the release of phenols in DJE.

3.2. *Identification of Polyphenols and their Decomposition Products after In Vitro Digestion*

In the PCA analysis (fig:antioxidants-2734457-f003), the first two principal components (PC1 and PC2) explained 95.17% of the total variance in the positive mode, while accounting for 83.49% of the total variance in the negative ion mode. The 2D scores plot indicated that the digested jackfruit pulp samples at the different gastric and intestinal digestion stages were clearly separated due to differences in the accumulation of metabolites.

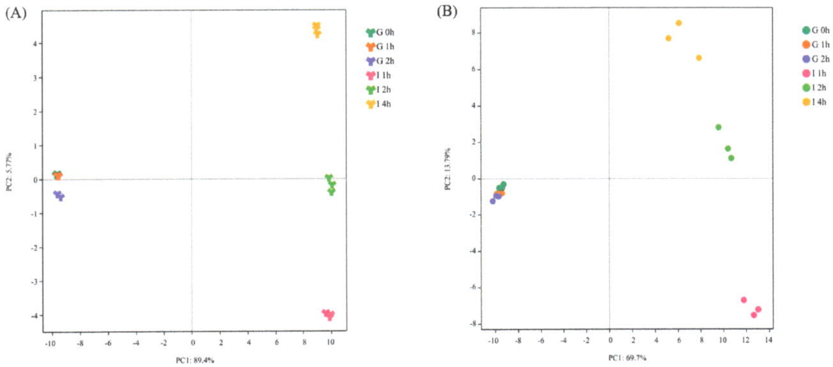

Figure 3. PCA score plot of the metabolites of DJE. (**A**) Positive mode and (**B**) negative mode. G0: Gastric digestion 0 h, G1: Gastric digestion 1 h, G2: Gastric digestion 2 h, I1: Intestinal digestion 1 h, I2: Intestinal digestion 2 h, I4: Intestinal digestion 4 h.

Preliminary identification was achieved by comparing the collected MS data with chemicals previously identified in the literature or registered in Massbank. A total of 30 substances were identified in DJE, including 22 flavonoids (8 flavonols, 5 flavones, and 4 flavanols), 6 hydroxycinnamic acids and derivatives, and 2 hydroxybenzoic acids and derivatives (tabref:antioxidants-2734457-t002). The total ion chromatogram (TIC) of MS spectral data is shown in Figure S1. The MS spectra and structural formulae of the monomeric substances are illustrated in Figure S2.

Hydroxybenzoic acids and derivatives. According to reference substances, compounds 1 and 2 at m/z 171.0291 $[M + H]^+$ and 257.0657 $[M + CH_3COO]^-$, with their resultant ions at m/z 125 and 153 attributed to the loss of CO_2, respectively, were identified as gallic and syringic acids, respectively [32].

Hydroxycinnamic acids and derivatives. Compound 3 (Rt = 7.522, m/z 337.0906) was determined as 5-p-coumaroylquinic acid with a main fragment at m/z 147 and 119, resulting from the loss of $[C_9H_8O_2\text{-}H]^-$ and $[C_8H_8O\text{-}H]^-$, respectively. Its fragmentation patterns aligned with previous reports [33,34]. Polyphenols often have multiple *cis*/*trans* isomers. Exposure to UV light leads to phytochemical isomerization of naturally occurring phenols that often appear in the *trans* conformation. O-caffeoylquinic is the main quinate derivative found in jackfruit polyphenols and is often esterified at positions 1, 3, 4, and 5 of quinic acid, yielding four positional isomers. In the negative mode, the MS/MS spectra of caffeoylquinic acid (CQA) typically display common product ions of m/z 191.06 ($C_7H_{12}O_6$), 173.05 ($C_7H_{10}O_5$), 179.0342 ($C_9H_8O_4$), and 135.0049 ($C_8H_8O_2$). These were attributed to the fragments of [quinic acid-H]$^-$, [quinic acid-H_2O-H]$^-$, [caffeic acid-H]$^-$, and [caffeic acid-CO_2-H]$^-$, respectively [35,36]. Comparing the retention times and product ion fragments with references from previous studies [37,38], compound 4 exhibited product ions at m/z 173 and 179, generated by the loss of [quinic acid-H_2O-H]$^-$ and [caffeic acid-H]$^-$, and identified as 1-O-caffeoylquinic acid. The $C_9H_7O_3$ and $C_{16}H_{15}O_7$ in compound 5 were cleaved to produce signals at m/z 163 and 319; the compound was assigned as (E,E)-3,5-di-O-caffeoylquinic acid based on a comparison with a previous report [39]. In the negative ionization mode, compound 6 (Rt = 1.674 min) had an [M-H]$^-$ of m/z 367.1056 and was proposed to be 3-O-caffeoylquinic acid methyl ester as it lost a hydrogen ion fragment with mass 1 under the negative mode conditions of mass spectrometry. Additionally, the product ion fragments at m/z 135 could be an adduct ion fragment of $C_8H_7O_2$ and a hydrogen ion, further supporting the hypothesis that compound 6 may be 3-O-caffeoylquinic acid methyl ester [37]. Similarly, based on the analysis of the fragmentation pattern and database search, compounds 7 and 8 were proposed as 4,5-di-O-caffeoylquinic acid ester and ethyl-3,4-dicaffeoylquinate, respectively.

Flavanols. Flavanols exist in plants as monomers (catechins, epicatechin, epigallocatechin, gallocatechin) or multimers (procyanidins or condensed tannins) [40]. Compound 9 (Rt = 1.563) showed a molecular ion ($[M + HCOO]^-$) at m/z of 353.0869. Based on the mass spectral information and comparison with a study [41], this fragment was considered as the adduct fragment of (+)-catechin and HCOO- ions. This compound generated a diagnostic ion at m/z 245 $[M\text{-}CH_2\text{-}CHOH\text{-}H]^-$, and so could be (+)-catechin hydrate. Compound 10 demonstrated the $[M-H]^-$ ion of m/z 441.0798 and fragments of m/z 167, 137, and 125, and so was regarded as (-)-epicatechin gallate. Compound 11 yielded a parent ion at m/z 459.0919, and the primary fragment ion at m/z 125 corresponded to a trihydroxybenzene moiety. It was identified as gallocatechin by comparison with a report by Liu et al (2020) [42]. Compound 12 yielded fragment ions of m/z 539, 407, and 285, respectively. m/z 539 is produced by the molecular ion shedding two molecules of H_2O (36 u). m/z 407 is produced by the molecular ion undergoing RDA cleavage (152 u) while shedding one more molecule of H_2O. m/z 285 corresponds to the molecular ion undergoing quinone methide fission (QM) cleavage (290 u). The molecular ion breakage and fragmentation are consistent with the A-type procyanidin dimer cleavage pattern [43,44]. By mass spectrometry database analysis, compound 12 was identified as procyanidin A1.

Flavonols. Flavonols are the most common flavonoid in food and are represented by kaempferol and quercetin [40]. Compounds 13 and 14 were found to be quercetin derivatives; compound 13 (Rt = 1.58 min, m/z 397.0781) corresponded to quercetin dihydrate. Meanwhile, precursor ions $[M\text{-}H]^-$ at m/z 789.2055 were detected in compound 14, which showed a major fragment ion ($[M\text{-}H\text{-}glycoside]^-$) at m/z 301. The compound was further identified by database search as quercetin-3-O-beta-D-glucose-7-O-beta-D-gentiobiosiden. Two kaempferol-diglycosides, namely, kaempferol-3-glucoside (compound 15, m/z 447) and kaempferol 3-O-robinobioside (compound 16, m/z 595.1632), were identified. Both compounds resulted in the dominant fragment ion at m/z 285, which was associated with the cleavage of a glycosidic linkage (glucoside or robinobioside) accompanied by a conformational change in H. In addition, compound 17 presented a precursor ion at 625.1758, fragment ions at 314 ($C_{28}H_{32}O_{16}$); it was then tentatively identified as isorhamnetin-3-O-neohesperidoside. The molecular formula of compound 18 was determined as $C_{21}H_{18}O_{14}$, by observing the secondary mass spectrometry ion fragmentation. The main fragment ion of the compound was m/z 493, and by comparison with previous study [45], the compound was identified as hibifolin. Analogously, compounds 19 ($C_{25}H_{26}O_7$) and 20 ($C_{33}H_{40}O_{15}$) were proposed as papyriflavonol A and sagittatoside A.

Flavanones. Flavanones are generally present as glycosides, and their aglycone form is released during digestion, mainly including hesperidin and naringenin. On the basis of the characteristic fragment ions at m/z 494 $[M + H\text{-}C_6H_{10}O_5]$, 465 $[M + H\text{-}C_6H_{10}O_4]$, 431 $[M + H\text{-}C_6H_{10}O_5\text{-}H_2O]$, and 303 $[M + H\text{-}C_6H_{10}O_4\text{-}C_6H_{10}O_5]$, compound 21 was proposed as hesperidin, which was in accordance with a relevant report [46]. Compound 22 was assigned as naringenin based on its forming deprotonated molecule at m/z 271.064 and product ions at m/z 177 $[M\text{-}H\text{-}C_6H_5OH]^-$, 151 (RDA fragmentation reaction cleaved at the C-ring of flavonoid aglycones) and 107 $[151\text{-}CO_2]^-$. Similarly, compounds 23 ($C_{18}H_{18}O_5$) and 24 ($C_{16}H_{14}O_5$) were easily assigned to naringenin trimethyl ether and 5-O-Methylnaringenin. The precursor ion $[C_{25}H_{26}O_6 + HCOO]^-$ at m/z 467.1715 and fragments at m/z 367 $[C_{21}H_{20}O_6\text{-}H]^-$, 45 $[C_2H_6O\text{-}H]^-$, and 123 $[C_7H_8O_2\text{-}H]^-$ were used to identify compound 25 as kuwanol C.

Table 2. UPLC-ESI-QTOF-MS characteristics of polyphenols and their metabolites in jackfruit pulp in vitro digestion.

Compd	Rt (min)	MS	MS/MS	Formula	Identification	Distribution					
						G0	G1	G2	I1	I2	I4
					Hydroxybenzoic acids and derivatives						
1	3.172	171.0291	125	$C_7H_6O_5$	Gallic acid			✓			
2	1.431	257.0657	153	$C_9H_{10}O_5$	Syringic acid			✓			
					Hydroxycinnamic acids and derivatives						
3	7.522	337.0906	147, 119	$C_{16}H_{18}O_8$	5-p-Coumaroylquinic acid			✓			✓
4	7.323	399.0939	173, 179	$C_{16}H_{18}O_9$	1-Caffeoylquinic acid			✓			
5	1.44	539.1212	163, 319	$C_{25}H_{24}O_{12}$	(E,E)-3,5-di-O-Caffeoylquinic acid				✓		
6	1.674	367.1056	135	$C_{17}H_{20}O_9$	3-O-Caffeoylquinic acid methyl ester				✓		
7	8.757	531.1475	204,163	$C_{26}H_{26}O_{12}$	4,5-di-O-Caffeoylquinic acid ester				✓		✓
8	4.503	543.1557	326, 163	$C_{27}H_{28}O_{12}$	Ethyl-3,4-dicaffeoylquinate					✓	
					Flavanols						
9	1.563	353.0869	245	$C_{15}H_{16}O_7$	(+)-Catechin hydrate			✓	✓		
10	8.304	441.0798	167, 137, 125	$C_{22}H_{18}O_{10}$	(−)-Epicatechin gallate		✓				
11	10.161	459.0919	125, 137,139	$C_{22}H_{18}O_{11}$	Gallocatechin			✓			
12	20.856	599.1153	539, 407, 285	$C_{30}H_{24}O_{12}$	Procyanidin A1			✓	✓		
					Flavonols						
13	1.58	397.0781	303, 301, 273	$C_{15}H_{14}O_9$	Quercetin dihydrate						✓
14	2.284	789.2055	591, 489	$C_{33}H_{40}O_{22}$	Quercetin-3-O-beta-D-glucose-7-O-beta-D-gentiobiosiden					✓	
15	1.567	447.4016	285	$C_{21}H_{18}O_{12}$	Kaempferol-3-glucuronide			✓	✓		
16	20.865	595.1632	287, 285, 449	$C_{27}H_{30}O_{15}$	Kaempferol 3-O-robinobioside						✓
17	15.968	625.1758	314	$C_{28}H_{32}O_{16}$	Isorhamnetin-3-O-neohesperidoside	✓				✓	
18	16.206	553.0848	493	$C_{21}H_{18}O_{14}$	Hibifolin	✓					
19	7.151	497.1851	438	$C_{25}H_{26}O_7$	Papyriflavonol A			✓			
20	1.679	677.2371	351	$C_{33}H_{40}O_{15}$	Sagittatoside A			✓			
					Flavanones						
21	3.37	633.1822	494, 465, 431	$C_{28}H_{34}O_{15}$	Hesperidin			✓	✓		
22	6.772	271.064	177,151, 107	$C_{15}H_{12}O_5$	Naringenin			✓			✓
23	7.3	373.1289	181, 161	$C_{18}H_{18}O_5$	Naringenin trimethyl ether			✓			
24	5.147	331.0828	193, 93	$C_{16}H_{14}O_5$	5-O-Methylnaringenin					✓	
25	29.446	467.1715	367, 45, 123	$C_{25}H_{26}O_6$	Kuwanol C			✓			
					Flavones						
26	15.971	615.1728	593	$C_{28}H_{32}O_{14}$	Fortunellin		✓				
27	7.301	395.1096	342, 357	$C_{20}H_{20}O_7$	Tangeretin				✓		
28	6.347	553.1154	521, 477	$C_{31}H_{20}O_{10}$	Bilobetin	✓		✓			✓
29	2.374	289.0703	271, 243, 153	$C_{15}H_{12}O_6$	Aromadendrin				✓		
30	2.469	463.0875	162	$C_{21}H_{20}O_{12}$	6-Hydroxyluteolin-7-glucoside						✓

G0: Gastric digestion 0 h, G1: Gastric digestion 1 h, G2: Gastric digestion 2 h, I1: Intestinal digestion 1 h, I2: Intestinal digestion 2 h, I4: Intestinal digestion 4.

Flavones. Compound 26, with the master ion $[M + H]^+$ at m/z 615.1728, infers that the fragment ion m/z 593 gives a sodium ion fragment in the positive mode, which is considered to be fortunellin since the fragmentation pattern is comparable to that reported previously [47]. Compound 27 presented a precursor ion at m/z 395.1096, easily identified as tangeretin. The molecular ion peak $[M\text{-}H]^-$ of compound 28 was at m/z 553.1154. In the high mass region, it loses one molecule of CH_3OH to form the fragment ion with the highest ionic strength m/z 521, and this fragment ion loses one molecule of CO_2 to acquire the fragment ion m/z 477. Therefore, the compound could be inferred to be bilobetin. Based on the analysis of its secondary mass spectrometry ion fragments 271, 243, and 153, compound 29 with $[M + H]^+$ at m/z 289.0703was identified as aromadendrin. Compound 30 (Rt = 2.469, $[M + H]^+$ at m/z 463.0875) confirmed that the fragment ion at m/z 303 was a 6-hydroxyaluminoenyl protein with a neutral loss of 162, corresponding to the disappearance of the hexose moiety. Hence, it was identified as 6-hydroxytyrosine-7-glucoside.

3.3. Antioxidant Activity

As shown in fig:antioxidants-2734457-f004A, in comparison with the initial gastric digestion stage (22.212 μg Trolox/mL), the DPPH radical scavenging ability of DJE increased by 3.54% and 17.73% after gastric digestion for 1 h and 2 h, respectively, which may be related to the acidic environment conducive to the release of antioxidants. DPPH radical scavenging ability was significantly reduced at 1 h during the intestinal digestion phase ($p < 0.05$), with a 12.07% decrease compared to after gastric digestion. It then started to rise and remained stable at 4 h of intestinal digestion. Pavan et al. [18] reported an increased antioxidant activity of jackfruit extracts after in vitro digestion using trolox equivalent antioxidant capacity and oxygen radical absorbance capacity methods. This may be because of the change in pH value from gastric digestion to intestinal digestion, leading to changes in polyphenolic compounds that affect the free radical scavenging ability of digested extracts [48]. Phenolic compounds are considered critical bioactive compounds in the fight against free radicals. However, no significant differences in ABTS radical scavenging ability were found, needing further investigation.

TPC is closely related to the antioxidant capacity of plants, and bioactive compounds can release monomers or glycosides during gastrointestinal digestion, thereby increasing the number of phenolic hydroxyl groups. This could be due to the interaction between phenolic hydroxyl groups as hydrogen donors and free radicals, boosting their free radical scavenging properties [31,49]. The results of correlation analysis (fig:antioxidants-2734457-f004C) showed that the TPC in this study correlated well with the antioxidant capability obtained by the DPPH assay, but not with the antioxidant capacity indicated by the ABTS assay. This is not entirely consistent with a report by Pods Dek et al. (2014) [50]. From this point of view, the link between TPC and antioxidant activity, except for the method of determining antioxidant activity [51], may also be related to the source of phenolic substances, main phenolic components, free or bound state, etc.

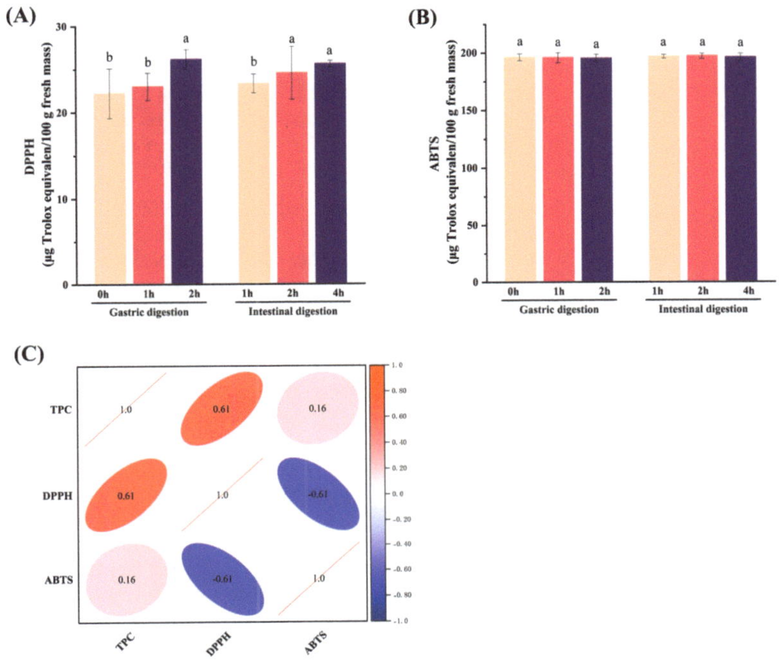

Figure 4. Antioxidant activity (**A**) DPPH radical scavenging ability, (**B**) ABTS radical scavenging ability and the correlation between total phenolic content and antioxidant activity (**C**) of DJE. Different letters indicate significant differences ($p < 0.05$). Positive correlation is indicated by red colors and negative correlation by blue colors.

4. Conclusions

This study focused on the impacts of simulated in vitro gastrointestinal digestion on the TPC, phenolic constituents, and antioxidant activity of DJE. The TPC in the fruit significantly increased after gastric digestion, while these values decreased and then increased during intestinal digestion. Overall, the TPC in the intestinal digest was higher than that in the gastrointestinal digest, suggesting that gastrointestinal digestion increased the TPC in DJE. In addition, 30 phenolic compounds were identified during in vitro simulated gastrointestinal digestion. The antioxidant activity of the digested samples, as determined by the DPPH assay, varied with TPC, and there was a correlation between them, but that correlation was not as strong in ABTS. Therefore, this study suggests that in vitro digestion can facilitate the release of polyphenols in DJE with antioxidant effects. These findings provide important references for the potential benefits of polyphenols in jackfruit pulp for gastrointestinal health. Considering that polyphenols may undergo significant transformation during the process of digestion and absorption, and that the altered forms may exhibit distinct biological properties and effects, future research should also take into account their intestinal flora and metabolic behavior, which may influence health and disease treatment outcomes.

Supplementary Materials: The following supporting information can be downloaded at: https://www.mdpi.com/article/10.3390/antiox13010037/s1, Table S1. Preparation of electrolyte solutions and simulated digestive fluids. Figure S1. Positive and negative total ion chromatogram of DJE. Figure S2. MS spectra and structural formulae of 30 compounds identified in JPE after in vitro simulated gastrointestinal digestion.

Author Contributions: M.C. designed the experiments, conducted the experiments, analyzed the data, and wrote and revised the manuscript. J.H. designed the experiments and conducted the experiments. Y.G. analyzed the data and revised the manuscript. G.W. analyzed the data. L.T. analyzed the data and revised the manuscript. C.L. wrote and revised the manuscript and gave the final approval. F.X. analyzed the data and revised the manuscript. K.Z. wrote and revised the manuscript and gave the final approval. All authors have read and agreed to the published version of the manuscript.

Funding: This work is financially supported by the Key Research and Development Project of Hainan Province (No. ZDYF2020218), the Chinese Academy of Tropical Agricultural Sciences for Science and Technology Innovation Team of National Tropical Agricultural Science Center (NO CATASCXTD202304), and the Central Public-interest Scientific Institution Basal Research Fund for the Chinese Academy of Tropical Agricultural Sciences (No. 1630142022009).

Institutional Review Board Statement: Not applicable.

Informed Consent Statement: Not applicable.

Data Availability Statement: Data is contained within the article.

Conflicts of Interest: The authors declare no conflicts of interest.

References

1. Shafiq, M.; Mehmood, S.; Yasmin, A.; Khan, S.J.; Khan, N.H.; Ali, S. Evaluation of phytochemical, nutritional and antioxidant activity of indigenously grown jackfruit (*Artocarpus heterophyllus* Lam). *J. Sci. Res.* **2017**, *9*, 135–143. [CrossRef]
2. Gupta, A.; Marquess, A.R.; Pandey, A.K.; Bishayee, A. Jackfruit (*Artocarpus heterophyllus* Lam.) in health and disease: A critical review. *Crit. Rev. Food Sci. Nutr.* **2022**, *63*, 6344–6378. [CrossRef] [PubMed]
3. Vazhacharickal, P.J.; Sajeshkumar, N.K.; Mathew, J.J.; Kuriakose, A.C.; Abraham, B.; Mathew, R.J.; Albin, A.N.; Thomson, D.; Thomas, R.S.; Varghese, N.; et al. Chemistry and medicinal properties of jackfruit (*Artocarpus heterophyllus*): A review on current status of knowledge. *Int. J. Innov. Res. Rev.* **2015**, *3*, 83–95.
4. Zhang, L.; Tu, Z.; Xie, X.; Wang, H.; Wang, H.; Wang, Z.; Sha, X.; Lu, Y. Jackfruit (*Artocarpus heterophyllus* Lam.) peel: A better source of antioxidants and α-glucosidase inhibitors than pulp, flake and seed, and phytochemical profile by HPLC-QTOF-MS/MS. *Food Chem.* **2017**, *234*, 303–313. [CrossRef] [PubMed]
5. Agunloye, O.M.; Oboh, G.; Ademiluyi, A.O.; Ademosun, A.O.; Akindahunsi, A.A.; Oyagbemi, A.A.; Omobowale, T.O.; Ajibade, T.O.; Adedapo, A.A. Cardio-protective and antioxidant properties of caffeic acid and chlorogenic acid: Mechanistic role of angiotensin converting enzyme, cholinesterase and arginase activities in cyclosporine induced hypertensive rats. *Biomed. Pharmacother.* **2019**, *109*, 450–458. [CrossRef]
6. Ali, F.; Rahul; Jyoti, S.; Naz, F.; Ashafaq, M.; Shahid, M.; Siddique, Y.H. Therapeutic potential of luteolin in transgenic Drosophila model of Alzheimer's disease. *Neurosci. Lett.* **2019**, *692*, 90–99. [CrossRef] [PubMed]
7. De Paulo Farias, D.; Neri-Numa, I.A.; de Araújo, F.F.; Pastore, G.M. A critical review of some fruit trees from the Myrtaceae family as promising sources for food applications with functional claims. *Food Chem.* **2020**, *306*, 125630. [CrossRef]
8. Elsayed, R.H.; Kamel, E.M.; Mahmoud, A.M.; El-Bassuony, A.A.; Bin-Jumah, M.; Lamsabhi, A.M.; Ahmed, S.A. *Rumex dentatus* L. phenolics ameliorate hyperglycemia by modulating hepatic key enzymes of carbohydrate metabolism, oxidative stress and PPARγ in diabetic rats. *Food Chem. Toxicol.* **2020**, *138*, 111202. [CrossRef]
9. Liu, F.; Li, L.; Lu, W.; Ding, Z.; Huang, W.; Li, Y.T.; Cheng, C.; Shan, W.S.; Xu, J.; He, W.; et al. Scutellarin ameliorates cartilage degeneration in osteoarthritis by inhibiting the Wnt/β-catenin and MAPK signaling pathways. *Int. Immunopharmacol.* **2020**, *78*, 105954. [CrossRef]
10. Alminger, M.; Aura, A.M.; Bohn, T.; Dufour, C.; El, S.N.; Gomes, A.; Karakaya, S.; Martínez-Cuesta, M.C.; McDougall, G.J.; Requena, T.; et al. In vitro models for studying secondary plant metabolite digestion and bioaccessibility. *Compr. Rev. Food Sci. Food Saf.* **2014**, *13*, 413–436. [CrossRef]
11. Hao, Y.; Yang, J.; Cui, J.; Fan, Y.; Li, N.; Wang, C.; Liu, Y.; Dong, Y. Stability and mechanism of phenolic compounds from raspberry extract under in vitro gastrointestinal digestion. *LWT Food Sci. Technol.* **2021**, *139*, 110552. [CrossRef]
12. Khochapong, W.; Ketnawa, S.; Ogawa, Y.; Punbusayakul, N. Effect of in vitro digestion on bioactive compounds, antioxidant and antimicrobial activities of coffee (*Coffea arabica* L.) pulp aqueous extract. *Food Chem.* **2021**, *348*, 129094. [CrossRef] [PubMed]
13. Lucas-González, R.; Viuda-Martos, M.; Pérez Álvarez, J.A.; Fernández-López, J. Changes in bioaccessibility, polyphenol profile and antioxidant potential of flours obtained from persimmon fruit (*Diospyros kaki*) co-products during in vitro gastrointestinal digestion. *Food Chem.* **2018**, *256*, 252–258. [CrossRef] [PubMed]
14. Aschoff, J.K.; Kaufmann, S.; Kalkan, O.; Neidhart, S.; Carle, R.; Schweiggert, R.M. In vitro bioaccessibility of carotenoids, flavonoids, and vitamin C from differently processed oranges and orange juices [*Citrus sinensis* (L.) Osbeck]. *J. Agric. Food Chem.* **2015**, *63*, 578–587. [CrossRef] [PubMed]

15. Konsue, N.; Bunyameen, N.; Donlao, N. Utilization of young jackfruit (*Artocarpus heterophyllus* Lam.) as a plant-based food ingredient: Influence of maturity on chemical attributes and changes during in vitro digestion. *LWT Food Sci. Technol.* **2023**, *180*, 114721. [CrossRef]
16. Zhu, K.; Yao, S.; Zhang, Y.; Liu, Q.; Xu, F.; Wu, G.; Dong, W.; Tan, L. Effects of in vitro saliva, gastric and intestinal digestion on the chemical properties, antioxidant activity of polysaccharide from *Artocarpus heterophyllus* Lam. (Jackfruit) pulp. *Food Hydrocoll.* **2019**, *87*, 952–959. [CrossRef]
17. Qu, D.; Liu, C.; Jiang, M.; Feng, L.; Chen, Y.; Han, J. After in vitro digestion, jackfruit flake affords protection against Acrylamide-induced oxidative damage. *Molecules* **2019**, *24*, 3322. [CrossRef]
18. Pavan, V.; Sancho, R.A.S.; Pastore, G.M. The effect of in vitro digestion on the antioxidant activity of fruit extracts (*Carica papaya, Artocarpus heterophillus* and *Annona marcgravii*). *LWT Food Sci. Technol.* **2014**, *59*, 1247–1251. [CrossRef]
19. Minekus, M.; Alminger, M.; Alvito, P.; Ballance, S.; Bohn, T.; Bourlieu, C.; Carrière, F.; Boutrou, R.; Corredig, M.; Dupont, D.; et al. A standardised static in vitro digestion method suitable for food—An international consensus. *Food Funct* **2014**, *5*, 1113–1124. [CrossRef]
20. Cheng, M.; He, J.; Wang, H.; Li, C.; Wu, G.; Zhu, K.; Chen, X.; Zhang, Y.; Tan, L. Comparison of microwave, ultrasound and ultrasound-microwave assisted solvent extraction methods on phenolic profile and antioxidant activity of extracts from jackfruit (*Artocarpus heterophyllus* Lam.) pulp. *LWT Food Sci. Technol.* **2023**, *173*, 114395. [CrossRef]
21. Dong, R.; Yu, Q.; Liao, W.; Liu, S.; He, Z.; Hu, X.; Chen, Y.; Xie, J.; Nie, S.; Xie, M. Composition of bound polyphenols from carrot dietary fiber and its in vivo and in vitro antioxidant activity. *Food Chem.* **2021**, *339*, 127879. [CrossRef] [PubMed]
22. Cheng, Y.; Wu, T.; Chu, X.; Tang, S.; Cao, W.; Liang, F.; Fang, Y.; Pan, S.; Xu, X. Fermented blueberry pomace with antioxidant properties improves fecal microbiota community structure and short chain fatty acids production in an in vitro mode. *LWT Food Sci. Technol.* **2020**, *125*, 109260. [CrossRef]
23. Bouayed, J.; Hoffmann, L.; Bohn, T. Total phenolics, flavonoids, anthocyanins and antioxidant activity following simulated gastro-intestinal digestion and dialysis of apple varieties: Bioaccessibility and potential uptake. *Food Chem.* **2011**, *128*, 14–21. [CrossRef] [PubMed]
24. Gutierrez-Grijalva, E.P.; Angulo-Escalante, M.A.; Leon-Felix, J.; Heredia, J.B. Effect of In Vitro Digestion on the Total Antioxidant Capacity and Phenolic Content of 3 Species of Oregano (*Hedeoma patens, Lippia graveolens, Lippia palmeri*). *J. Food Sci.* **2017**, *82*, 2832. [CrossRef] [PubMed]
25. Tischer Seraglio, S.K.; Valese, A.C.; Daguer, H.; Bergamo, G.; Azevedo, M.S.; Nehring, P.; Gonzaga, L.V.; Fett, R.; Oliveira Costa, A.C. Effect of in vitro gastrointestinal digestion on the bioaccessibility of phenolic compounds, minerals, and antioxidant capacity of Mimosa scabrella Bentham honeydew honeys. *Food Res. Int.* **2017**, *99*, 670–678. [CrossRef] [PubMed]
26. Liu, X.; Shi, J.; Yi, J.; Zhang, X.; Ma, Q.; Cai, S. The effect of in vitro simulated gastrointestinal digestion on phenolic bioaccessibility and bioactivities of Prinsepia utilis Royle fruits. *LWT Food Sci. Technol.* **2021**, *138*, 110782. [CrossRef]
27. Tu, F.; Xie, C.; Li, H.; Lei, S.; Li, J.; Huang, X.; Yang, F. Effect of in vitro digestion on chestnut outer-skin and inner-skin bioaccessibility: The relationship between biotransformation and antioxidant activity of polyphenols by metabolomics. *Food Chem.* **2021**, *363*, 130277. [CrossRef]
28. Ma, Y.; Gao, J.; Wei, Z.; Shahidi, F. Effect of in vitro digestion on phenolics and antioxidant activity of red and yellow colored pea hulls. *Food Chem.* **2021**, *337*, 127606. [CrossRef]
29. Peña-Vázquez, G.I.; Dominguez-Fernández, M.T.; Camacho-Zamora, B.D.; Hernandez-Salazar, M.; Urías-Orona, V.; De Peña, M.; de la Garza, A.L. In vitro simulated gastrointestinal digestion impacts bioaccessibility and bioactivity of Sweet orange (*Citrus sinensis*) phenolic compounds. *J. Funct. Foods* **2022**, *88*, 104891. [CrossRef]
30. Velderrain-Rodríguez, G.; Quirós-Sauceda, A.; Mercado-Mercado, G.; Ayala-Zavala, J.F.; Astiazarán-García, H.; Robles-Sánchez, R.M.; Wall-Medrano, A.; Sayago-Ayerdi, S.; González-Aguilar, G.A. Effect of dietary fiber on the bioaccessibility of phenolic compounds of mango, papaya and pineapple fruits by an in vitro digestion model. *Food Sci. Technol.* **2016**, *36*, 188–194. [CrossRef]
31. Gil-Izquierdo, A.; Gil, M.I.; Ferreres, F.; Tomás-Barberán, F. In Vitro Availability of Flavonoids and Other Phenolics in Orange Juice. *Food Chem.* **2001**, *49*, 1035–1041. [CrossRef] [PubMed]
32. Kumar, S.; Singh, A.; Kumar, B. Identification and characterization of phenolics and terpenoids from ethanolic extracts of Phyllanthus species by HPLC-ESI-QTOF-MS/MS. *J. Pharm. Anal.* **2017**, *7*, 214–222. [CrossRef] [PubMed]
33. De Bellis, R.; Chiarantini, L.; Potenza, L.; Gorassini, A.; Verardo, G.; De Marco, R.; Benayada, L.; Stocchi, V.; Cristina Albertini, M.; Fraternale, D. High production of secondary metabolites and biological activities of *Cydonia oblonga* Mill. pulp fruit callus. *J. Funct. Foods* **2022**, *94*, 105133. [CrossRef]
34. Kiselova-Kaneva, Y.; Galunska, B.; Nikolova, M.; Dincheva, I.; Badjakov, I. High resolution LC-MS/MS characterization of polyphenolic composition and evaluation of antioxidant activity of *Sambucus ebulus* fruit tea traditionally used in Bulgaria as a functional food. *Food Chem.* **2022**, *367*, 130759. [CrossRef] [PubMed]
35. Alves Filho, E.G.; Sousa, V.M.; Rodrigues, S.; de Brito, E.S.; Fernandes, F.A.N. Green ultrasound-assisted extraction of chlorogenic acids from sweet potato peels and sonochemical hydrolysis of caffeoylquinic acids derivatives. *Ultrason. Sonochem.* **2020**, *63*, 104911. [CrossRef]
36. Yang, L.; Marney, L.; Magana, A.A.; Choi, J.; Wright, K.; Mcferrin, J.; Gray, N.E.; Soumyanath, A.; Stevens, J.F.; Maier, C.S. Quantification of caffeoylquinic acids and triterpenes as targeted bioactive compounds of *Centella Asiatica* in extracts and formulations by liquid chromatography mass spectrometry. *J. Chromatogr. Open* **2023**, *4*, 100091. [CrossRef]

37. Acero, N.; Gradillas, A.; Beltran, M.; Garcia, A.; Mingarro, D.M. Comparison of phenolic compounds profile and antioxidant properties of different sweet cherry (*Prunus avium* L.) varieties. *Food Chem.* **2019**, *279*, 260–271. [CrossRef]
38. Daud, M.; Fatanah, D.N.; Abdullah, N.; Ahmad, R. Evaluation of antioxidant potential of *Artocarpus heterophyllus* L. J33 variety fruit waste from different extraction methods and identification of phenolic constituents by LCMS. *Food Chem.* **2017**, *232*, 621–632. [CrossRef]
39. Torres, A.; Aguilar-Osorio, G.; Camacho, M.; Basurto, F.; Navarro-Ocana, A. Characterization of polyphenol oxidase from purple sweet potato (*Ipomoea batatas* L. Lam) and its affinity towards acylated anthocyanins and caffeoylquinic acid derivatives. *Food Chem.* **2021**, *356*, 129709. [CrossRef]
40. Leonard, W.; Zhang, P.; Ying, D.; Adhikari, B.; Fang, Z. Fermentation transforms the phenolic profiles and bioactivities of plant-based foods. *Biotechnol. Adv.* **2021**, *49*, 107763. [CrossRef]
41. Abu-Reidah, I.M.; Arráez-Román, D.; Lozano-Sánchez, J.; Segura-Carretero, A.; Fernández-Gutiérrez, A. Phytochemical Characterisation of Green Beans (*Phaseolus vulgaris* L.) by Using High-performance Liquid Chromatography Coupled with Time-of-flight Mass Spectrometry. *Phytochem. Anal.* **2013**, *24*, 105–116. [CrossRef] [PubMed]
42. Liu, Y.; Zhang, X.; Zhan, L.; Xu, C.; Sun, L.; Jiang, H.; Sun, C.; Li, X. LC-Q-TOF-MS characterization of polyphenols from white bayberry fruit and its antidiabetic effect in KK-A^y mice. *ACS Omega* **2020**, *5*, 17839–17849. [CrossRef] [PubMed]
43. Hsu, C.; Lin, G.; Lin, H.; Chang, S. Characteristics of proanthocyanidins in leaves of *Chamaecyparis obtusa* var. formosana as strong α-glucosidase inhibitors. *J. Sci. Food Agric.* **2018**, *98*, 3806–3814. [CrossRef] [PubMed]
44. Li, S.; Xiao, J.; Chen, L.; Hu, C.; Chen, P.; Xie, B.; Sun, Z. Identification of A-series oligomeric procyanidins from pericarp of Litchi chinensis by FT-ICR-MS and LC-MS. *Food Chem.* **2012**, *135*, 31–38. [CrossRef]
45. Xue, C.; Guo, J.; Qian, D.; Duan, J.; Shang, E.; Shu, Y.; Lu, Y. Identification of the potential active components of Abelmoschus manihot in rat blood and kidney tissue by microdialysis combined with ultra-performance liquid chromatography/quadrupole time-of-flight mass spectrometry. *J. Chromatogr. B* **2011**, *879*, 317–325. [CrossRef] [PubMed]
46. Tejada, S.; Pinya, S.; Martorell, M.; Capó, X.; Tur, J.A.; Pons, A.; Sureda, A. Potential anti-inflammatory effects of hesperidin from the genus Citrus. *Curr. Med. Chem.* **2017**, *24*, 4929. [CrossRef] [PubMed]
47. Ge, Y. Chemical constituents of Fortunella margarita fruits. *J. Chin. Med. Mater.* **2014**, *37*, 435–438. (In Chinese)
48. Quail, W.; Tao, Y.; Lu, M.; Yuan, B.; Chen, J.; Zeng, M.; Qin, F.; Guo, F.; He, Z. Stability of the phenolic compounds and antioxidant capacity of five fruit (apple, orange, grape, pomelo and kiwi) juices during in vitro: Imulated gastrointestinal digestion. *Int. J. Food Sci. Technol.* **2018**, *53*, 1131–1139.
49. Barreira, J.C.M.; Ferreira, I.C.F.R.; Oliveira, M.B.P.P.; Pereira, J.A. Antioxidant activities of the extracts from chestnut flower, leaf, skins and fruit. *Food Chem.* **2008**, *107*, 1106–1113. [CrossRef]
50. Pods Dek, A.; Majewska, I.; Redzynia, M.; Sosnowska, D.; Kozio Kiewicz, M. In vitro inhibitory effect on digestive enzymes and antioxidant potential of commonly consumed fruits. *J. Agric. Food Chem.* **2014**, *62*, 4610–4617. [CrossRef]
51. Jayawardena, N.; Watawana, M.I.; Waisundara, V.Y. The total antioxidant capacity, total phenolics content and starch hydrolase inhibitory activity of fruit juices following pepsin (gastric) and pancreatin (duodenal) digestion. *J. Consum. Prot. Food Saf.* **2015**, *10*, 349–357. [CrossRef]

antioxidants

Article

Elevating Cereal-Based Nutrition: *Moringa oleifera* Supplemented Bread and Biscuits

Teresa Ferreira [1], Sandra M. Gomes [2,3] and Lúcia Santos [2,3,*]

1 FEUP—Faculty of Engineering, University of Porto, Rua Dr. Roberto Frias, 4200-465 Porto, Portugal; up201806792@edu.fe.up.pt

2 ALiCE—Associate Laboratory in Chemical Engineering, Faculty of Engineering, University of Porto, Rua Dr. Roberto Frias, 4200-465 Porto, Portugal; scgomes@fe.up.pt

3 LEPABE—Laboratory for Process Engineering, Environment, Biotechnology and Energy, Faculty of Engineering, University of Porto, Rua Dr. Roberto Frias, 4200-465 Porto, Portugal

* Correspondence: lsantos@fe.up.pt; Tel.: +351-225-081-682; Fax: +351-225-081-440

Abstract: Enhancing the nutritional value of commonly consumed, cost-effective staple foods, such as bread and biscuits, by fortifying them with *Moringa oleifera* leaf powder (MOLP) and its phenolic-rich extract holds substantial potential for addressing malnutrition. This study evaluated the phenolic extract from MOLP obtained through Soxhlet extraction, focusing on its antioxidant, antibacterial, and antidiabetic properties. The resulting extract exhibited a total phenolic content (TPC) of 138.2 mg of gallic acid equivalents/g. The ABTS and DPPH assays presented IC_{50} values of 115.2 mg/L and 544.0 mg/L, respectively. Furthermore, the extract displayed notable α-amylase inhibition and no cytotoxicity towards human fibroblasts. The primary phenolic compounds identified were catechin, epicatechin, and caffeic acid. Subsequently, MOLP and its extract were incorporated into bread and biscuits, replacing 5% of wheat flour, resulting in fortified functional foods. The fortified products exhibited improved TPC and antioxidant activity compared to the non-fortified foods. Furthermore, they displayed the ability to inhibit microbial growth, leading to an extended shelf life. Sensory analysis indicated that the products incorporated with the extract were preferred over those with MOLP. These results have demonstrated the viability of using MOLP and its phenolic-rich extract as an environmentally sustainable strategy for enhancing the quality of cereal-based products.

Keywords: food fortification; cereal-based products; functional foods; bioactive compounds; antioxidants

Citation: Ferreira, T.; Gomes, S.M.; Santos, L. Elevating Cereal-Based Nutrition: *Moringa oleifera* Supplemented Bread and Biscuits. *Antioxidants* **2023**, *12*, 2069. https://doi.org/10.3390/antiox12122069

Academic Editors: Baojun Xu and Bin Du

Received: 29 October 2023
Revised: 18 November 2023
Accepted: 23 November 2023
Published: 1 December 2023

1. Introduction

In low- to middle-income countries, predominantly Africa and Southern Asia, monotonous diets consisting mainly of starchy foods are consumed since these are more easily accessible and affordable compared to other more nutritious sources that include fruits, vegetables, and animal-sourced foods. In these countries, food subsidies mostly involve starchy staples, including wheat flour, bread, and rice. On average, the cost of a healthy diet is 60% more than a diet that just meets the minimum dietary energy needs through a starchy staple; however, such diets are deficient in dietary diversity and can lead to micronutrient deficiencies—a significant public health concern [1]. Moreover, the limited shelf life of foods exacerbates food security and hunger issues in developing countries, where inadequate transportation and storage infrastructure result in significant food loss. This leads to food waste, economic losses for farmers, and limited access to nutritious food, negatively impacting communities [2].

With a growing urge to address these concerns and an increasing interest in leading a healthy lifestyle, functional foods have gained popularity and are now vastly studied and included in human diets [3,4]. Consequently, food fortification techniques have become vital in the food industry. Many functional ingredients are derived from natural sources, such as plants, which contain a high number of nutrients, vitamins, and compounds with beneficial properties for human health known as bioactive compounds (BACs) [5].

Moringa oleifera is a plant nutraceutical, given that it serves both as a nutritional and medicinal plant and can be incorporated as a natural food ingredient to produce functional foods. It belongs to the Moringacea family, which comprises 13 other species and is extensively cultivated across the Middle East, Africa, and Asia. Although originally from India, it has been widely cultivated and naturalised in many countries around the world, including tropical and subtropical regions of Asia, Africa, and South America [3,6]. Remarkably, the geographical distribution of this plant aligns with regions affected by malnutrition [7].

M. oleifera is commonly known as the miracle plant or tree of life and is believed to hold beneficial properties in all its parts, including leaves, roots, pods, seeds, and flowers, due to containing a high number of essential nutrients such as proteins, minerals, vitamins (A, B1, B2, B3, C, and E), and fibre [8]. Furthermore, the plant is known to possess a significant quantity of BACs, specifically phenolic compounds (PCs), tannins, and carotenoids, giving this plant antioxidant, antimicrobial, anti-inflammatory, hypoglycaemic, hepatoprotective, and anticancer properties [9,10]. The most common PCs present in *M. oleifera* include phenolic acids (e.g., caffeic acid, chlorogenic acid, gallic acid) and flavonoids (e.g., catequin, epicatequin, quercetin, kaempferol, rutin) [11].

PCs are secondary metabolites of plants that function as bio-antioxidants able to reduce reactive oxygen species (ROS) by donating hydrogen atoms of phenolic hydroxyls and by transferring electrons from such phenolic hydroxyls [12]. When incorporated into functional foods, besides helping prevent numerous diseases such as heart-related disorders and certain cancers, they also aid in prolonging their shelf life [13]. They achieve it by inhibiting oxidation, neutralising free radicals, delaying oxidative reactions, chelating metal ions, and regenerating other antioxidants, thus preserving the freshness, flavour, texture, and nutritional value of foods for a longer period of time [14]. This helps reduce food waste and enhances food security. PCs can be extracted from *M. oleifera* leaf powder (MOLP) using polar solvents such as ethanol since these compounds are relatively polar; however, the phenolic content in MOLP extracts varies depending on several factors, such as the plant's origin and extraction conditions [15].

Furthermore, *M. oleifera* leaf and seed extracts have been shown to possess antidiabetic properties due to their inhibitory activity towards α-amylase and α-glucosidase enzymes responsible for breaking down carbohydrates into simple sugars [16]. Inhibiting these enzymes is a common approach to regulating blood glucose levels by slowing down the digestion of carbohydrates. Elevated blood glucose levels can trigger glycation, a process that causes secondary complications of diabetes, including impaired vision, heart disease, kidney failure, and more. Compounds that inhibit glycation have the potential to alleviate these complications by reducing the formation of advanced glycation end products and mitigating the associated problems related to oxidative stress, inflammation, and compromised immune function. Currently, diabetes management primarily involves oral hypoglycaemic medications, which can produce side effects such as lactic acidosis, gastrointestinal issues, and anaemia, therefore, alternatives (especially from food and vegetable sources) are necessary [17].

The interesting composition of *M. oleifera*, allied with the fact that it is a fast-growing tree that can thrive in poor soils and dry lands requiring little agricultural care, presents a sustainable approach to tackle malnutrition and food insecurity, as in providing access to safe, nutritious foods to meet their dietary needs for an active and healthy life by food fortification technique.

To preserve consumer safety, it is crucial to ensure that the consumption of fortified products does not lead to any negative effects. Several studies were conducted in vitro and in vivo to explore the safety of MOLP and extracts, with most of them showing a high degree of safety. In vivo studies on rats showed that *M. oleifera* aqueous extracts were safe, presenting acute toxicity only at supra-supplementation levels (3 g/kg), being considered safe at levels ≤1 g/kg b.wt. (body weight), a dose that is still higher than the ones commonly consumed [18]. The same study indicated that concentrations of aqueous leaf extracts exceeding 20 mg/mL were found to be toxic to human peripheral blood mononuclear cells. It is worth noting, though, that achieving such concentrations through oral ingestion is not feasible. In another study, it was determined that ethanolic leaf extracts exhibited no adverse effects on fibroblasts within the concentration spectrum of 0.02 g/mL to 100 g/mL [19]. Lastly, no negative effects were observed in any human trials involving MOLP to date. Therefore, *M. oleifera* extracts seem to be safe for human consumption in the amounts commonly used [20]. Despite the existing findings, there is still a need for more toxicity studies to guarantee that supplements made from *M. oleifera* leaves are harmless for human health. There is a growing interest in directly including powders extracted from *M. oleifera* leaves or seeds as a natural food fortification in a range of cereal-based products, including bread, biscuits, cake, pasta, and various snacks. Furthermore, there is a rising awareness of the topic of food preservation, which warrants the usefulness of antioxidants to improve the quality of foods. The commercial availability of synthetic antioxidant compounds, such as butylated hydroxytoluene (BHT) and butylated hydroxyanisole (BHA), is limited due to their potential toxicity towards vital organs in the body, which pushes interest in the ongoing study and consumer preference for natural antioxidant compounds [21]. These natural alternatives, such as PCs, are generally considered safer and have garnered significant attention for their ability to effectively deliver antioxidant and antimicrobial properties to preserve food products and prevent the occurrence of rancidity caused by oxidation [21].

Most of the research on the fortification of bakery foods has been conducted using MOLP. Table 1 summarises the methods and findings of some studies conducted on the incorporation of MOLP in cereal-based products.

Table 1. Studies on the incorporation of *Moringa oleifera* in cereal-based food products.

Product	Study Methods	Findings	Ref.
Dried pasta	Formulation of pasta containing semolina and 10%, 20%, 30%, and 40% of MOLP. The moisture, protein, and ash content of the samples were investigated. A sensory evaluation and a texture profile analysis were also conducted.	The pasta fortified with MOLP exhibited higher protein and ash contents in comparison to the control pasta and a lower moisture content. Regarding textural parameters and consumer acceptability, the pasta with 20% MOLP inclusion was found to be the best.	[22]
Bread	Assessment of the impact of MOPL addition at levels ranging from 0% to 10% on the proximate, mineral, antioxidant, and sensory attributes of WWF-leavened bread.	The MOLP-supplemented bread showed noticeable improvements in proximate and mineral profiles. The TPC of MOLP in comparison to WWF was much higher. Additionally, a 5% MOLP-based value-added bread demonstrated significantly higher antioxidant activities. The overall acceptability scores for WWF-leavened bread decreased progressively as MOLP addition levels increased.	[23]
Gluten-free bread	Fortification of rice semolina gluten-free bread with different amounts of MOLP (2.5%, 5%, 7.5%, and 10%). The TPC and antioxidant activity were determined. A texture and sensory analysis were also performed.	The addition of MOLP resulted in a significant decrease in the volume of the bread samples, except for the 2.5% MOLP. Additionally, a slight decrease in hardness and chewiness was observed with the addition of 2.5% and 10% MOLP. The TPC and antioxidant activity increased as the amount of MOLP increased. Among all MOLP-containing bread samples, the most acceptable bread was the one containing 2.5% MOLP in comparison to the control.	[24]

Table 1. *Cont.*

Product	Study Methods	Findings	Ref.
Biscuits	Production of biscuits with WWF substituted with MOLP (2.5%, 5%, and 10%). Evaluated the nutritional content, bioactive compounds, antioxidant, physical, and α-amylase inhibitory properties. The sensory attributes of the cookies were also determined.	MOLP-supplemented cookies had a significant enhancement in their bioactive compounds, antioxidant, and α-amylase inhibitory properties. Protein, ash, fat, and fibre contents were significantly increased in MOLP-substituted cookies. The sensory acceptance of the cookies decreased with increasing levels of WWF substitution.	[25]
Biscuits	Preparation of biscuits by substituting WWF with 5%, 10%, and 15% MOLP. The effect of MOLP on the rheological, microstructural, nutritional, textural, and organoleptic characteristics of biscuits was tested.	The addition of MOLP led to higher water absorption, softer dough, and an altered cookie texture. Sensory evaluation favoured cookies with 10% MOLP. Nutritional components like protein, iron, calcium, β-carotene, and dietary fibre increased with a higher MOLP content (0–15%).	[26]

MOLP—*Moringa oleifera* leaf powder; TPC—total phenolic content; WWF—whole wheat flour.

Numerous studies have confirmed the effectiveness of *M. oleifera* in enhancing the nutritional content of various cereal-based food products. Nevertheless, as far as the authors are aware, there is a notable absence of studies that have explored the utilisation of *M. oleifera* extracts, which contain a significantly higher concentration of BACs compared to the powdered form. Furthermore, the existing studies tend to focus more on the nutritional value of the plant and less on their antioxidant effectiveness, which aids in decreasing the rancidity process of foods and increasing their shelf life, a particularly important aspect in the case of bread since it presents a short shelf life of around 3–7 days.

Therefore, the aim of this study was to produce bread and biscuits fortified with MOLP and its respective extract and assess their effect on the physicochemical properties of these foods, such as antioxidant capacity and microbial stability. A sensory assessment of the fortified bread and biscuits was conducted to evaluate the consumers' demands and preferences for these novel products.

2. Materials and Methods

2.1. Samples and Reagents

The *Moringa oleifera* leaf powder (MOLP) used during this work was provided by Agostinho Neto University (UAN). The leaves were collected in Luanda, Angola (8°57′24.9″ S, 13°13′02.9″ E), washed with running tap water, dried until obtaining a constant weight, and grounded to obtain a homogeneous powder (particle size < 250 μm).

To produce the bread and biscuits, wheat flour (type 55), yeast, salt, eggs (about 60 g each), butter, and sugar were purchased from a supermarket in Porto, Portugal.

Acetonitrile (Ref. 45983, C_2H_3N, CAS 75-05-8), ethanol (Ref. 83813.360, C_2H_6O, CAS 64-17-5), and methanol (Ref. 20834.291, C_3H_2O, CAS 67-56-1) were obtained from VWR (Radnor, PA, USA). Merck Millipore Mill-Q water purification equipment from Billerica (MA, USA), with 18.2 Ω of electric resistance, was used to obtain ultrapure water.

The sodium carbonate (Ref. 1.06392, CNa_2O_3, CAS 497-19-8) acquired from Merck (Darmstadt, Germany) and the Folin–Ciocalteu reagent (Ref. 47641) obtained from Sigma-Aldrich (St. Louis, MO, USA) were utilised for analysing the total phenolic content.

To assess the antioxidant capacity, Sigma-Aldrich reagents were used: 2,2-diphenyl-1-picrylhydrazyl (DPPH) (Ref. D9132, $C_{18}H_{12}N_5O_6$, CAS 1898-66-4) and 2,2′-azino-bis(3-ethylbenzothiazoline-6-sulfonic acid) (ABTS) (Ref. A1888, $C_{18}H_{24}N_6O_6S_4$, CAS 30931-67-0).

For antibacterial evaluation, agar (Ref. J637, CAS 9002-18-0) and Plate Count Agar (Ref. 84608.0500) were obtained from VWR, m-Lauryl Sulfate Broth (Ref. 0734) was purchased from Sigma-Aldrich, and Rose-Bengal Chloramphenicol Agar (Ref. 1.00467.0500) was acquired from Merck. Sorbic acid (Ref. S1626, $C_6H_8O_2$, CAS 110-44-1), which served as a positive control, was also acquired from Sigma-Aldrich. Dimethyl sulfoxide (Ref. 41640, C_2H_6OS, CAS 67-68-5) was obtained from Honeywell (Charlotte, NC, USA).

The α-amylase inhibition capacity was evaluated using α-amylase from porcine pancreas (Ref. A3176) and starch from corn (Ref. S4180, $C_6H_{10}O_5$, CAS 9005-25-8) acquired from Sigma-Aldrich.

For the cytotoxicity evaluation, Foetal Bovine Serum (FBS), Dulbecco's Modified Eagle's Medium (DMEM), trypsin, and Penicillin-Streptomycin (PenStrep) were purchased from Gibco (Cambridge, MA, USA).

The analytical standards used for phenolic compound quantification were acquired from Sigma-Aldrich: chlorogenic acid (Ref. 1115545, $C1_6H_{18}O_9$, CAS 327-97-9), gallic acid (Ref. 147915, $C_7H_6O_5$, CAS 149-91-7), catechin (Ref. 43412, $C_{15}H_{14}O_6$, CAS 154-23-4), kaempferol (Ref. 60010, $C1_5H_{10}O_6$, CAS 520-18-3), caffeic acid (Ref. C0625, $C_9H_8O_4$, CAS 331-39-5), epicatechin (Ref. E1753, $C_{15}H_{14}O_6$, CAS 490-46-0), and quercetin (Ref. Q4951, $C_{15}H_{10}O_7$, CAS 117-39-5).

2.2. *Extraction of Bioactive Compounds from Moringa oleifera*

Phenolic compounds from *Moringa oleifera* leaf powder (MOLP) were obtained through a solid–liquid extraction method utilising a Soxhlet apparatus, as reported in the literature [27]. The extraction extended over a 2 h period, employing ethanol as the solvent, with a sample-to-solvent ratio of 1:40 (*m*/*v*). Following this, solvent evaporation took place using a rotary evaporator (Rotavapor R-200, BUCHI Laboratories, Flawil, Switzerland), followed by a continuous nitrogen stream at 2 mbar.

2.3. *Characterisation of Moringa oleifera Extract*

2.3.1. Total Phenolic Content and Antioxidant Capacity

The assessment of the total phenolic content (TPC) in the extract was conducted using the Folin–Ciocalteu method, following the procedure outlined in the literature [28]. A solution of the extract at a concentration of 1000 mg/L was incubated with the Folin–Ciocalteu reagent and a sodium carbonate solution (333.3 g/L) for a duration of 2 h at room temperature. Following that, the absorbance was recorded at 750 nm, and the results were expressed in milligrams of gallic acid equivalents (GAE) per gram of the extract. The antioxidant capacity of the MOLP extract was assessed through the DPPH and ABTS assays, according to literature procedures with minor modifications [29]. In the ABTS test, the extract solutions (ranging from 100 mg/L to 2500 mg/L) were incubated with the ABTS solution for 15 min, and the absorbance was measured at 734 nm. For the DPPH assay, the extract solutions (ranging from 1500 mg/L to 8000 mg/L) were incubated for 40 min with the DPPH solution, after which the absorbance was measured at 515 nm. In both assays, a curve of the percentage inhibition of free radicals vs. the extract concentration was constructed, and the IC_{50} values were determined. The results were also expressed in milligrams of Trolox equivalents (TE) per gram of the extract. All measurements were performed in triplicate.

2.3.2. Antibacterial Capacity

The antibacterial activity of the extract was tested against *Escherichia coli* and *Staphylococcus aureus* bacteria using the disk diffusion test in the biological culture medium Plate Count Agar (PCA). Two different concentrations of the extract (500 mg/mL and 1000 mg/mL) were prepared in 2% aqueous DMSO. A common food preservative, sorbic acid, was used as the positive control (500 mg/mL and 1000 mg/mL) while ultrapure water was used as the negative control. The test was performed according to the protocol described by Ferreira and Santos [28]. Bacterial suspensions of each microorganism, with an optical density of 0.1 at 610 nm, were plated on PCA medium. Afterwards, 7 μL of the extract or ascorbic acid solution (at both concentrations) were added to a sterile 5 mm diameter disc in triplicate. The same was performed for the negative control. After incubating the plates for 24 h at 37 °C, the inhibition levels were measured in triplicate.

2.3.3. α-Amylase Inhibition

The extract was tested for its inhibitory potential using a modified α-amylase assay, following the procedures described in reference [30]. In summary, 250 μL of the extract solution (1 mg/mL in ethanol) were mixed with 250 μL of α-amylase solution (100 mg/L in ultrapure water) and incubated at 37 °C for 10 min. Subsequently, 250 μL of starch solution (1% *w*/*v*) were added, followed by another 10 min incubation at 37 °C. Then, 500 μL of 3,5-dinitrosalicylic acid (DNS) reagent were added, and the samples were heated in boiling water for 10 min. After cooling, 5 mL of ultrapure water was added, and the absorbance was measured at 540 nm. The measurements were performed in triplicate.

2.3.4. Phenolic Compounds Quantification

This study used high-performance liquid chromatography with a diode array detector (HPLC-DAD) to identify and quantify phenolic compounds in the *M. oleifera* extract. An Elite LaChrom HPLC system (Hitachi, Tokyo, Japan) with a Puroshper® STAR RP-18 endcapped LiChroCART® column (Merck, Germany) was utilised. Sample solutions were prepared using an acetonitrile:water:ethanol solvent mixture (2:1:1 *v*/*v*/*v*). The mobile phase consisted of two eluents: Milli-Q water with 0.5% orthophosphoric acid (A) and methanol:acetonitrile (80:20 *v*/*v*) (B). The gradient ranged from 10% to 70% B over 60 min [31]. Compounds were detected at specific wavelengths: catechin and epicatechin were detected at 222 nm, gallic acid at 275 nm, and caffeic acid, along with chlorogenic acid, at 322 nm. Additionally, kaempferol and quercetin were identified at 365 nm. Quantification relied on calibration curves and the external standard method, with spike samples aiding peak identification and quantification. All measurements were performed in triplicate.

2.3.5. Cytotoxicity

The cytotoxicity of the obtained extract was evaluated against HFF-1 human fibroblasts using the resazurin assay. The cells were cultured in DMEM supplemented with 10% FBS (*v*/*v*) and 1% PenStrep (*v*/*v*) at 37 °C in a 5% CO_2 atmosphere. Briefly, 200 μL of HFF-1 cells were seeded in a 96-well plate at a density of 13,000 cells/well and incubated for 24 h at 37 °C, 5% CO_2. Afterwards, 40 μL of the medium were replaced by the extract solution (5000 mg/L in ultrapure water), and the plate was incubated overnight at 37 °C with 5% CO_2. Non-treated cells (cells without the addition of any compound) were used as a negative control, while hydrogen peroxide was used as a positive control since it is an oxidising agent capable of causing cell death. After incubation, the medium was replaced by 200 μL of resazurin (20% *v*/*v*), and the plate was incubated for 4 h, protected from the light, at 37 °C with 5% CO_2. Finally, the fluorescence was assessed using the microplate reader Synergy HT (BioTek Instruments Inc., Winooski, VT, USA) at an excitation wavelength of 530 nm and an emission wavelength of 590 nm. The cell viability percentage was then calculated in reference to the untreated cells. Four independent measurements were performed in triplicate.

2.4. *Incorporation of Moringa oleifera Leaf Extract in Bread and Biscuits*

2.4.1. Production of Breads and Biscuits

To evaluate the effect of the incorporation of MOLP and extract in bread (Br) and biscuits (Bi), three formulations of each food were prepared: a negative control (NC) with no additional ingredients; a formulation with MOLP at a wheat flour substitution level of 5% *w*/*w* (MP); and a formulation with the extract at a wheat flour substitution level of 5% *w*/*w* given the extraction efficiency (ME). The ingredients of each formulation are detailed in Tables 2 and 3.

Table 2. List of ingredients for each bread formulation.

Ingredient	NC-Br	MP-Br	ME-Br
Water (g)	210	210	210
Salt (g)	4	4	4
Wheat flour (g)	360	342	352.55
Yeast (g)	2	2	2
MOLP (g)	-	18	-
M. oleifera extract (g)	-	-	7.45

MOLP—*Moringa oleifera* leaf powder; NC-BR—Negative control bread; MP-Br—Bread with 5% flour substitution with MOLP; ME-Br—Bread with *M. oleifera* extract flour substitution corresponding to 5% of MOLP.

Table 3. List of ingredients for each biscuit formulation.

Ingredient	NC-Bi	MP-Bi	ME-Bi
Butter (g)	125	125	125
Sugar (g)	125	125	125
No. Eggs	1	1	1
Wheat flour (g)	250	237.5	244.82
Yeast (g)	2	2	2
MOLP (g)	-	12.5	-
M. oleifera extract (g)	-	-	5.18

MOLP—*Moringa oleifera* leaf powder; NC-Bi—Negative control biscuit; MP-Bi—Biscuit with 5% flour substitution with MOLP; ME-Bi—Biscuit with *M. oleifera* extract flour substitution corresponding to 5% of MOLP.

Breads were produced using the Home Bread machine from Tefal (Sarcelles, France) using program 3 (French bread). After production, each bread loaf was sliced and sealed in separate plastic bags. Biscuits were produced using the Thermomix machine (Vorwerk, Wuppertal, Germany). The ingredients were mixed following the instructions, and afterwards the biscuits were left to bake for 8 min at 200 °C in a convection oven from Gaggenau (Gaggenau, Germany). The biscuits were also stored in separate plastic bags. The bags were stored at room temperature and kept in the dark to avoid the breakdown of the bioactive compounds present in *M. oleifera* since they are photosensitive.

2.4.2. Antioxidant Capacity

Subsequently, the antioxidant properties of the bread and biscuits were examined by extracting the phenolic compounds present in each formulation using ethanol. Briefly, 2 g of sample were subjected to an extraction process with the addition of 4 mL of ethanol, followed by three repetitions of vortexing (1 min) and ultrasonic bathing (5 min). The resulting solutions underwent centrifugation for 20 min at 3000 rpm using a Rotofix 32 A centrifuge (Hettich, Tuttlingen, Germany). The collected supernatant was then subjected to an additional round of 4 mL of ethanol addition and a repetition of the entire process. Finally, the resulting supernatant was utilised for the analysis of the antioxidant capacity (ABTS and DPPH) of the various formulations, following the methodologies delineated in Section 2.3.1. The bread was analysed the day after production (t0), four days after production (t1), and one week after production (t2). The biscuits were analysed the day after production (t0), two weeks after production (t1), four weeks after production (t2), and seven weeks after production (t3). All measures were performed in triplicate.

2.4.3. Microbial Analysis

For this assay, two different mediums were used: Lauryl Sulphate Agar (LSA), selective to coliform microorganisms, and Rose Bengal Chloramphenicol Agar (RBC), selective to yeast and moulds. The analysis was performed according to the literature [27]. The bread crumb/biscuits were mixed in 9 mL of saline solution (0.9% NaCl), followed by 1 min of vortexing. Subsequently, a volume of 100 μL from each solution was inoculated onto the respective medium. The LSA plates underwent a 37 °C incubation for 24 h, while the RBC plates were incubated for 7 days at 25 °C. Subsequently, the plates were examined for the presence of microorganisms. The analysis was performed in duplicate.

2.4.4. Sensory Assessment

For the sensory evaluation, the freshly made breads were sliced into individual portions, one for each participant, who were habitual bread and biscuit consumers and ranged from 20 to 63 years of age. The panel consisted of 20 untrained panellists: 12 females and 8 males. Using a 5-point hedonic scale (1: dislike extremely, 3: neither like nor dislike, and 5: like extremely), the bread's visual appeal, colour, flavour, aroma, texture, and overall acceptability were assessed. The same reasoning was applied to the biscuits for their sensory assessment.

2.5. Statistical Analysis

The statistical analysis was conducted using GraphPad Prism 8.0.2, employing an analysis of variance (ANOVA) and Tukey's multiple comparisons test. Statistically significant differences were considered for values with $p < 0.05$ at a 95% confidence interval. The results were expressed as mean ± standard deviation.

3. Results and Discussion

3.1. Characterisation of Moringa oleifera Extract

In this study, the phenolic compounds (PCs) from M. oleifera leaf powder (MOLP) were extracted using a solid–liquid extraction technique with a Soxhlet apparatus. Soxhlet extraction offers several appealing advantages over other extraction methods. Its simplicity requires only low-cost equipment, facilitating the extraction of large quantities of material without the need for subsequent filtration. Despite its downsides, such as the prolonged extraction time and solvent usage, this extraction method continues to be extensively employed in many laboratories [32]. Recent advancements in extraction techniques have shifted the focus towards more environmentally friendly and sustainable solutions. Techniques such as supercritical extraction and ultrasound-assisted extraction (UAE) have gained popularity [33], together with microwave-assisted extraction (MAE) and pressurised liquid extraction (PLE). These techniques often yield very rich phenolic extracts due to the possibility of using moderate temperatures and short extraction times, reducing the probability of degradation of PCs and lowering the environmental footprint.

Table 4 comprises the extraction yields obtained in different studies regarding the extraction of PCs from MOLP using various extraction techniques. Various factors, including the extraction method, solvent, time, temperature, and sample mass-to-solvent volume ratio, can influence the extraction yield. In the current study, an extraction yield of 41.4 ± 11.8% was obtained, a value superior to most of the previously reported studies, showing the efficiency of the Soxhlet method in extracting phenolic compounds.

Table 4. Extraction yields reported in the literature for the extraction of phenolic compounds from *M. oleifera* leaf powder.

Extraction Method	Extraction Conditions	Extraction Yield (%)	Ref.
UAE	Solvent: 70% ethanol *w*/*v* ratio: 1:40 Time: 30 min + 2.5 h Temperature: RT + 50 °C	34.1 ± 0.9	[28]
Sonication	Solvent: 80% ethanol *w*/*v* ratio: 1:50 Time: 30 min (×3)	56.44 ± 0.82	[34]

Table 4. *Cont.*

Extraction Method	Extraction Conditions	Extraction Yield (%)	Ref.
Maceration	Solvent: 50% or 70% ethanol *w/v* ratio: 1:40 Time: 72 h Temperature: RT	Et50: 38.34 ± 1.17 Et70: 40.50 ± 1.24	[35]
Percolation	Solvent: 50% or 70% ethanol	Et50: 34.47 ± 1.41 Et70: 32.75 ± 1.93	
Soxhlet	Solvent: 50% or 70% ethanol *w/v* ratio: 1:50 Time: 20 h	Et50: 33.58 ± 1.58 Et70: 35.87 ± 1.12	

UAE—ultrasound-assisted extraction; *w/v*—weight of sample/volume of solvent (g/mL); Et50—50% ethanol solvent; Et70—70% ethanol solvent.

The phenolic-rich extract obtained in the present work was characterised regarding its total phenolic content (TPC) and antioxidant, antimicrobial, and anti-diabetic properties. The results are presented in Table 5.

Table 5. Results from the bioactive characterisation of *Moringa oleifera* extract.

TPC ($mg_{GAE}/g_{extract}$)	Antioxidant Capacity (IC_{50}—$mg_{extract}/L$) (TEAC—$mg_{TE}/g_{extract}$)		Antibacterial Capacity (d_{halo}—mm)		α-Amylase Inhibition Capacity (%)
	DPPH	ABTS	*E. coli*	*S. aureus*	
138.2 ± 17.0	544.0 ± 7.9 12.8 ± 0.2	115.2 ± 4.9 32.8 ± 1.4	ND	ND	94.1 ± 0.4

ABTS—2,2′-azino-bis(3-ethylbenzothiazoline-6-sulfonic acid); DPPH—2,2-diphenyl-1-picrylhydrazyl; GAE—gallic acid equivalents; IC_{50}—concentration of extract needed to inhibit 50% of the free radicals; ND—not detected; TE—Trolox equivalents; TEAC—Trolox equivalent antioxidant capacity; TPC—total phenolic content. The results are expressed as means ± standard deviations of three independent measurements.

The TPC and antioxidant capacity are highly influenced by many parameters, such as the plant source, geographical location, climate, and growing conditions. The timing of harvest, extraction method, storage conditions, and chemical composition of PCs also play a role [35]. From Table 5, it is possible to observe that the extract obtained in the current work presented an average TPC value of 138.2 $mg_{GAE}/g_{extract}$. Concerning the antioxidant capacity, the results were expressed both in IC_{50} values, which represent the extract concentration required to inhibit 50% of the free radicals (DPPH or ABTS), and in comparison with Trolox (TEAC), a standard antioxidant compound. The smaller the value of IC_{50}, the less amount of extract is needed to inhibit the radicals to the same extent, therefore the stronger the antioxidant capacity. In the case of the TEAC, the higher the value, the stronger the antioxidant capacity. The *M. oleifera* extract was able to inhibit both radicals; however, the extract presented a higher antioxidant capacity towards ABTS since its IC_{50} value is much lower compared to DPPH (115.2 ± 4.9 mg/L vs. 544.0 ± 7.9 mg/L) and its TEAC is higher. This higher capacity to inhibit ABTS compared to DPPH was also described by other researchers [28,36]. In another study, the exact same extraction method and conditions were used to extract PCs from MOLP. The results of the IC_{50} of DPPH and ABTS assays were quite similar to those of this study: 636.0 ± 9.2 mg/L and 205.2 ± 4.6 mg/L, respectively; however, the TPC (79 $mg_{GAE}/g_{extract}$) was almost half of the one obtained in the present study, suggesting that a higher TPC does not necessarily mean a higher antioxidant capacity [27]. A similar pattern was noted in another study employing ultrasound-assisted solid–liquid extraction, where MOLP extract exhibited a reduced TPC at 54.5 ± 16.8 mg_{GAE}/g and higher antioxidant capacity, reflected in lower IC_{50} values (133.4 ± 12.3 mg/L for DPPH and 60.0 ± 9.9 mg/L for ABTS) [28]. In a different study, similar extraction conditions as the ones selected for the present work were used, with the exception of the extraction solvent (70% acetone). The extract obtained had a TPC of 26.07 ± 0.80 mg_{GAE}/g. These results suggest that using ethanol as compared to acetone might be more fitting to extract phenolic compounds from *M. oleifera* leaves [37]. Other studies have reported that using a mixture of ethanol and water as the extraction solvent usually tends to yield a higher number of TPC in the extracts when compared to using ethanol alone [38,39]. These findings suggest that the choice of extraction solvents may impact both the phenolic composition and antioxidant properties of *M. oleifera* extracts.

Regarding the antibacterial activity, there were no inhibition halos identified in the *M. oleifera* leaf extract sample. These results should not be interpreted as the extract's inability to hinder the growth of *E. coli* and *S. aureus*, as the halo may still be present below the disk. While other research highlights the antibacterial efficacy of *M. oleifera* leaf extract against these microorganisms, it is crucial to acknowledge that the concentrations investigated in those studies significantly exceeded the ones examined in our research. This divergence in concentration levels might account for the discrepancies observed in the obtained results [40]. Furthermore, the source of the *M. oleifera* leaves, the extraction method, the inhibition medium, and the reagents used also have an impact.

In this study, the mean percentage of α-amylase inhibition was 94.1 ± 0.4%. Other studies have also proven the efficacy of *M. oleifera* leaf extract in inhibiting this enzyme. A particular study found that the leaves have the highest percentage of α-amylase inhibition when methanolic solvents are used (65.6 ± 4.93%), followed by other extraction solvents (hexane: 52.3 ± 2.5%, distilled water: 43.3 ± 2.3%, ethanol: 33 ± 2.6%) [17]. In another report, the authors stated that the methanolic leaf extract (5 mg/mL) presented a percentage enzyme inhibition of 50.6 ± 5.9%, followed by hexane (47.3 ± 2.8%) at the same concentration [41]. The α-amylase inhibitory effects may be attributed to the extracts' richness in a diverse number of phytochemical and antinutritional factors, inducing an optimistic path for creating innovative cereal-based products that can help control chronic illnesses such as type II diabetes [13,41].

To have a better understanding of the biological characteristics demonstrated by the *M. oleifera* extract, an HPLC-DAD examination was conducted to identify and quantify the principal phenolic compounds in the obtained extract. As indicated in Table 6, it is evident that catechin emerged as the predominant phenolic compound in the extract, trailed by other flavonoids like epicatechin and caffeic acid. Residual amounts of chlorogenic acid and quercetin were also found. Gallic acid and kaempferol were not detected.

Table 6. Main phenolic compounds present in the *M. oleifera* extract quantified by HPLC-DAD.

Compounds	RT (min)	Calibration Curves	R^2	IDL (mg/L)	IQL (mg/L)	Phenolic Concentration ($mg_{compound}/g_{extract}$)
Catechin	24.38	$A = 1.57 \times 10^5 C - 7.93 \times 10^5$	0.9861	41.50	138.34	1.24
Epicatechin	30.34	$A = 4.15 \times 10^5 C - 1.56 \times 10^6$	0.9983	5.82	19.39	0.09
Caffeic acid	29.23	$A = 5.56 \times 10^5 C - 1.56 \times 10^6$	0.9992	4.03	13.43	0.07
Chlorogenic acid	26.62	$A = 1.91 \times 10^5 C - 1.16 \times 10^5$	0.9999	2.84	9.48	0.01
Quercetin	52.79	$A = 7.37 \times 10^5 C - 2.68 \times 10^5$	0.9994	1.37	4.58	0.01

A—peak area; C—standard concentration (mg/L); IDL—instrumental detection limit; IQL—instrumental quantification limit; R^2—coefficient of determination; RT—retention time.

Previous studies have already reported the presence of these compounds in *M. oleifera* leaf extract [42]. The literature reports show some variability in the values obtained due to factors such as the origin and cultivation conditions of *M. oleifera* trees, as well as the extraction method used to isolate phenolic compounds, which can influence their concentration. In a recent study, the same PCs as this study were identified and quantified. Catequin, epicatechin, caffeic acid, chlorogenic acid, and quercetin were found at concentrations of 19.83, 0.67, 0.16, 1.04, and 0.06 $mg_{compound}/g_{extract}$, respectively [43]. Other studies have also reported higher concentrations of PCs. For example, in a separate study that examined seven phenolic compounds, the concentrations extended from 19.65 mg/g for kaempferol to 65.83 mg/g for chlorogenic acid [44]. Likewise, in another study, the concentrations of chlorogenic acid, kaempferol, epicatechin, quercetin, and gallic acid were also higher than those obtained in this research [45]. Moreover, it is crucial to highlight that while some authors acknowledged gallic acid as a primary compound in *M. oleifera* leaves [46], this compound was not found in the analysed extract. This aligns with the observations of other researchers who similarly did not detect gallic acid [47]. One possible explanation for this could be that gallic acid is a very polar phenolic compound and, thus, potentially challenging to extract using ethanol or hydroalcoholic solutions with low water content. Nevertheless, the results obtained in the present work suggest that the presence of the phenolic compounds identified in the *M. oleifera* leaf extract can explain its antioxidant and antidiabetic properties.

Finally, cytotoxicity studies were performed to confirm the *M. oleifera* extract's safety for human consumption. An extract's concentration of 1000 mg/L was selected, taking into consideration that it is a concentration that presents great biological properties (as seen in Table 5). The results are presented in Figure 1.

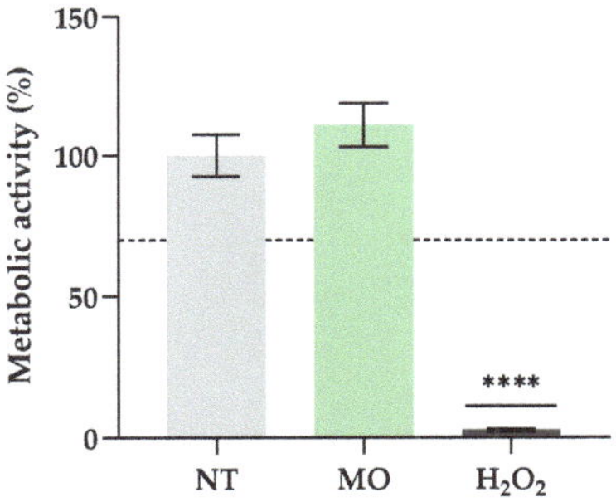

Figure 1. Cytotoxicity results of *Moringa oleifera* extract towards HFF-1 cells (human fibroblasts). The dashed line corresponds to 70% of cell viability. The results are expressed as means ± standard deviations of 4 independent measurements. NT—non-treated cells (negative control); MO—cells incubated with *M. oleifera* extract; H_2O_2—cells incubated with hydrogen peroxide (positive control). ****—statistically different values ($p < 0.0001$) in comparison to NT cells.

When in contact with cells, a substance can be considered non-cytotoxic when the cells present a viability superior to 70% [48]. As shown in Figure 1, the *M. oleifera* extract presented similar metabolic activity (not statistically different) compared to the non-treated cells and was well above the limit of 70%. This result demonstrated that, at the concentration tested, the extract is safe to be used in human applications, such as its incorporation in cereal-based products. Other studies have also accessed the safety of *M. oleifera* leaf extracts in fibroblasts, proving that the ethanolic extract is safe at concentrations from 0.02 g/mL to 100 g/mL [19], which are much higher than the one tested in this study (0.001 g/mL). According to another report, fibroblasts remained unaffected by aqueous and methanolic extracts when concentrations below 0.5 mg/mL were examined [49]. Regarding the present work, further analysis is required, using higher concentrations, to determine the limit of incorporation that is safe and does not pose any risk to human health.

3.2. *Characterisation of Fortified Breads and Biscuits*

3.2.1. Physical Characterisation

To analyse the feasibility of using MOLP and its extract to create value-added cereal-based products, three different formulations of bread and biscuits were prepared. A negative control bread/biscuit without *M. oleifera* incorporation (NC-Br/NC-Bi), a bread/biscuit with a 5% substitution level of wheat flour for MOLP (MP-Br/MP-Bi), and a third bread/biscuit with substituted wheat flour for *M. oleifera* extract corresponding to a 5% MOLP substitution taking into consideration the extraction yield (ME-Br/ME-Bi). The visual appearance of the different formulations produced is shown in Figures 2 and 3.

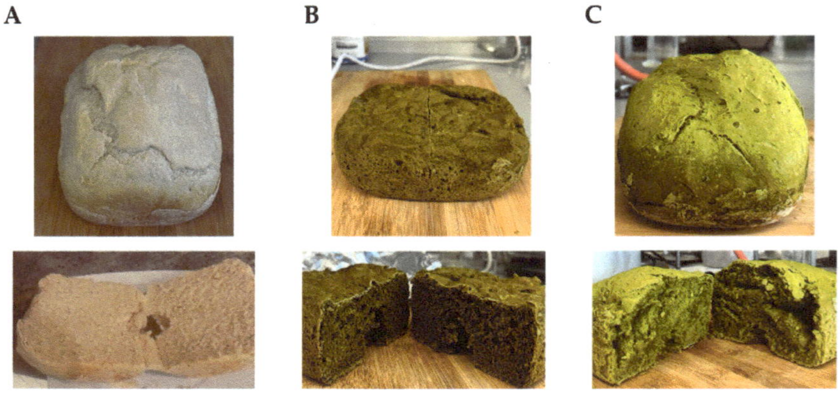

Figure 2. Visual appearance of bread's crust (**top**) and crumb (**down**). (**A**) NC-Br—bread with no additives (negative control); (**B**) MP-Br—bread with 5% MOLP; and (**C**) ME-Br—bread with extract corresponding to 5% MOLP.

Figure 3. Visual appearance of biscuits' formulations. (**A**) NC-Bi—biscuits with no additives (negative control); (**B**) MP-Bi—biscuits with 5% MOLP; and (**C**) ME-Bi—biscuits with extract corresponding to 5% MOLP.

The fortified breads and biscuits presented a greenish hue due to the dark green colour of the MOLP and its respective extract, with the bread fortified with MOLP (MP-Br) appearing to be darker (Figure 2B). It was also shown to be denser, shorter, and less voluminous compared to the regular bread (NC-Br) (Figure 2A) and the bread fortified with MOLP extract (ME-Br) (Figure 2C). These findings were also noted by other authors, who attributed these observations to the powder's binding properties and moisture retention. Compounds in the powder can interact with gluten, hindering dough expansion during fermentation and leading to a more compact crumb structure. Additionally, the powder's moisture absorption competes with yeast activity, affecting the rise. The reduction in both height and volume might also result from the antimicrobial effects of *M. oleifera* leaves on the yeast's leavening activities while the dough ferments [50].

Regarding the biscuits, there was no clear visible difference in colour between MP-Bi and ME-Bi (Figure 3), neither the volume nor density of the cookies after baking. However, other researchers have noted that the progressive inclusion of MOLP from 0 to 10% resulted in a reduction in the spread ratio in biscuits due to reduced diameter and increased thickness, linked to gluten dilution and decreased water available for gluten hydration. The authors argued that adding composite flour increased dough viscosity and formed aggregates by competing for limited free water in the cookie dough [51].

3.2.2. Total Phenolic Content and Antioxidant Capacity

To understand the impact of the incorporation of MOLP and its phenolic-rich extract on the breads' and biscuits' properties, the TPC and antioxidant capacities of the foods were analysed over time. The results obtained for the TPC of the formulations at different timepoints are presented in Figure 4.

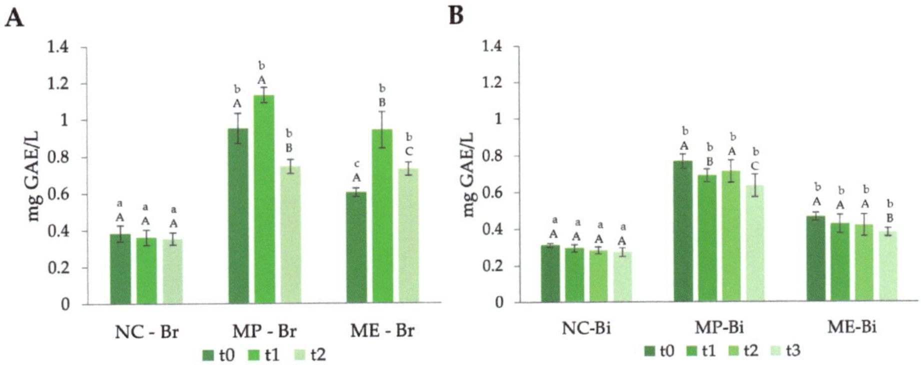

Figure 4. Total phenolic content variation in the breads (**A**) and biscuits (**B**) throughout the period of the study. NC-Br—bread with no additives (negative control); MP-Br—bread with 5% MOLP; ME-Br—bread with extract corresponding to 5% MOLP. NC-Bi—biscuits with no additives (negative control); MP-Bi—biscuits with 5% MOLP; ME-Bi—biscuits with extract corresponding to 5% MOLP. The analysis was performed at different timepoints: for bread, t0—day after production, t1—four days after production, and t2—one week after production; for biscuits, t0—day after production, t1—two weeks after production, t2—four weeks after production, and t3—seven weeks after production. The results are expressed as means ± standard deviations of 4 independent measurements. Different lowercase letters (a–c) represent statistically different values ($p < 0.05$) for the same timepoint. Different capital letters (A–C) represent statistically different values ($p < 0.05$) for the same formulation.

It is clear that the fortification of both the bread and biscuits with *M. oleifera* led to a substantial rise in the phenolic content of the formulations throughout the entire duration of this study. This outcome was anticipated, given the rich abundance of phenolic compounds (PCs) present in *M. oleifera* leaves. Other authors had similar findings [23,51]. Moreover, the TPC of the foods fortified with MOLP (MP-Br and MP-Bi) presented a higher value at all the timepoints of this study in comparison to the extract-incorporated formulations (ME-Br and ME-Bi). This finding can be due to the fact that the MOLP extraction might have led to the degradation or loss of some of the PCs, thereby reducing their concentration in the extract compared to the original powder. This could have led to a dilution effect where the PCs in the extract were spread out over a larger volume of bread, potentially resulting in lower measured TPC compared to the powder.

Phenolic compounds are susceptible to degradation from factors like light, heat, and oxygen, which can result in a decrease in their content over time [52]. It is noticeable from Figure 4 that both the fortified breads and biscuits TPC decreased over time. Regarding the bread (Figure 4A), when comparing MP-Br and ME-Br, the first one had a significantly higher phenolic content right after production. However, with time, this initial difference in phenolic content between MP-Br and ME-Br seemed to diminish, and after one week, both formulations exhibited nearly identical (non-significantly different) TPC values. This suggests that ME-Br has the interesting property of retaining its PCs more effectively over time. The same happened with biscuits (Figure 4B).

Regarding the antioxidant properties of the breads and biscuits, the results for both the DPPH and ABTS assays for both foods at the different timepoints are displayed in Figure 5.

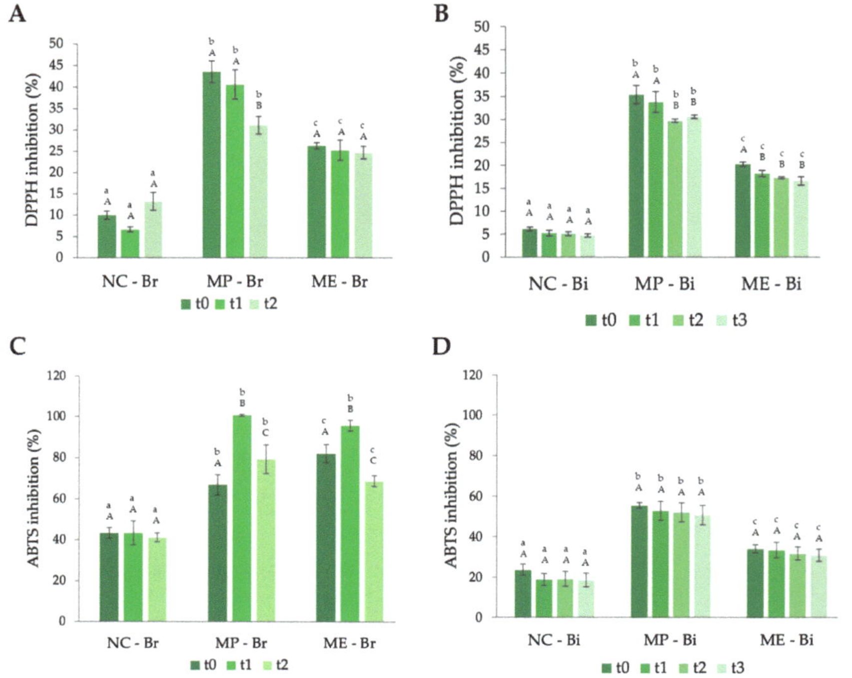

Figure 5. Antioxidant property variation in the breads and biscuits throughout the period of this study. (**A**) DPPH assay results for bread; (**B**) DPPH assay results for biscuits; (**C**) ABTS assay results for bread; and (**D**) ABTS assay results for biscuits. NC-Br—bread with no additives (negative control); MP-Br—bread with 5% MOLP; ME-Br—bread with extract corresponding to 5% MOLP. NC-Bi—biscuits with no additives (negative control); MP-Bi—biscuits with 5% MOLP; ME-Bi—biscuits with extract corresponding to 5% MOLP. The analysis was performed at different timepoints: for bread, t0—day after production, t1—four days after production, and t2—one week after production; for biscuits, t0—day after production, t1—two weeks after production, t2—four weeks after production, and t3—seven weeks after production. The results are expressed as means ± standard deviations of 4 independent measurements. Different lowercase letters (a–c) represent statistically different values ($p < 0.05$) for the same timepoint. Different capital letters (A–C) represent statistically different values ($p < 0.05$) for the same formulation.

From Figure 5, it is clear that the fortification of bread and biscuits increased their antioxidant capacity throughout the entire period of this study due to showing a substantially higher percentage inhibition for both radicals, with the ABTS assay having higher values, as expected (Figure 5C,D). Similar to the TPC assay, the MP-Br and MP-Bi formulations presented a higher antioxidant capacity in comparison to the extract-incorporated foods. While there is a lack of existing literature on the inclusion of *M. oleifera* extracts into bread, previous studies, as mentioned before, have explored the enhancement of bread through the addition of MOLP. These studies consistently noted a boost in the antioxidant capacity of the bread with increasing levels of fortification [23,24]. Nevertheless, an analysis of these properties over time was not conducted.

Analysing Figure 5A,C, it is noticeable that for the three breads, the antioxidant capacity had the tendency to decrease over time; however, in the DPPH assay, ME-Br showed once again that it can retain the PCs more efficiently than MP-Br. Regarding the biscuits (Figure 5B,D), the values of their antioxidant capacity were lower in comparison to the breads due to their containing a larger number of different ingredients, which in turn creates a dilution effect on the PCs present in the samples. Other authors have also proven the increasing level of antioxidant capacity of biscuits with increasing levels of MOLP substitution. In the study conducted by Fapetu et al., the biscuit with a 5% wheat flour substitution for MOLP presented a DPPH inhibition of 40.42%, which is close to the value presented in this study of 35.37% [25]. Although it is not so common to incorporate MOLP extract in biscuits, some authors have tested the inclusion of 1% MOLP extract and also stated that it caused an increase in the antioxidant activity of the biscuits [53].

3.2.3. Microbial Contamination

The presence of microorganisms, more specifically coliforms (using LSA medium), yeast, and moulds (using RBC medium), was investigated in each food formulation at different time periods. The results are presented in Table 7.

Table 7. Results from microbial contamination analysis of breads and biscuits during the study period.

Formulation	Timepoint	LSA	RBC
		(CFU/mL)	
NC-Br	t0	ND	ND
	t1	>30 × 10^5	>30 × 10^5
MP-Br	t0	ND	ND
	t1	>30 × 10^5	ND
ME-Br	t0	ND	ND
	t1	3.05 × 10^5	ND
NC-Bi	t0	ND	ND
	t1	ND	ND
	t2	ND	ND
	t3	ND	ND
MP-Bi	t0	ND	ND
	t1	ND	ND
	t2	ND	ND
	t3	ND	ND
ME-Bi	t0	ND	ND
	t1	ND	ND
	t2	ND	ND
	t3	ND	ND

ND—not detected; LSA—Lauryl Sulfate Agar medium; RBC—Rose-Bengal Chloramphenicol Agar medium. NC-Br—bread with no additives (negative control); MP-Br—bread with 5% MOLP; ME-Br—bread with extract corresponding to 5% MOLP. NC-Bi—biscuits with no additives (negative control); MP-Bi—biscuits with 5% MOLP; ME-Bi—biscuits with extract corresponding to 5% MOLP. The analysis was performed at different timepoints: for bread, t0—day after production and t1—four days after production; for biscuits, t0—day after production, t1—two weeks after production, t2—four weeks after production, and t3—seven weeks after production.

As shown in Table 7, the three bread formulations had no significant contamination the day after production (t0) in both LSA and RBC mediums. Considering the typical 4-day shelf life of bread at room temperature (t1), it was expected to find microbial contamination by the fourth day. As observed, the negative control bread (NC-Br) displayed a substantial number of colony-forming units (CFU) in both mediums, aligning with expectations. However, in the LSA medium, the bread with MOLP (MP-Br) presented a high degree of microbial contamination, which was somewhat unexpected since MOLP-enriched bread had the highest TPC and should theoretically delay oxidative spoilage, which in turn hinders microbial growth. In contrast, ME-Br exhibited impressive results, with only 3.05 × 10^5 CFU/mL, which is lower than the unsatisfactory limit established by the Centre of Food Safety, Department of Food and Environmental Hygiene, categorising an upper limit of microbial concentration equal to $\geq 10^6$ CFU/mL for bakery and confectionery products [54]. This outcome underscores the extract's effectiveness as a potent food preservative, validating its role in extending shelf life and maintaining product quality. A possible explanation for these findings is the fact that although MP-Br contained a higher phenolic content value than ME-Br, the MOLP extract might contain other compounds that have been proven to have an antimicrobial effect (such as alkaloids and glycosides) in a higher concentration, which aid in delaying the microbial growth in the bread [55].

Analysing yeast and mould growth results (RBC medium), there was a substantial growth of colonies for the NC-Br four days after production (t1) (>30 $\times$ 10^5 CFU/mL) and no significant growth of colonies for the fortified bread samples at any timepoint. These results show that both the MOLP and its extract were able to efficiently inhibit microbial contamination in the food samples owing to their remarkable anti-microbial effect and antioxidant capacity, therefore having the ability to extend their shelf life. Other studies have also proven the efficacy of the incorporation of MOLP in inhibiting the growth of microorganisms, yeasts, and moulds in bread [56].

Regarding the biscuits, since they typically have an extended shelf life, there was no significant contamination in any of the mediums or timepoints, which suggests that this experiment should be conducted in a prolonged time frame in order to capture the potential growth of microbials. Other authors, however, have noted an increase in shelf life with MOLP-incorporated biscuits and crackers [57,58].

3.2.4. Sensory Assessment of the Breads and Biscuits

A sensory assessment was conducted to measure consumers' acceptability of bread and biscuits fortified with MOLP and its respective extracts. The obtained results are expressed in Figure 6.

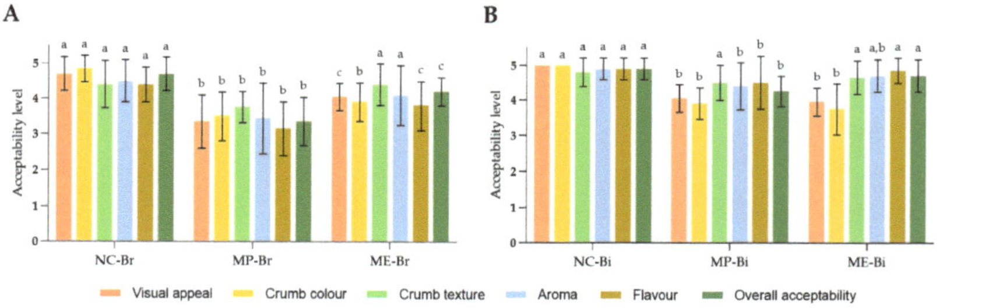

Figure 6. Results from the sensory assessment of bread (**A**) and biscuits (**B**). NC-Br—bread with no additives (negative control); MP-Br—bread with 5% MOLP; ME-Br—bread with extract corresponding to 5% MOLP. NC-Bi—biscuits with no additives (negative control); MP-Bi—biscuits with 5% MOLP; ME-Bi—biscuits with extract corresponding to 5% MOLP. The results are expressed as means $\pm$ standard deviations. Different lowercase letters (a–c) represent statistically different values ($p < 0.05$) for the same sensory parameter.

The sensory evaluation showed that the control bread and biscuits, as expected, had the highest scores in all categories. The decrease in scores of the fortified foods was likely due to the herbaceous taste of the leaves and noticeable colour changes in the foods, which differed from their familiar golden-brown appearance. Previous studies have noted that as MOLP levels increased in bread and biscuits (at 5% and 10% replacement levels), overall acceptability decreased, with brown bread being less affected than white bread [24,57,59,60].

The bread with *M. oleifera* extract (ME-Br) received a more positive response from consumers compared to the one incorporated with MOLP (MP-Br), as observed in Figure 6A. This preference could be attributed to the extract's milder taste and lighter green colour. Regarding crumb texture, NC-Br and ME-Br received high scores, while MP-Br had decreased texture acceptability since MOLP increases bread hardness and gumminess due to its high fibre content [61]. Fibre absorbs water and swells during baking, resulting in a firmer and denser texture. Regarding the biscuits, the extract-incorporated biscuit (ME-Bi) was also better accepted than the powder-incorporated biscuit (MP-Bi), as seen in Figure 6B. Moreover, in general, the biscuits were better received than the bread in all categories due to the fact that the biscuits contain sweeter ingredients in their formulation that are able to mask the herbaceous taste and aroma of *M. oleifera* more effectively.

In summary, the control bread and biscuits had the highest overall acceptability, the extract-enriched bread and biscuits scored slightly lower but still favourably, and the MOLP-incorporated foods had a moderate level of acceptability, noting that the biscuits had a higher score in all categories in comparison to the bread. Therefore, this study suggests that incorporating *M. oleifera* leaf extract in cereal-based foods is a more promising approach to offering the numerous benefits of this plant while still keeping in mind consumers' taste preferences and satisfaction.

4. Conclusions

This work is intended to evaluate the feasibility of utilising *Moringa oleifera* leaf powder (MOLP) and its phenolic-rich extract, renowned for their antioxidant capabilities, to enhance the quality of cereal-based products, with a specific focus on bread and biscuits. The results suggested that the extract derived from MOLP exhibited remarkable antioxidant and antidiabetic properties, highlighting its potential as a functional food ingredient for managing blood sugar levels. Regarding the cytotoxicity studies, these demonstrated that an extract's concentration of 1000 mg/L is safe for human consumption. Furthermore, the fortified breads and biscuits, enriched with MOLP and its extract, consistently demonstrated an elevated total phenolic content and antioxidant capacity, outperforming the negative control over the study duration. Moreover, the application of both MOLP and its extract effectively limited microbial growth during the storage of food products, emphasising their dual role in enhancing food preservation. Moreover, there was an evident preference for extract-infused foods over powder-infused alternatives among consumers. The results with a 5% substitution level were quite promising, and as future work, it would be interesting to study higher levels to understand what the technological limit for the supplementation level is as well as the effect on the sensory properties. In conclusion, the potentiality of food fortification using MOLP and its phenolic-rich extract is a novel approach to the food industry to produce value-added products that offer richer nutrient composition, potential disease-preventing benefits, and an extended shelf life to tackle food insecurity and malnutrition.

Despite being promising, the use of nutraceuticals such as *Moringa oleifera* in foods must be carefully analysed to define a safe dosage for consumers. Although a cytotoxicity analysis was performed in the present work, it is important to perform a more detailed analysis on the safety of the extract, for example, to evaluate potential allergens. Moreover, conducting a nutritional assessment of bread and biscuits, both prior to and following the inclusion of the powder and extract, would provide insights into their impact on nutritional content. Lastly, an evaluation of the effect of *M. oleifera* addition on the physical parameters of the products (such as colour and texture) would contribute to a thorough understanding of the sensory attributes of these products.

Author Contributions: Conceptualization, S.M.G. and L.S.; methodology, T.F. and S.M.G.; validation, T.F.; formal analysis, T.F.; investigation, T.F.; resources, L.S.; data curation, T.F.; writing—original draft preparation, T.F. and S.M.G.; writing—review and editing, S.M.G. and L.S.; visualisation, S.M.G. and L.S.; supervision, L.S.; project administration, L.S.; funding acquisition, L.S. All authors have read and agreed to the published version of the manuscript.

Funding: This work was financially supported by: Project MORfood (Ref[a]. 541163254/2019), funded by the Foundation for Science and Technology (FCT) and Aga Khan Development Network (AKDN), LA/P/0045/2020 (ALiCE), UIDB/00511/2020, and UIDP/00511/2020 (LEPABE), funded by national funds through FCT/MCTES (PIDDAC).

Institutional Review Board Statement: Not applicable.

Informed Consent Statement: Not applicable.

Data Availability Statement: The data presented in this study are available on request from the corresponding author.

Acknowledgments: The authors would like to thank Anabela Leitão from Agostinho Neto University for providing the *Moringa oleifera* leaf powder used in this study.

Conflicts of Interest: The authors declare no conflict of interest.

References

1. FAO; IFAD; UNICEF; WFP; WHO. *The State of Food Security and Nutrition in the World. Repurposing Food and Agricultural Policies to Make Healthy Diets More Affordable*; FAO: Rome, Italy, 2022.
2. Paraschivu, M.; Cotuna, O.; Matei, G.; Sarateanu, V. Are Food Waste and Food Loss a Real Threat for FOOD Security? *Sci. Pap. Ser. Manag. Econ. Eng. Agric. Rural Dev.* **2022**, *22*, 479–484.
3. Hodas, F.; Zorzenon, M.R.T.; Milani, P.G. *Moringa oleifera* potential as a functional food and a natural food additive: A biochemical approach. *An. Acad. Bras. Ciênc.* **2021**, *93*, e20210571. [CrossRef] [PubMed]
4. Farag, M.A.; Abdelwareth, A.; Sallam, I.E.; El Shorbagi, M.; Jehmlich, N.; Fritz-Wallace, K.; Schäpe, S.S.; Rolle-Kampczyk, U.; Ehrlich, A.; Wessjohann, L.A.; et al. Metabolomics reveals impact of seven functional foods on metabolic pathways in a gut microbiota model. *J. Adv. Res.* **2020**, *23*, 47–59. [CrossRef] [PubMed]
5. Llorent-Martinez, E.J.; Gordo-Moreno, A.I.; Fernandez-de Cordova, M.L.; Ruiz-Medina, A. Preliminary Phytochemical Screening and Antioxidant Activity of Commercial *Moringa oleifera* Food Supplements. *Antioxidants* **2023**, *12*, 110. [CrossRef] [PubMed]
6. Hisam, E.E.A.; Rofiee, M.S.; Khalid, A.M.; Jalaluddin, A.F.; Yusof, M.I.M.; Idris, M.H.; Ramli, S.; James, R.J.; Yoeng, W.J.; Kek, T.L.; et al. Combined extract of *Moringa oleifera* and Centella asiatica modulates oxidative stress and senescence in hydrogen peroxide-induced human dermal fibroblasts. *Turk. J. Biol.* **2018**, *42*, 33–44. [CrossRef] [PubMed]
7. Rathnayake, A.R.M.H.A.; Navaratne, S.B.; Uthpala, T.G.G. *Moringa oleifera* Plant and the Nutritional and Medicinal Properties of *Moringa olifera* Leaves. In *Trends & Prospects in Processing of Horticultural Crops*; Chakraborty, I., Paul, P.K., Mani, A., Tiwary, A.K., Prasad, K., Eds.; Today and Tomorrow's Printers and Publishers: New Delhi, India, 2019; pp. 251–268.
8. Oyeyinka, A.T.; Oyeyinka, S.A. *Moringa oleifera* as a food fortificant: Recent trends and prospects. *J. Saudi Soc. Agric. Sci.* **2018**, *17*, 127–136. [CrossRef]
9. Das, A.K.; Rajkumar, V.; Verma, A.K.; Swarup, D. Moringa oleiferia leaves extract: A natural antioxidant for retarding lipid peroxidation in cooked goat meat patties. *Int. J. Food Sci. Technol.* **2012**, *47*, 585–591. [CrossRef]
10. Stohs, S.J.; Hartman, M.J. Review of the Safety and Efficacy of *Moringa oleifera*. *Phytother. Res.* **2015**, *29*, 796–804. [CrossRef]
11. Vonghirundecha, P.; Chusri, S.; Meunprasertdee, P.; Kaewmanee, T. Microencapsulated functional ingredients from a *Moringa oleifera* leaf polyphenol-rich extract: Characterization, antioxidant properties, in vitro simulated digestion, and storage stability. *LWT* **2022**, *154*, 112820. [CrossRef]
12. Singh, D.; Arya, P.V.; Aggarwal, V.P.; Gupta, R.S. Evaluation of Antioxidant and Hepatoprotective Activities of *Moringa oleifera* Lam. Leaves in Carbon Tetrachloride-Intoxicated Rats. *Antioxidants* **2014**, *3*, 569–591. [CrossRef]
13. Vergara-Jimenez, M.; Almatrafi, M.M.; Fernandez, M.L. Bioactive Components in *Moringa oleifera* Leaves Protect against Chronic Disease. *Antioxidants* **2017**, *6*, 91. [CrossRef] [PubMed]
14. Shahidi, F.; Zhong, Y. Novel antioxidants in food quality preservation and health promotion. *Eur. J. Lipid Sci. Technol.* **2010**, *112*, 930–940. [CrossRef]
15. Vongsak, B.; Sithisarn, P.; Gritsanapan, W. Bioactive contents and free radical scavenging activity of *Moringa oleifera* leaf extract under different storage conditions. *Ind. Crops Prod.* **2013**, *49*, 419–421. [CrossRef]
16. Giuberti, G.; Rocchetti, G.; Lucini, L. Interactions between phenolic compounds, amylolytic enzymes and starch: An updated overview. *Curr. Opin. Food Sci.* **2020**, *31*, 102–113. [CrossRef]
17. Karray, A.; Alonazi, M.; Jallouli, R.; Alanazi, H.; Ben Bacha, A. A Proteinaceous Alpha-Amylase Inhibitor from *Moringa oleifera* Leaf Extract: Purification, Characterization, and Insecticide Effects against *C. maculates* Insect Larvae. *Molecules* **2022**, *27*, 4222. [CrossRef] [PubMed]
18. Asare, G.A.; Gyan, B.; Bugyei, K.; Adjei, S.; Mahama, R.; Addo, P.; Otu-Nyarko, L.; Wiredu, E.K.; Nyarko, A. Toxicity potentials of the nutraceutical *Moringa oleifera* at supra-supplementation levels. *J. Ethnopharmacol.* **2012**, *139*, 265–272. [CrossRef] [PubMed]
19. Ashraf, M.; ALAM, S.S.; Fatima, M.; Altaf, I.; Khan, F.; Afzal, A. Comparative Anti-Influenza Potential of *Moringa oleifera* Leaves and Amantadine Invitro. *Pak. Postgrad. Med. J.* **2017**, *28*, 127–131. [CrossRef]
20. Chhikara, N.; Kaur, A.; Mann, S.; Garg, M.; Sofi, S.A.; Panghal, A. Bioactive compounds, associated health benefits and safety considerations of *Moringa oleifera* L.: An updated review. *Nutr. Food Sci.* **2020**, *51*, 255–277. [CrossRef]
21. Siddhuraju, P.; Becker, K. Antioxidant properties of various solvent extracts of total phenolic constituents from three different agroclimatic origins of drumstick tree (*Moringa oleifera* Lam.) leaves. *J. Agric. Food Chem.* **2003**, *51*, 2144–2155. [CrossRef]
22. Kakde, S.B.; Masih, D.; Sonkar, C. Utilization of Moringa leaves powder as valuable food ingredients in pasta preparation. *J. Pharmacogn. Phytochem.* **2018**, *7*, 1053–1056.
23. Khan, M.A.; Shakoor, S.; Ameer, K.; Farooqi, M.A.; Rohi, M.; Saeed, M.; Asghar, M.T.; Irshad, M.B.; Waseem, M.; Tanweer, S.; et al. Effects of Dehydrated Moringa (*Moringa oleifera*) Leaf Powder Supplementation on Physicochemical, Antioxidant, Mineral, and Sensory Properties of Whole Wheat Flour Leavened Bread. *J. Food Qual.* **2023**, *2023*, 4473000. [CrossRef]

24. Bourekoua, H.; Różyło, R.; Gawlik-Dziki, U.; Benatallah, L.; Zidoune, M.N.; Dziki, D. Evaluation of physical, sensorial, and antioxidant properties of gluten-free bread enriched with *Moringa oleifera* leaf powder. *Eur. Food Res. Technol.* **2018**, *244*, 189–195. [CrossRef]
25. Fapetu, A.P.; Karigidi, K.O.; Akintimehin, E.S.; Olawuwo, T.; Adetuyi, F.O. Effect of partial substitution of wheat flour with *Moringa oleifera* leaf powder on physical, nutritional, antioxidant and antidiabetic properties of cookies. *Bull. Natl. Res. Cent.* **2022**, *46*, 46–53. [CrossRef]
26. Dachana, K.B.; Rajiv, J.; Indrani, D.; Prakash, J. Effect of Dried Moringa (*Moringa oleifera* Lam) Leaves on Rheological, Microstructural, Nutritional, Textural and Organoleptic Characteristics of Cookies. *J. Food Qual.* **2010**, *33*, 660–677. [CrossRef]
27. Gomes, S.M.; Albuquerque, D.; Santos, L. Innovative Approaches for Food: Using Natural Phenolic-Rich Extracts to Produce Value-Added Fresh Pasta. *Int. J. Mol. Sci.* **2023**, *24*, 12451. [CrossRef]
28. Ferreira, S.M.; Santos, L. Incorporation of phenolic extracts from different by-products in yoghurts to create fortified and sustainable foods. *Food Biosci.* **2023**, *51*, 102293. [CrossRef]
29. Bobo-García, G.; Davidov-Pardo, G.; Arroqui, C.; Vírseda, P.; Marín-Arroyo, M.R.; Navarro, M. Intra-laboratory validation of microplate methods for total phenolic content and antioxidant activity on polyphenolic extracts, and comparison with conventional spectrophotometric methods. *J. Sci. Food Agric.* **2015**, *95*, 204–209. [CrossRef]
30. Minekus, M.; Alminger, M.; Alvito, P.; Ballance, S.; Bohn, T.; Bourlieu, C.; Carrière, F.; Boutrou, R.; Corredig, M.; Dupont, D.; et al. A standardised static in vitro digestion method suitable for food—An international consensus. *Food Funct.* **2014**, *5*, 1113–1124. [CrossRef]
31. Gomes, S.M.; Miranda, R.; Santos, L. Sustainable Cosmetics: Valorisation of Kiwi (*Actinidia deliciosa*) By-Products by Their Incorporation into a Moisturising Cream. *Sustainability* **2023**, *15*, 14059. [CrossRef]
32. Bimakr, M.; Rahman, R.A.; Taip, F.S.; Ganjloo, A.; Salleh, L.M.; Selamat, J.; Hamid, A.; Zaidul, I. Comparison of different extraction methods for the extraction of major bioactive flavonoid compounds from spearmint (*Mentha spicata* L.) leaves. *Food Bioprod. Process.* **2011**, *89*, 67–72. [CrossRef]
33. Santana, C.M.; Ferrera, Z.S.; Padron, M.E.T.; Rodriguez, J.J.S. Methodologies for the extraction of phenolic compounds from environmental samples: New approaches. *Molecules* **2009**, *14*, 298–320. [CrossRef] [PubMed]
34. Tai, T.K.; Thongklay, J.; Meunprasertdee, P.; Kornthattalim, P.; Kaaewmanee, T. A Comparison of Three Extraction Methods for Phenolic Compounds and Antioxidant Activities from *Moringa oleifera* Leaves. *Chiang Mai J. Sci.* **2018**, *45*, 2779–2789.
35. Vongsak, B.; Sithisarn, P.; Mangmool, S.; Thongpraditchote, S.; Wongkrajang, Y.; Gritsanapan, W. Maximizing total phenolics, total flavonoids contents and antioxidant activity of *Moringa oleifera* leaf extract by the appropriate extraction method. *Ind. Crops Prod.* **2013**, *44*, 566–571. [CrossRef]
36. Putra, N.R.; Rizkiyah, D.N.; Zaini, A.S.; Yunus, M.A.C.; Machmudah, S.; Idham, Z.B.; Ruslan, M.S.H. Effect of particle size on yield extract and antioxidant activity of peanut skin using modified supercritical carbon dioxide and soxhlet extraction. *J. Food Process. Preserv.* **2018**, *42*, e13689. [CrossRef]
37. Yen, N.T.H.; Quoc, L.P.T. Optimization of ultrasound-assisted extraction of phenolic compounds from fresh *Moringa oleifera* leaves with a response surface methodology and comparison with the Soxhlet extraction method. *Bull. Chem. Soc. Ethiop.* **2022**, *36*, 261–275. [CrossRef]
38. Laskar, R.A.; Sk, I.; Roy, N.; Begum, N.A. Antioxidant activity of Indian propolis and its chemical constituents. *Food Chem.* **2010**, *122*, 233–237. [CrossRef]
39. Mello, B.C.B.S.; Petrus, J.C.C.; Hubinger, M.D. Concentration of flavonoids and phenolic compounds in aqueous and ethanolic propolis extracts through nanofiltration. *J. Food Eng.* **2010**, *96*, 533–539. [CrossRef]
40. Jahan, S.; Shahjahan, M.; Rasna, S.S.; Aktar, M.; Sultana, S.; Ahmed, S.M.; Sabrin, F.; Nahar, S. Antibacterial Effect of Moringa (*Moringa oleifera*) Leaf Ethanolic Extract Against Staphylococcus aureus and Escherichia coli. *Mymensingh Med. J.* **2022**, *31*, 976–982.
41. Magaji, U.F.; Sacan, O.; Yanardag, R. Alpha amylase, alpha glucosidase and glycation inhibitory activity of *Moringa oleifera* extracts. *S. Afr. J. Bot.* **2020**, *128*, 225–230. [CrossRef]
42. Niziol-Lukaszewska, Z.; Furman-Toczek, D.; Bujak, T.; Wasilewski, T.; Hordyjewicz-Baran, Z. *Moringa oleifera* L. Extracts as Bioactive Ingredients That Increase Safety of Body Wash Cosmetics. *Dermatol. Res. Pract.* **2020**, *2020*, 8197902. [CrossRef]
43. Gomes, S.M.; Leitão, A.; Alves, A.; Santos, L. Incorporation of *Moringa oleifera* Leaf Extract in Yoghurts to Mitigate Children's Malnutrition in Developing Countries. *Molecules* **2023**, *28*, 2526. [CrossRef]
44. Ijarotimi, O.S.; Adeoti, O.A.; Ariyo, O. Comparative study on nutrient composition, phytochemical, and functional characteristics of raw, germinated, and fermented *Moringa oleifera* seed flour. *Food Sci. Nutr.* **2013**, *1*, 452–463. [CrossRef] [PubMed]
45. Oboh, G.; Ademiluyi, A.O.; Ademosun, A.O.; Olasehinde, T.A.; Oyeleye, S.I.; Boligon, A.A.; Athayde, M.L. Phenolic Extract from *Moringa oleifera* Leaves Inhibits Key Enzymes Linked to Erectile Dysfunction and Oxidative Stress in Rats' Penile Tissues. *Biochem. Res. Int.* **2015**, *2015*, 175950. [CrossRef] [PubMed]
46. Singh, B.N.; Singh, B.R.; Singh, R.L.; Prakash, D.; Dhakarey, R.; Upadhyay, G.; Singh, H.B. Oxidative DNA damage protective activity, antioxidant and anti-quorum sensing potentials of *Moringa oleifera*. *Food Chem. Toxicol.* **2009**, *47*, 1109–1116. [CrossRef]
47. Rodríguez-Pérez, C.; Quirantes-Piné, R.; Fernández-Gutiérrez, A.; Segura-Carretero, A. Optimization of extraction method to obtain a phenolic compounds-rich extract from *Moringa oleifera* Lam leaves. *Ind. Crops Prod.* **2015**, *66*, 246–254. [CrossRef]
48. *ISO 10993-12:2021*; Biological Evaluation of Medical Devices. Part 12: Sample Preparation and Reference Materials. International Organization for Standardization: Geneva, Switzerland, 2021.

49. Landázuri, A.C.; Gualle, A.; Castañeda, V.; Morales, E.; Caicedo, A.; Orejuela-Escobar, L.M. *Moringa oleifera* Lam. leaf powder antioxidant activity and cytotoxicity in human primary fibroblasts. *Nat. Prod. Res.* **2021**, *35*, 6194–6199. [CrossRef]
50. Bukar, A.; Uba, A.; Oyeyi, T. Antimicrobial profile of *Moringa oleifera* lam. Extracts against some food—Borne microorganisms. *Bayero J. Pure Appl. Sci.* **2010**, *3*, 43–48. [CrossRef]
51. Ajibola, C.F.; Oyerinde, V.O.; Adeniyan, O.S. Physicochemical and Antioxidant Properties of Whole-Wheat Biscuits Incorporated with *Moringa oleifera* Leaves and Cocoa Powder. *J. Sci. Res. Rep.* **2015**, *7*, 195–206. [CrossRef]
52. Jimenez-Moreno, N.; Volpe, F.; Moler, J.A.; Esparza, I.; Ancin-Azpilicueta, C. Impact of Extraction Conditions on the Phenolic Composition and Antioxidant Capacity of Grape Stem Extracts. *Antioxidants* **2019**, *8*, 597. [CrossRef]
53. Abdel-Samie, M.A.S.; Abdulla, G. Effect of Moringa Leaves (*Moringa oleifera* Lam.) on Some Physico-Chemical and Sensory Properties of Wheat Flour Cookies. *Zagazig J. Agric. Res.* **2014**, *41*, 305–314.
54. Srinivasamurthy, S.; Yadav, U.; Sahay, S.; Singh, A. Development of muffin by incorporation of dried *Moringa oleifera* (Drumstick) leaf powder with enhanced micronutrient content. *Int. J. Food Sci. Nutr.* **2017**, *2*, 173–178.
55. Akinyeye, A.J. Phytochemical and antimicrobial evaluation of leaf and seed of *Moringa oleifera* extracts. *Int. J. Res. Med. Health Sci.* **2014**, *4*, 1–10.
56. Aly, A.A.; Zaky, E.A.; Khatab, N.R.; Hameed, A.M.; Kadasah, S. The Biological and Chemical Ameliorative Effects of Bread Substituted with Dried Moringa Leaves. *Arab. J. Chem.* **2022**, *15*, 103875. [CrossRef]
57. Govender, L.; Siwela, M. The Effect of *Moringa oleifera* Leaf Powder on the Physical Quality, Nutritional Composition and Consumer Acceptability of White and Brown Breads. *Foods* **2020**, *9*, 1910. [CrossRef]
58. Ismawati, R.; Wahini, M.; Romadhoni, I.F.; Aina, Q. Sensory Preference, Nutrient Content, and Shelf Life of Moringa Oliefera Leaf Crackers. *Int. J. Adv. Sci. Eng. Inf.* **2019**, *9*, 489–494. [CrossRef]
59. El-Gammal, R.; Ghoneim, G.; ElShehawy, S. Effect of Moringa Leaves Powder (*Moringa oleifera*) on Some Chemical and Physical Properties of Pan Bread. *J. Food Dairy Sci.* **2016**, *7*, 307–314. [CrossRef]
60. Sengev, A.I.; Abu, J.O.; Gernah, D.I. Effect of *Moringa oleifera* Leaf Powder Supplementation on Some Quality Characteristics of Wheat Bread. *Food Nutr. Sci.* **2013**, *4*, 270–275. [CrossRef]
61. Madane, P.; Das, A.K.; Pateiro, M.; Nanda, P.K.; Bandyopadhyay, S.; Jagtap, P.; Barba, F.J.; Shewalkar, A.; Maity, B.; Lorenzo, J.M. Drumstick (*Moringa oleifera*) Flower as an Antioxidant Dietary Fibre in Chicken Meat Nuggets. *Foods* **2019**, *8*, 307. [CrossRef]

antioxidants

MDPI

Article

Chlorogenic Acid Alleviated AFB1-Induced Hepatotoxicity by Regulating Mitochondrial Function, Activating Nrf2/HO-1, and Inhibiting Noncanonical NF-κB Signaling Pathway

Qianqian Wang [1], Tianxu Liu [1], Matthew Koci [2], Yanan Wang [1], Yutong Fu [1], Mingxin Ma [1], Qiugang Ma [1] and Lihong Zhao [1,*]

[1] State Key Laboratory of Animal Nutrition and Feeding, Poultry Nutrition and Feed Technology Innovation Team, College of Animal Science and Technology, China Agricultural University, No. 2. West Road Yuanming Yuan, Beijing 100193, China; wqq@cau.edu.cn (Q.W.); liutianx@cau.edu.cn (T.L.); wyn@cau.edu.cn (Y.W.); yutong_fu@cau.edu.cn (Y.F.); mamingxin@cau.edu.cn (M.M.); maqiugang@cau.edu.cn (Q.M.)

[2] Prestage Department of Poultry Science, North Carolina State University, Raleigh, NC 27695, USA; mdkoci@ncsu.edu

* Correspondence: zhaolihongcau@cau.edu.cn; Tel.: +86-10-62731019; Fax: +86-10-62732774

Abstract: Aflatoxin B1 (AFB1), a kind of mycotoxin, imposes acute or chronic toxicity on humans and causes great public health concerns. Chlorogenic acid (CGA), a natural phenolic substance, shows a powerful antioxidant and anti-inflammatory effect. This study was conducted to investigate the effect and mechanism of CGA on alleviating cytotoxicity induced by AFB1 in L-02 cells. The results showed that CGA (160 μM) significantly recovered cell viability and cell membrane integrity in AFB1-treated (8 μM) cells. Furthermore, it was found that CGA reduced AFB1-induced oxidative injury by neutralizing reactive oxygen species (ROS) and activating the nuclear factor erythroid 2-related factor 2 (Nrf2)/heme oxygenase-1 (HO-1) signaling pathway. In addition, CGA showed anti-inflammatory effects as it suppressed the expression of inflammation-related genes (*IL-6*, *IL-8*, and *TNF-α*) and AFB1-induced noncanonical nuclear factor kappa-B (NF-κB) activation. Moreover, CGA mitigated AFB1-induced apoptosis by maintaining the mitochondrial membrane potential (MMP) and inhibiting mRNA expressions of *Caspase-3*, *Caspase-8*, *Bax*, and *Bax/Bcl-2*. These findings revealed a possible mechanism: CGA prevents AFB1-induced cytotoxicity by maintaining mitochondrial membrane potential, activating Nrf2/HO-1, and inhibiting the noncanonical NF-κB signaling pathway, which may provide a new direction for the application of CGA.

Keywords: chlorogenic acid; aflatoxin B1; hepatotoxicity; antioxidant; anti-inflammatory; mitochondrial function

Citation: Wang, Q.; Liu, T.; Koci, M.; Wang, Y.; Fu, Y.; Ma, M.; Ma, Q.; Zhao, L. Chlorogenic Acid Alleviated AFB1-Induced Hepatotoxicity by Regulating Mitochondrial Function, Activating Nrf2/HO-1, and Inhibiting Noncanonical NF-κB Signaling Pathway. *Antioxidants* **2023**, *12*, 2027. https://doi.org/10.3390/antiox12122027

Academic Editors: Baojun Xu and Bin Du

Received: 23 October 2023
Revised: 10 November 2023
Accepted: 11 November 2023
Published: 22 November 2023

1. Introduction

Aflatoxins are a kind of metabolite produced by the fungi *Aspergillus parasiticus* and *A. flavus* [1]. Among the known aflatoxin derivatives (B1, B2, G1, and G2), aflatoxin B1 (AFB1) possesses the highest toxicity, and it was classified as a Group 1 (carcinogenic to humans) carcinogen by the International Agency for Research on Cancer (IARC) [2]. Several studies suggested that exposure to AFB1 causes liver oxidative injury by the excessive production of reactive oxygen species (ROS) in organs [3]. In general, low levels of ROS play an important role in maintaining physiological conditions, while high levels of ROS are associated with pathological conditions due to their damage to subcellular structures and biomolecules [4]. Similarly, ROS induced by AFB1 caused an imbalance in the antioxidant system and damaged biological molecules including lipids, proteins, and DNA in cellular systems [5]. Moreover, AFB1 induced mitochondrial damage as it could uncouple mitochondrial oxidative phosphorylation and increase the opening of mitochondrial permeability transition pores [6]. The combination of these negative effects induced cells to produce potentially catastrophic genetic alterations, eventually resulting in apoptosis or autophagy [7]. Therefore, it is important to explore effective methods for alleviating AFB1-induced damage to human health.

Nutritional antioxidants, including polyphenols, antioxidant vitamins, and trace elements, were targeted for mitigating AFB1-induced toxin damage. For example, it was reported that a Se supplement in broilers' diet alleviates AFB1-induced liver damage by increasing the activities of antioxidant enzymes like catalase (CAT), glutathione peroxidase (GSH-Px), superoxide dismutase (SOD), and the molecular regulation of the death receptors pathway [8]. Vitamin E was shown to ameliorate AFB1-induced nephrotoxicity in rats, probably by the correction of the resultant oxidative stress parameters as well as the inhibition of the protein expression of caspase-3 [9]. In addition, the growing body of studies has shown that polyphenols such as carotenoids and quercetin are able to mitigate AFB1-induced hepatic damage by mediating the expression of antioxidant enzymes [10,11]. However, it is noted that many existing studies focus on phenotypic measures, while the underlying molecular mechanisms related to nutritional antioxidants, including chlorogenic acid (CGA), are still poorly understood.

CGA, a major family of phenolic acids derived from the esterification of cinnamic acids, is abundant in coffee, fruit, and vegetables. It was found that CGA has various biological activities such as free radical scavenging and antioxidant, antibacterial, anti-inflammatory, and cardiovascular protection [12–14]. Previous studies reported that CGA could protect the liver from damage caused by chemicals or lipopolysaccharide [15]. Further research showed that CGA has a protective effect on the liver of rats by modulating inflammation and redox homeostasis [16]. In addition, researchers found that CGA protects against aluminum-induced cytotoxicity by enhancing the activities of antioxidant enzymes (SOD and CAT) and increasing the expression of nuclear factor erythroid 2-related factor 2 (Nrf2) in primary hippocampal neuronal cells [17]. Based on these results, CGA may be a potential preventive agent for alleviating AFB1-induced hepatotoxicity. However, the protective effect of CGA on the cytotoxicity induced by AFB1 and its underlying mechanism remain elusive. Therefore, this study aims to investigate the antioxidant effect and the underlying action mechanism of CGA on a hepatic cell exposed to AFB1.

2. Materials and Methods

2.1. Materials and Reagents

CGA (purity, 98.0%) was purchased from Solarbio (Beijing, China). AFB1 was obtained from Pribolab (Singapore City, Singapore). Fetal bovine serum (FBS) and Roswell Park Memorial Institute (RPMI) 1640 medium were purchased from GIBCO (Grand Island, NE, USA). Antibodies were obtained from Proteintech (Wuhan, China).

2.2. Cell Culture

Human liver cells L-02 were purchased from Tongpai biotechnology co, ltd (Shanghai, China). Cells were maintained in RPMI 1640 containing 10% fetal bovine serum (FBS), 100 U/mL penicillin, and 100 μg/mL streptomycin (GIBCO, Grand Island, NE, USA) at 37 °C with 5% CO_2.

2.3. Cell Viability Assay

To select suitable AFB1 and CGA concentrations, the cell viability was determined with a CCK-8 assay. Briefly, cells (1×10^4 cells per well) were seeded in 96-well plates for 24 h, and then cells were treated with different concentrations of AFB1 (from 1 to 32 μM) for 24 h or different concentrations of CGA (from 10 to 320 μM) for 6 h. For the CGA + AFB1 experiment, after being pre-incubated with culture medium (with or without CGA) for 6 h, the supernatant was removed, and then the cells were treated with culture medium (with or without AFB1) for another 24 h. Thereafter, the CCK-8 assay was performed by adding 110 μL of fresh medium containing 10 μL of CCK-8 solution to the cells and incubating them for 1 h at 37 °C. The absorbance was measured at 450 nm using a microplate reader

(Thermo Labsystems, Pittsburgh, PA, USA). Cell viability was calculated as follows: Cell viability (100%) = (OD treatment/OD control) × 100%.

2.4. Determination of Lactate Dehydrogenase (LDH), Aspartate Aminotransferase (AST), and Alanine Aminotransferase (ALT) Activity

The L-02 cells were seeded in 6-well plates (5×10^5 cells per well) for 24 h; after four treatments (Control, CGA, AFB1, CGA + AFB1), the supernatant was collected and centrifuged at 3000 r/min for 10 min. The activities of LDH, AST, and ALT were measured with assay kits (Nanjing Jiancheng Bioengineering Institute, Nanjing, China) according to manufacturer's protocol.

2.5. RNA-Seq Analysis

Total RNA of cells was extracted with TRIzol reagent (Vazyme, Shanghai, China) according to the manufacturer's instructions. Samples were submitted to Shanghai Majorbio Bio-pharm Technology (Shanghai, China) for mRNA purification, library preparation, and sequencing. The data were analyzed on the free online platform of Majorbio Cloud Platform (URL: www.majorbio.com, accessed on 20 June 2020).

2.6. Oxidative Stress Analysis

Cells were collected with 300 μL Dulbecco's Phosphate-Buffered Saline (DPBS) and crushed 10 times continuously with ultrasonic cell crusher (60%) after treatment. Then, the cell homogenate centrifuged at 12,000× g for 10 min at 4 °C to collect the supernatant. Total protein concentrations of parallel samples were measured using a BCA Protein Assay kit (Beyotime, Shanghai, China). Malondialdehyde (MDA) level, total antioxidant capacity (T-AOC), and SOD activity were measured with assay kits according to the manufacturer's protocol (Nanjing Jiancheng Bioengineering Institute, Nanjing, China). The production of cellular ROS was measured with a reactive oxygen species assay kit (Beyotime, Shanghai, China) according to the manufacturer's instructions. Briefly, the treated cells were incubated with DCF-DA (10 mM) for 30 min at 37 °C. Then, cells were observed with a fluorescence microscope.

2.7. Mitochondrial Membrane Potential Determination

Cells were incubated with JC-1 solution for 20 min at 37 °C in the murk environment and were aspirated with the staining solution. The fluorescence intensity of cells in each well was tested by an inverted fluorescence microscope.

2.8. Quantitative Real-Time PCR

Total RNA of cells was extracted with TRIzol reagent (Vazyme, Shanghai, China) according to the manufacturer's instructions. cDNA was synthesized using PrimeScript RT Master Mix (Vazyme, Shanghai, China). mRNA expression levels of genes were determined by quantitative real-time PCR (RT-qPCR) using SYBR Premix Ex Taq (Vazyme, Shanghai, China). Oligonucleotide primers of apoptosis-related genes (Caspase-3, Caspase-8, Caspase-9, Bax and Bcl-2) were designed based on databases of National Center for Biotechnology Information (NCBI) using Oligo (V 7.0) and synthesized by Sango Biotech Co., Ltd. (Shanghai, China). The relative mRNA expression levels of each target gene were calculated based on the expression of the housekeeping gene GAPDH using the $2^{-\Delta\Delta Ct}$ method [18]. Table 1 lists the primers of apoptosis-related genes used in this study.

2.9. Western Blotting

Total protein of cells was collected and determined using BCA protein detection kit (Beyotime, Shanghai, China) after treatment. Equal amounts of proteins were separated by sodium dodecyl sulfate polyacrylamide gel (gel concentration, 10%; intensity of electricity, 120 V; time, 60 min) electrophoresis (SDS-PAGE) and transferred onto the polyvinylidene difluoride (PVDF) membranes. The membranes were blocked with TBST buffer containing 5% skim milk powder for 1 h, followed by incubating with the primary antibodies (included anti-TRAF2 (concentration, 1:1000), anti-RelB (concentration, 1:3000), anti-p52 (concentration, 1:1000), anti-Nrf2 (concentration, 1:6000), anti-Keap-1 (concentration,1:5000), anti-HO-1 (concentration,1:3000), and anti-GAPDH (concentration, 1:8000)) at 4 °C overnight (Proteintech, Wuhan, China), and then further incubated with the secondary antibodies (Proteintech, Wuhan, China; concentration, 1:8000) for 1 h at room temperature. The immunoreactive bands were visualized by enhanced chemiluminescence and quantitatively analyzed by ImageJ2 software.

Table 1. The primers of apoptosis-related genes used in this study.

Gene	Forward Primer (5′-3′)	Reverse Primer (5′-3′)	Gene Accession Number
Caspase-3	CCAAAGATCATACATGGAAGCG	CTGAATGTTTCCCTGAGGTTTG	XM_054350958.1
Caspase-8	CAAACTTCACAGCATTAGGGAC	ATGTTACTGTGGTCCATGAGTT	NM_033355.4
Caspase-9	TCCAGGAAGGTTTGAGGACC	CCCTTTCACCGAAACAGCAT	XM_011542273.4
Bax	CCCGAGAGGTCTTTTTCCGAG	CCAGCCCATGATGGTTCTGAT	XM_047439168.1
Bcl-2	GACTTCGCCGAGATGTCCAG	GAACTCAAAGAAGGCCACAATC	XM_054318967.1
GAPDH	CTCTGCTCCTCCTGTTCGAC	TTAAAAGCAGCCCTGGTGAC	NM_001357943.2

2.10. Molecular Docking of CGA with the KELCH-like ECH-Associated Protein1 (Keap-1)-Nrf2 Complex

The structure data file (SDF) of CGA was downloaded from the PubChem website (Compound CID: 1794427) and converted to PDB format on OpenBabel 3.1.1. The structure of Keap-1-Nrf2 complex was downloaded from the RCSB Protein Data Bank (PDB ID: 2FLU). AutoDockTools1.5.7 software was used to modify the receptor protein, such as water removal and hydrogenation. Next, AutoDock Vina 1.1.2 was used to perform the molecular docking of the best ligand with each protein. Eventually, ligand binding flexibilities with the binding pocket residues were drawn using Discovery Studio 2019.

2.11. Statistical Analyses

All experiments were performed at least three times with similar results and the data obtained are expressed as the mean ± standard deviation (SD) unless otherwise specified. Analyses were performed using GraphPad Prism Version 7 software (GraphPad, San Diego, CA, USA). Statistical differences between control and the other groups were detected by one-way analysis of variance (ANOVA) followed by multiple comparisons. Differences were considered statistically significant when the p-values were $\leq$0.05.

3. Results

3.1. CGA Attenuated AFB1-Induced L-02 Cell Cytotoxicity

We first investigated the cytotoxic effect of AFB1 on L-02 cells. After a 24 h incubation, the viability of hepatocytes was significantly decreased with 8, 16, and 32 μM AFB1 treatment compared to the cells without AFB1 treatment. (Figure 1A). Therefore, 8 μM AFB1 was used in the following experiments. The cell viability in L-02 cells with the increasing concentrations of CGA is shown in Figure 1B. The cell viability after 6 h incubation was minorly variational when the concentration of CGA was lower than 160 μM. However, it was hazardous to L-02 cells when the concentration of CGA reached 320 μM. Consequently, 160 μM CGA was used in the following treatments. To investigate whether CGA could protect L-02 cells from AFB1-induced toxicity, the cells were pretreated with CGA for 6 h, and then exposed to AFB1 for 24 h. Remarkably, 160 μM CGA completely abolished the decrease in cell viability induced by 8 μM AFB1 (Figure 1C). Pretreatment of cells with CGA also reduced LDH release as well as intracellular levels of ALT compared with only AFB1-treated cells, indicating CGA could help maintain the cell membrane integrity (Figure 1D,E). These data demonstrate that CGA ameliorates the survival and function of L-02 cells with AFB1 exposure.

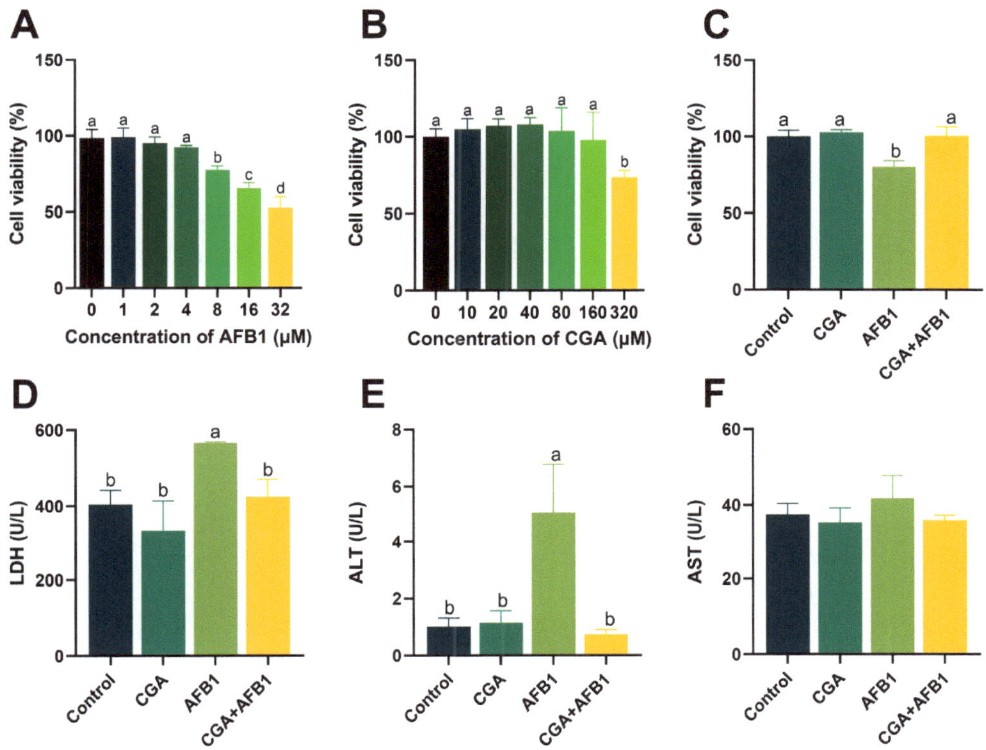

Figure 1. CGA attenuated AFB1-induced L-02 cell cytotoxicity. (**A**) the cell viability of L-02 cells after different concentrations of AFB1 incubations for 24 h, $n = 6$. (**B**) the cell viability of L-02 cells after different concentrations of CGA incubations for 6 h, $n = 6$. (**C**) effect of CGA on AFB1-induced cells viability in L-02 cells, $n = 6$. (**D**) the LDH concentration of L-02 cells in different treatments. (**E**) the ALT concentration of L-02 cells in different treatments. (**F**) the AST concentration of L-02 cells in different treatments. Note: No shoulder marker or same shoulder marker in column means the difference is not significant ($p > 0.05$), different shoulder markers mean the difference between groups was significant ($p \leq 0.05$).

3.2. RNA-Seq Analysis

To investigate the protective mechanisms of CGA on AFB1-induced L-02 cells, the RNA-seq was used to determine the whole gene expression pattern of L-02 cells. After AFB1 was treated, a total of 1716 genes were changed, including 1172 upregulated genes and 544 downregulated genes compared to the control group (Figure 2A,C). There were 28 upregulated and 117 downregulated genes in the CGA + AFB1 group compared with the AFB1 group (Figure 2B,C). However, there were minor differences in genes between the control group and CGA group at a fold change of 1.8 per gene ($p \leq 0.05$). According to the GO enrichment analysis (Figure 2D,E), the DEGs were classified into three main categories: biological process, cellular component, and molecular function. The most obvious difference was found in the biological process category and the mainly changed GO terms included "intrinsic apoptotic signaling pathway in response to DNA damage by p53 class mediator", "apoptotic signaling pathway", "molecular function regulator", and "inflammatory response". Among these changes, we summarized the functions of the main changes as "oxidative stress", "inflammatory response", and "apoptosis" (Figure 2F).

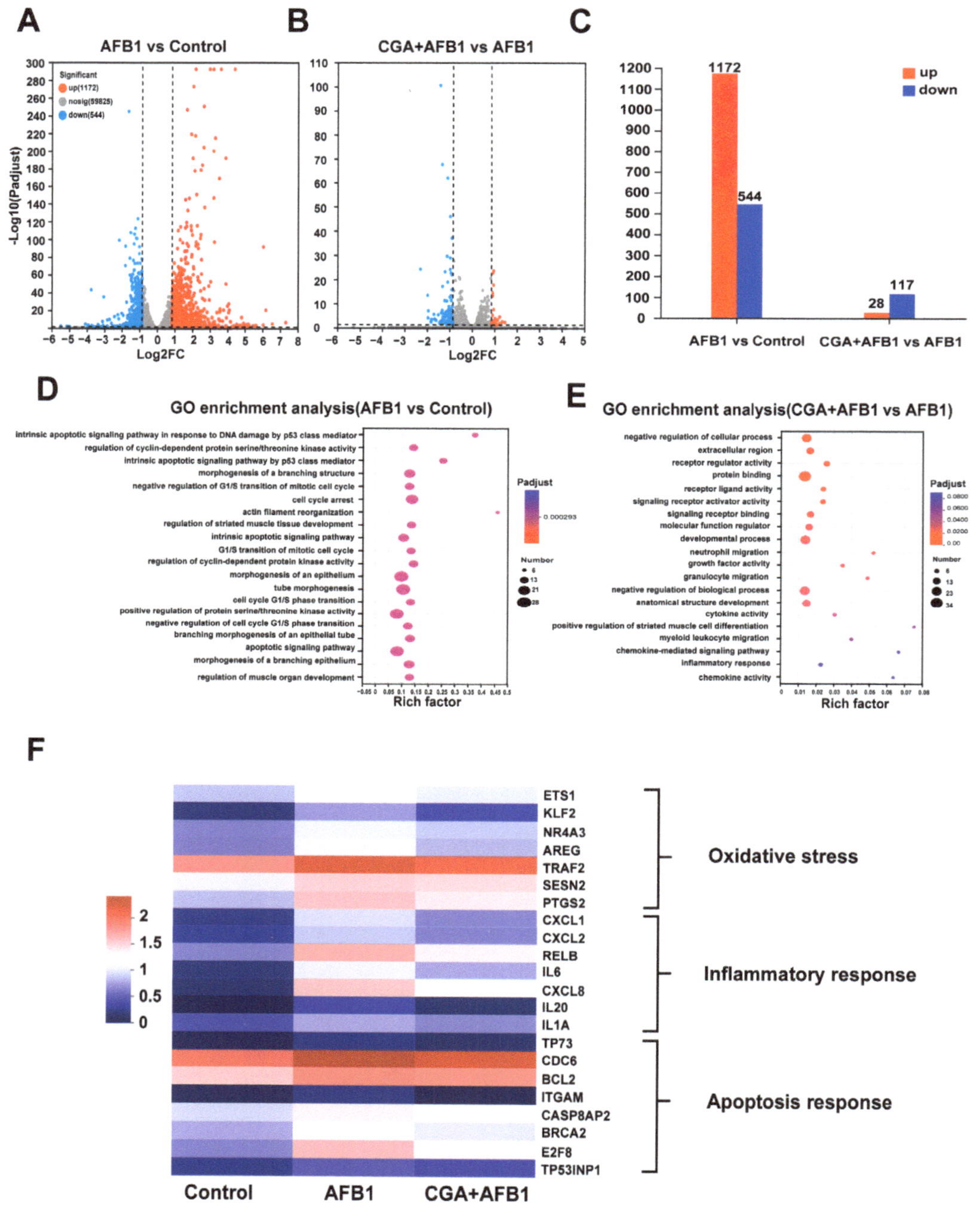

Figure 2. The RNA-seq of L-02 cells was used to determine the whole gene expression pattern. (**A**) the volcano plot shows the distribution of genes and the results of significant differences in genes in AFB1 group vs. control group. (**B**) the volcano plot shows the distribution of genes and the results of significant differences in genes in CGA + AFB1 group vs. AFB1 group. (**C**) the genes of up-regulation and down-regulation in this study, $p \leq 0.05$. (**D**) the GO enrichment analysis of AFB1 group vs. control group. (**E**) the GO enrichment analysis of CGA + AFB1 group vs. AFB1 group. (**F**) heat map of the representative function.

3.3. *CGA Alleviated AFB1-Induced Oxidative Damage by Activating Nrf2/Heme Oxygenase-1 (HO-1) Signaling Pathway in L-02 Cells*

To investigate the effect of CGA on the oxidative stress reaction to cells induced by AFB1, the ROS levels of L-02 cells were measured in using a fluorescent probe. Imaging indicated that the percentage of DCF positive cells increased after AFB1 treatment (Figure 4A). Interestingly, the AFB1-induced enhancement in ROS content was remarkedly inhibited by CGA pretreatment. The high production of ROS is usually accompanied by a decrease in enzyme activity and membrane fluidity [19]. In the present study, cells in AFB1 group exhibited lower activities of T-AOC and SOD as well as higher concentrations of MDA ($p \leq 0.05$) compared with control cells. However, pretreatment with CGA in AFB1-treated L-02 cells markedly promoted the activities of T-AOC and SOD but decreased the level of MDA compared with AFB1-treated L-02 cells ($p \leq 0.05$) (Figure 4B–D). Thus, these results indicate that CGA could alleviate AFB1-induced oxidative damage to L-02 cells. Increasing evidence has indicated that the Nrf2/HO-1 signaling pathway plays a pivotal role in the protection of oxidative stress [20]. In order to explore the protective mechanisms of CGA on AFB1-induced oxidative damage to L-02 cells, a computational molecular docking analysis was performed in this work to evaluate whether there is any affinity between CGA and the Keap-1-Nrf2 complex. The result showed that there were high-affinity (9.4 kcal/mol) hydrogen binding events between the residues of ARG415, VAL512, VAL418, VAL465, ILE559, VAL606, and GLY367 in the Keap-1-Nrf2 complex and CGA (Figure 4G), which meant that CGA interacted with the Keap-1-Nrf2 complex. Moreover, the levels of Nrf2 and HO-1 were detected using Western blot analysis to further confirm the effects of CGA in the Nrf2/HO-1 signaling pathway. The results revealed that pretreatment with CGA significantly increased the protein expression levels of HO-1 and nuclear Nrf2 in L-02 cells compared with AFB1 treatment, without effects on the expression levels of Keap-1 and total Nrf2 (Figure 4H–M). Therefore, it indicated that CGA probably alleviated AFB1-induced oxidative damage by interacting with the Keap-1-Nrf2 complex, promoting nuclear translocation of Nrf2 and upregulating the expression of HO-1.

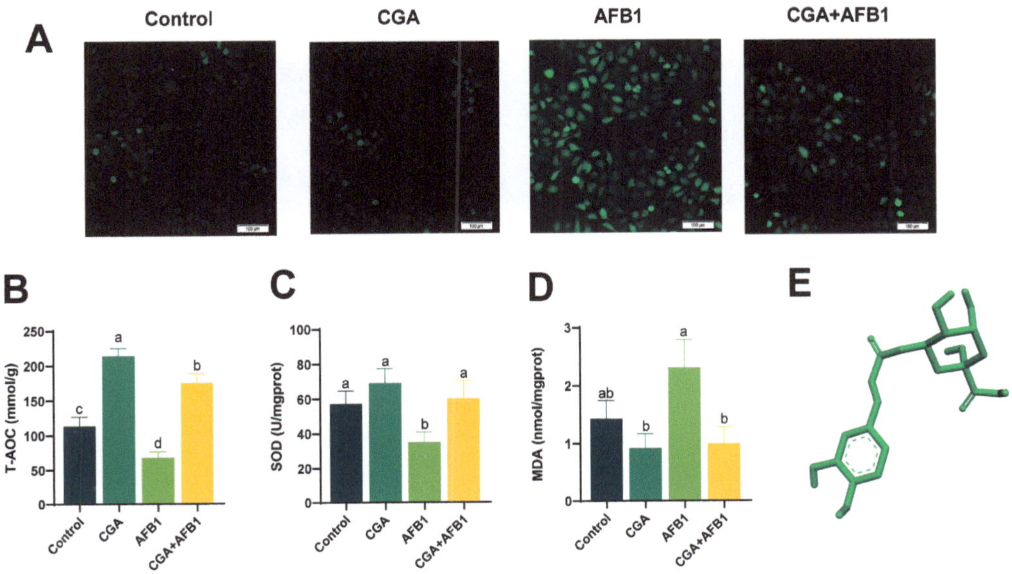

Figure 3. *Cont.*

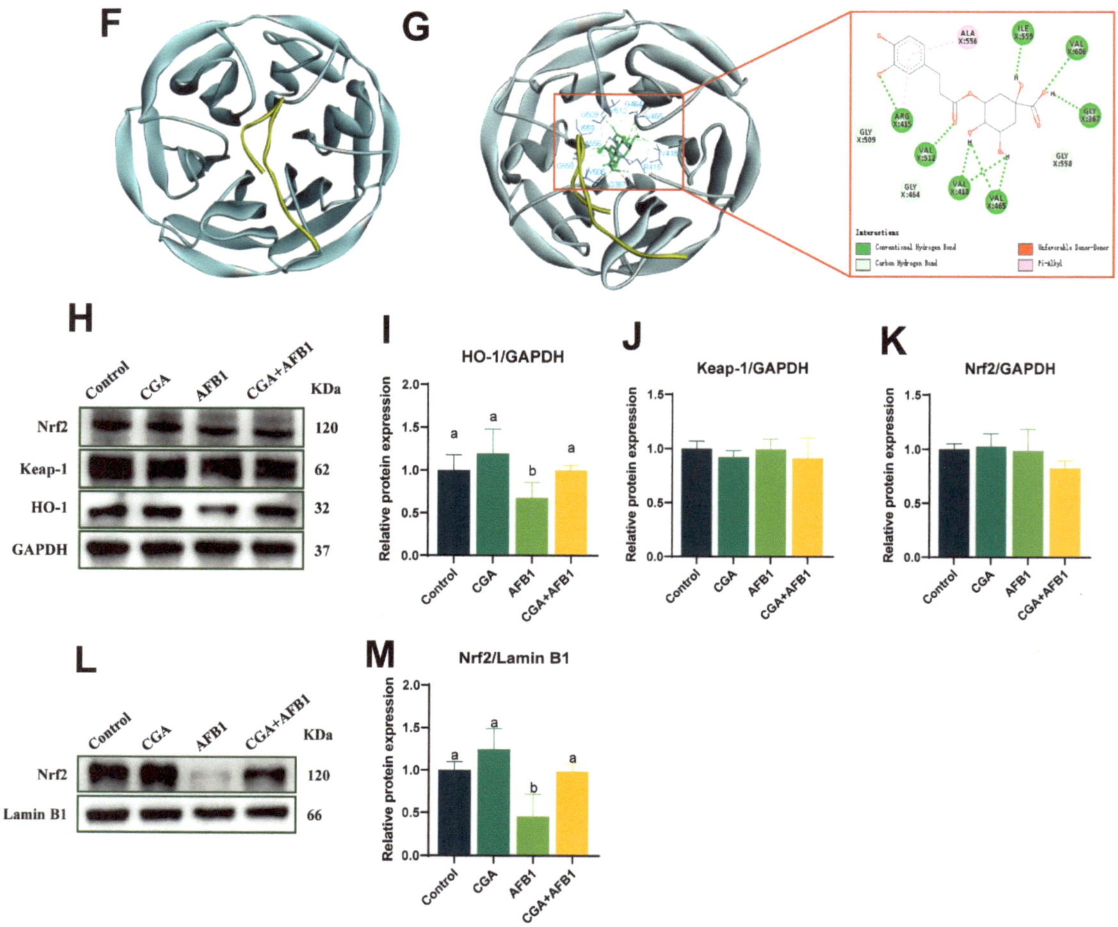

Figure 3. CGA alleviated AFB1-induced oxidative damage by activating Nrf2/HO-1 signaling pathway in L-02 cells. (**A**) effects of CGA on the ROS production of AFB1-induced L-02 cells. (**B**) the T-AOC activity of L-02 cells in different treatments. (**C**) the SOD activity of L-02 cells in different treatments. (**D**) the MDA concentration of L-02 cells in different treatments. (**E**) the modular structure of CGA. (**F**) the modular structure of Keap-1-Nrf2 complex. (**G**) the high interaction energy (−9.4 kcal/mol) between CGA and Keap-1-Nrf2 complex. (**H–K**) Western blots and quantification of total Nrf2, Keap-1, and HO-1. (**L**,**M**) Western blots and quantification of Nrf2 in the nuclear. Note: No shoulder marker or same shoulder marker in the same row means the difference is not significant ($p > 0.05$), different shoulder markers mean the difference between groups was significant ($p \leq 0.05$).

3.4. CGA Alleviated AFB1-Induced Inflammatory Response by Preventing Noncanonical Nuclear Factor Kappa-B (NF-κB) Pathway in L-02 Cells

Inflammatory response is an important mechanism in the toxic effects of AFB1. In this research, clustering analysis showed that AFB1 notably elevated expressions of inflammation-related genes in L-02 cells, which were significantly suppressed by CGA pretreatment. In addition, noncanonical NF-κB signaling pathway-related genes including *TRAF2*, *RelB*, and *p52* were significantly increased in the AFB1 group compared with those in the control group. However, these changes were substantially recovered by CGA pretreatment (Figure 4A). The gene set enrichment analysis (GSEA) also supported that AFB1 exposure significantly activated the NF-κB complex, while it was inhibited by pretreatment with CGA (Figure 4B,C). RT-qPCR results revealed that AFB1 exposure caused up-regulations on the mRNA expressions of *IL-6*, *IL-8*, and *TNF-α* in L-02 cells, whereas the mRNA abundances of these genes were down-regulated in AFB1-treated cells with CGA pretreatment ($p \leq 0.05$) (Figure 4D–F). KEGG pathway enrichment analysis identified the pathway connected with "NF-κB signaling pathway" as well (Figure 4G,H). Based on this, we speculated that CGA may regulate AFB1-induced inflammatory response via the noncanonical NF-κB signaling pathway. Therefore, the expression levels of TRAF2, RelB, and p52 were detected using Western blot analysis to verify the bioinformatics analysis results. The results revealed that the expression levels of TRAF2, RelB, and p52 were significantly increased in the AFB1 group, but these changes were effectively alleviated in AFB1-treated cells with CGA pretreatment (Figure 4I–L). Collectively, these results suggested that CGA could prevent noncanonical NF-κB pathway activation induced by AFB1 in L-02 cells.

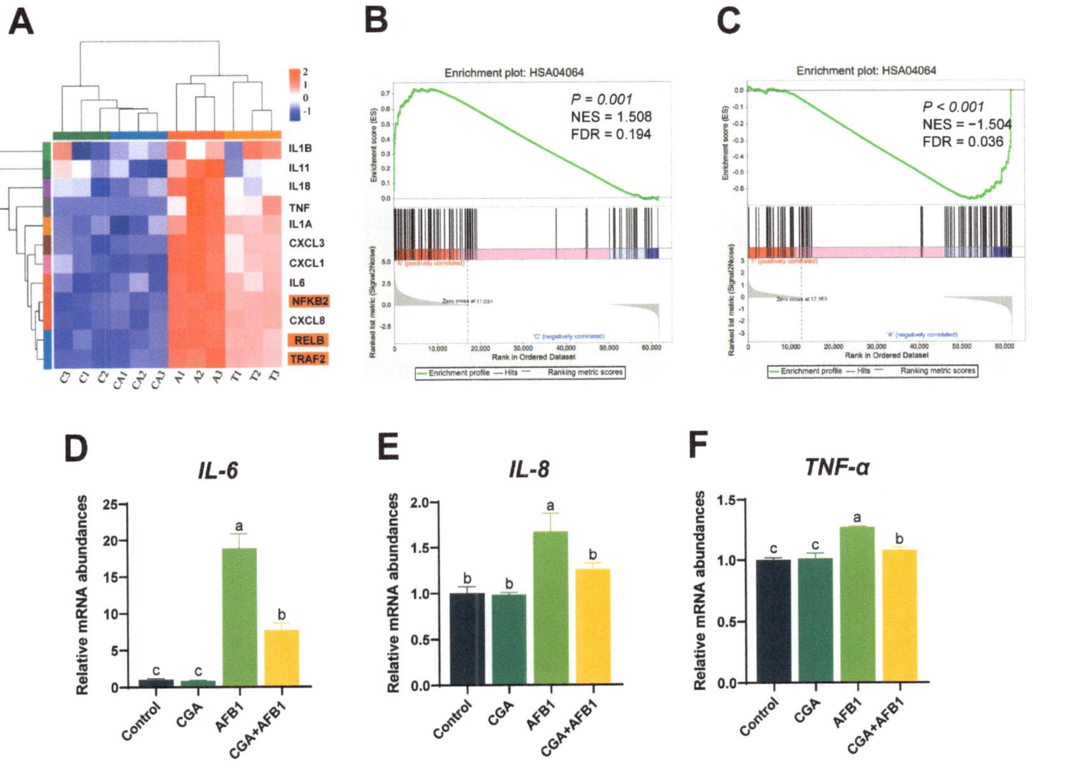

Figure 4. *Cont.*

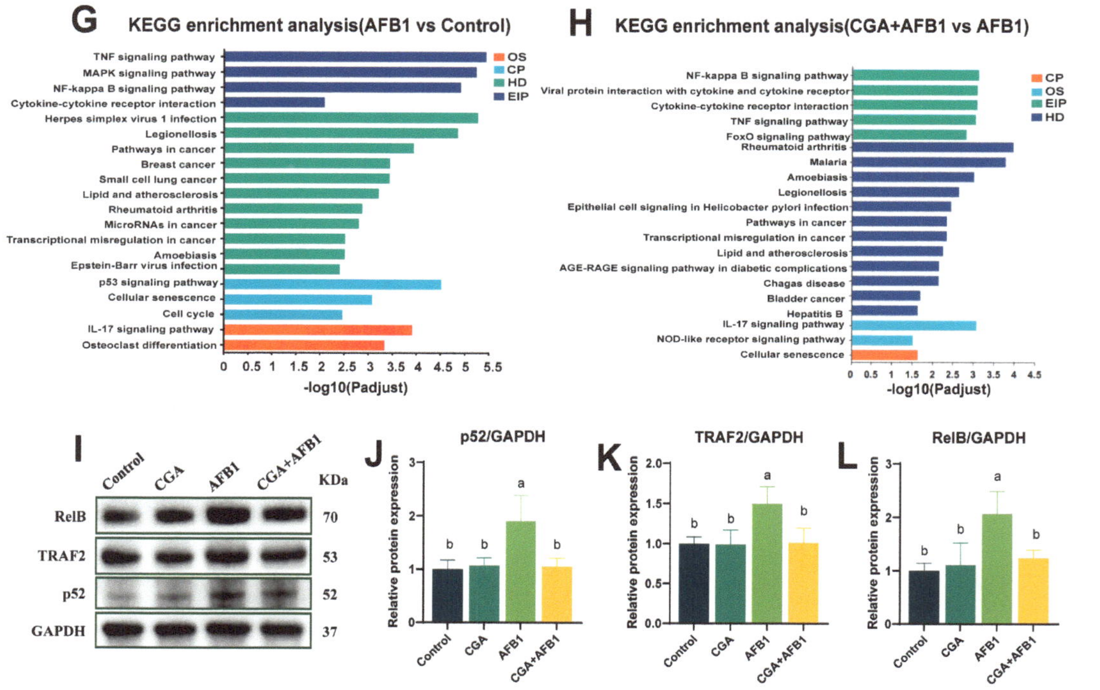

Figure 4. CGA prevented NF-κB pathway activation induced by AFB1 in L-02 cells. (**A**) clustering analysis of inflammation-related genes. (**B**) GSEA plot of a significant gene set associated with NF-κB-mediated signaling (Control vs. AFB1). FDR, false discovery rate; NES, normalized enrichment score. (**C**) GSEA plot of a significant gene set associated with NF-κB-mediated signaling (AFB1 vs. CGA + AFB1). (**D–F**) relative mRNA expressions of *IL-6*, *IL-8*, and *TNF-α* in L-02 cells. (**G**) KEGG pathway (AFB1 vs. Control) enrichment. "NF-κB signaling pathway" is shown with shadow red. (**H**) KEGG pathway (CGA + AFB1 vs. AFB1) enrichment. "NF-κB signaling pathway" is shown with shadow red. (**I–L**) Western blots and quantification of RelB, TRAF2, and p52. Note: No shoulder marker or same shoulder marker in the same row means the difference is not significant ($p > 0.05$), different shoulder markers mean the difference between groups was significant ($p \leq 0.05$).

3.5. CGA Alleviated AFB1-Induced Apoptosis in L-02 Cells

Alteration of mitochondrial membrane potential (MMP) is an important link in oxidative stress-induced apoptosis [21]. When MMP is higher than the cell membrane spot, the probe aggregates at the cell membrane and emits red fluorescence. However, the probe excites green fluorescence when the membrane potential drops. We observed that green fluorescence in AFB1-treated cells was increased, suggesting that AFB1 treatment triggered mitochondrial damage. After pretreatment with CGA, the fluorescence of AFB1-treated cells was nearly recovered to the normal group (Figure 5A). In addition, apoptosis is usually associated with proapoptotic and antiapoptotic proteins of the Bcl-2 family, initiator caspases (Caspase-8, -9, and -10), and effector caspases (Caspase-3, -6, and -7) [22]. In our study, L-02 cells treated with AFB1 increased the mRNA expressions of *Caspase-3*, *Caspase-8*, *Bax*, and *Bax/Bcl-2* compared with control cells. Interestingly, pretreatment with CGA in AFB1-treated L-02 cells decreased the mRNA expressions of *Caspase-3*, *Caspase-8*, *Bax*, and *Bax/Bcl-2* compared with the AFB1-treated cells (Figure 5B–G). These data indicated that CGA could alleviate AFB1-induced apoptosis by maintaining MMP and inhibiting the expressions of apoptosis-related genes in L-02 cells.

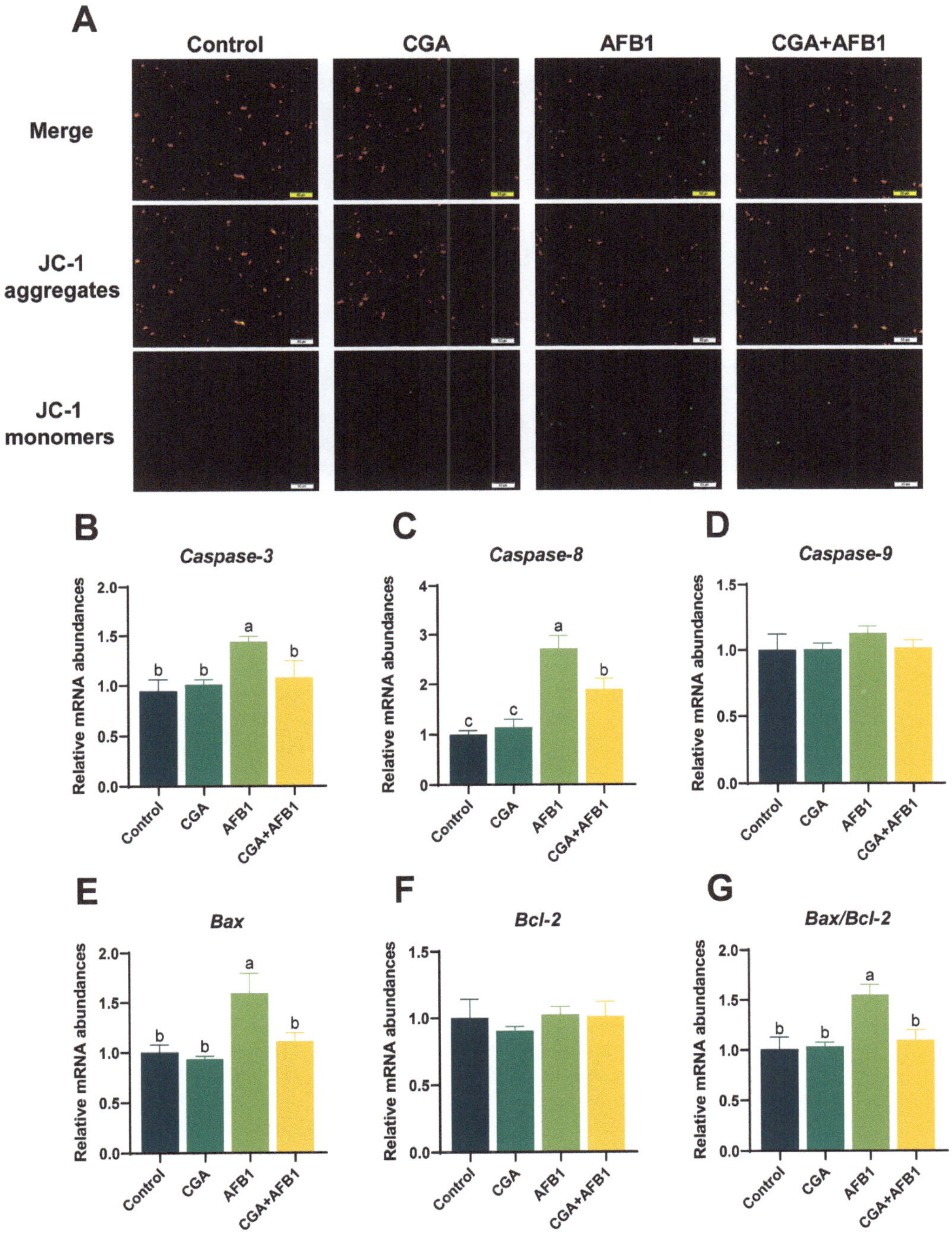

Figure 5. CGA alleviated AFB1-induced apoptosis in L-02 cells. (**A**) effects of CGA pretreatment on the mitochondrial membrane potential in AFB1-induced L-02 cells. (**B–G**) relative mRNA expressions of *Caspase-3*, *Caspase-8*, *Caspase-9*, *Bax*, *Bcl-2*, and *Bax/Bcl-2* in L-02 cells. Note: No shoulder marker or same shoulder marker in the same row means the difference is not significant ($p > 0.05$), different shoulder markers mean the difference between groups was significant ($p \leq 0.05$).

4. Discussion

AFB1, a kind of mycotoxin produced by fungi, is always detected from grains such as groundnuts and maize without suitable storage [23]. This toxic metabolism is recognized as the most hazardous because it imposes acute or chronic toxicities on humans and thus causes great public health concerns [24,25]. It has been reported that AFB1 has a variety of adverse health effects such as immunotoxicity, neurotoxicity, and hepatotoxicity, causing childhood stunting, liver and renal injury, immunosuppression, and even cancer [26,27]. Several studies have shown that AFB1-exo-8,9-epoxide (AFBO), a metabolite of AFB1 in the liver, could induce mutations in the oncogene *p53*, which is associated with cell cycle, DNA repair, apoptosis, and autophagy in human hepatocytes [28,29]. Hence, how to alleviate AFB1-induced hepatocyte injury remains to be further studied. CGA, a kind of dietary polyphenol, shows powerful antioxidant and anti-inflammatory effects on animals' bodies [30]. Moreover, numerous studies have confirmed that CGA could protect cells against oxidative stress by activating the cellular antioxidant defense system and counterbalancing the production of ROS. However, the protective mechanism of CGA on AFB1-induced cytotoxicity has not yet been clearly elucidated. Therefore, the effect and related mechanism of CGA on AFB1-induced hepatic cytotoxicity were explored in this study. We found that CGA addition could alleviate AFB1-induced cell damage with the increase in cell viability and the decrease in LDH release, confirming the protective effect of this polyphenol on cell membrane integrity. This is in accordance with a previous study which revealed that CGA addition could alleviate cytotoxicity of DON in IPEC-J2 cells with a decrease in LDH release [31].

To further explore the protective mechanism of CGA treatment on the AFB1-treated L-02 cells, RNA-seq and bioinformatics analyses were carried out in this work. GO analysis showed that CGA alleviated AFB1-induced cell damage by regulating oxidative stress, apoptosis, and inflammatory response. The imbalance between oxidation and anti-oxidation is one of the landmark events in the development of AFB1-induced hepatotoxicity. Oxidative stress is mainly manifested as excessive production of ROS and a decrease in the concentrations of antioxidant enzymes and an increase in the concentration of MDA [32]. Several studies have revealed that the mitigation of oxidative stress is conducive to curing damaged liver cells and a variety of therapeutic agents have significant antioxidant effects [33,34]. Cho et al. (2017) suggested that CGA ameliorated the incidence of photo-aging in UV radiation–exposed fibroblasts through decreasing the ROS production in mitochondria [35]. CGA has also been shown to attenuate LPS-induced oxidative stress through increasing the activities of GSH and SOD, and the GSH/GSSG ratio [36]. Accordingly, CGA could alleviate AFB1-induced oxidative injury in L-02 cells by directly neutralizing ROS production and indirectly modulating the activities of the antioxidant enzymes.

Nrf2 is the key regulator of intracellular ROS and proinflammatory cytokine expression [37]. In the physiological state, Nrf2 is ubiquitinated through binding with Keap-1 in the cytoplasm. In the presence of oxidative stresses, Nrf2 then translocates into the nucleus and forms a coactivator complex to promote transcription and mRNA expressions of antioxidant and detoxifying genes, including *HO-1*, *GSH*, *NQO1*, and *GCLC*. Under homeostatic conditions, *HO-1* expression is low or absent in most cells. However, it is highly upregulated by most cells in response to a vast number of pro-oxidant stimuli and provides protection against oxidative damage [38]. A previous study reported that CGA reduced radiation-induced apoptosis and DNA damage via Nrf2 activation in hepatocellular carcinoma [39]. CGA has also been shown to prevent diabetic nephropathy by suppressing oxidative stress and inflammation through activation of the Nrf2/HO-1 signaling pathway and inhibition of the NF-κB signaling pathway [40]. Additionally, CGA has been found to prevent acute liver injury via the upregulation of Nrf2 and the inhibition of NLRP3 inflammasome activation [41]. In the present study, it was found that there were high-affinity (9.4 kcal/mol) hydrogen binding events between the Keap-1-Nrf2 complex and CGA. CGA interacting with Keap-1 might lead to the release of Nrf2 from the Keap-1-Nrf2 complex, which is essential for Nrf2 activation. In addition, the protein levels of Nrf2 in nucleus and its corresponding downstream HO-1 were significantly increased after CGA treatment, suggesting that the protective role of CGA in AFB1-induced oxidative injury may be in part related to its activation of the Nrf2/HO-1 pathway through facilitating Nrf2 nuclear translocation and up-regulating the expression of HO-1.

Inflammation is always considered to interact with oxidative stress. Inflammation increases oxidative damage, which in turn aggravates inflammation [42]. Numerous studies have indicated that over-expressions of cytokines such as IL-1, IL-6, and TNF-α may promote liver inflammation during chronic liver injury [43]. In this study, AFB1 significantly activated the relative mRNA abundances of *IL-6*, *IL-8*, and *TNF-α* in L-02 cells, which was in agreement with the previous study [44]. However, CGA addition was able to remit cell inflammation caused by AFB1 exposure, proving CGA possess anti-inflammatory functions. Considerable evidence suggested that CGA can regulate the inflammation caused by some factors. In the model of mouse inflammatory bowel diseases (IBD), CGA showed a dose-dependent decrease in expression and nuclear translocation of the NF-κB subunit, which was accompanied by the suppression of systemic inflammation in mice with colitis [45]. CGA was also reported to have a protective effect on myocardial infarction by reducing the inflammatory response and exerting antioxidant activity [46]. Another study indicated that CGA exerted protective effects against hypoxanthine and potassium oxonate-induced hyperuricemia in mice, which was attributed to its ability to suppress NF-κB activation [47]. In this study, GSEA profiling showed that NF-κB signatures were remarkably suppressed by CGA treatment. KEGG pathway enrichment analysis also identified that the related pathways connected with the "NF-κB signaling pathway". Precise regulation of NF-κB signaling is crucial for regulating the expressions of many inflammatory factors and it is activated through the classical and noncanonical pathways [48]. Therein, the non-canonical NF-κB signaling cascade relies on the inducible processing of the NF-κB2 precursor protein (p100) and mediates the activation of the p52/RelB NF-κB complex [49]. Moreover, during the inflammatory response, the activation of NF-κB signaling pathways can induce the release of a variety of pro-inflammatory cytokines [50]. In the present study, the expression levels of TRAF2, RelB, and p52 were significantly increased in the AFB1 group compared with those in the normal group, while CGA supplementation reversed those changes. Collectively, these results suggested that CGA could alleviate AFB1-induced inflammatory response in L-02 cells by inhibiting the noncanonical NF-κB signaling pathway.

It has been shown that mitochondria are the primary site to produce ROS and excessive ROS accumulation could disrupt the MMP, leading to mitochondrial dysfunction, thus triggering apoptosis [51]. Polyphenol has been, in many contexts, shown to directly affect mitochondrial activity and maintain MMP [52]. A previous study reported that CGA pretreatment improved $H_2O_2^{\bullet-}$ induced intestinal mitochondrial damage, increased ROS production, and decreased cytochrome C release [53]. In line with this outcome, our study found that CGA can improve AFB1-induced hepatic mitochondrial damage and protect the integrity of mitochondrion effectively. Moreover, it has been reported that mitochondria-dependent apoptosis is usually associated with caspase activation [54]. Increased mRNA expressions of *Caspase-3*, *Caspase-8*, and *Bax* have been observed in response to AFB1 treatment, which is consistent with a previous study suggesting that AFB1 exposure induced ROS-dependent caspase-mediated apoptosis in normal human cells [55]. Interestingly, the pretreatment of CGA markedly decreased the mRNA expressions of *Caspase-3*, *Caspase-8*, *Bax*, and *Bax/Bcl-2* in AFB1-treated cells. Consequently, these results suggested that CGA might alleviate AFB1-induced apoptosis by maintaining MMP of cells and inhibiting the expressions of apoptosis-related genes (Figure 6).

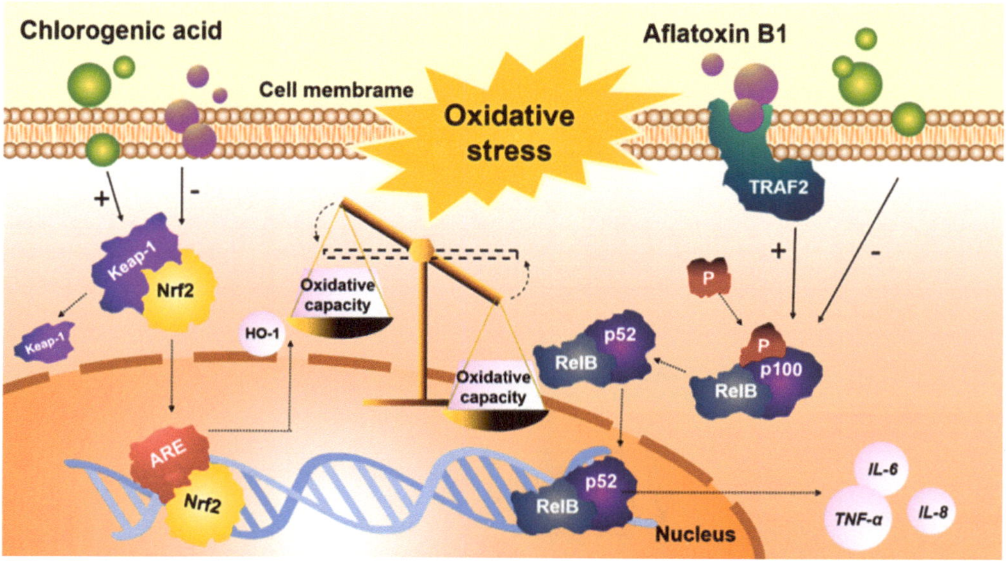

Figure 6. Schematic diagram of the protective effects of CGA on AFB1-induced cytotoxicity in L-02 cells.

5. Conclusions

This study demonstrated that AFB1 exposure induced oxidative stress, inflammatory response, and apoptosis, leading to a decrease in cell viability eventually. Moreover, CGA prevented AFB1-induced oxidative damage by neutralizing ROS production and modulating anti-oxidative enzyme activity, as well as activating the Nrf2/HO-1 signaling pathway. CGA alleviated AFB1-induced inflammatory response by inhibiting the noncanonical NF-κB signaling pathway and CGA alleviated AFB1-induced apoptosis by maintaining MMP and inhibiting the expressions of apoptosis-related genes, ultimately restoring cell viability.

Author Contributions: Q.W.: Conceptualization, Resources, Writing—original draft. T.L.: Data curation, Writing—review and editing, Visualization. M.K.: Conceptualization, Methodology, Writing—review and editing, Validation. Y.W.: Methodology, Software. Y.F.: Writing—review and editing. M.M.: Writing—review and editing. Q.M.: Writing—review and editing, Methodology. L.Z.: Conceptualization, Methodology, Writing—review and editing, Supervision, Funding acquisition. All authors have read and agreed to the published version of the manuscript.

Funding: This work was supported by the National Natural Science Foundation of China (Grant No. 31972604), the National Key Research and Development Program of China (Program No. 2021YFC2103003), and the Youth Elite Development Program (Grade A) of the College of Animal Science and Technology, China Agricultural University.

Institutional Review Board Statement: Not applicable.

Informed Consent Statement: Not applicable.

Data Availability Statement: Data are contained within the article. The transcriptomic data have been deposited into the NCBI sequence Read (SRA) with the dataset identifier SRP470990.

Conflicts of Interest: The authors declare that they have no known competing financial interests or personal relationships that could have appeared to influence the work reported in this paper.

References

1. Song, C.; Yang, J.; Wang, Y.; Ding, G.; Guo, L.; Qin, J. Mechanisms and transformed products of aflatoxin B1 degradation under multiple treatments: A review. *Crit. Rev. Food Sci. Nutr.* **2022**, 1–13. [CrossRef]
2. Zhu, Y.; Xu, Y.; Yang, Q. Antifungal properties and AFB(1) detoxification activity of a new strain of Lactobacillus plantarum. *J. Hazard. Mater.* **2021**, *414*, 125569. [CrossRef]
3. Abdel-Daim, M.M.; Abdeen, A.; Jalouli, M.; Abdelkader, A.; Megahed, A.; Alkahtane, A.; Almeer, R.; Alhoshani, N.M.; Al-Johani, N.S.; Alkahtani, S.; et al. Fucoidan supplementation modulates hepato-renal oxidative stress and DNA damage induced by aflatoxin B1 intoxication in rats. *Sci. Total Environ.* **2021**, *768*, 144781. [CrossRef]
4. Bodega, G.; Alique, M.; Puebla, L.; Carracedo, J.; Ramirez, R.M. Microvesicles: ROS scavengers and ROS producers. *J. Extracell. Vesicles* **2019**, *8*, 1626654. [CrossRef]
5. Liu, Y.; Wang, W. Aflatoxin B1 impairs mitochondrial functions, activates ROS generation, induces apoptosis and involves Nrf2 signal pathway in primary broiler hepatocytes. *Anim. Sci. J.* **2016**, *87*, 1490–1500. [CrossRef]
6. Cheng, L.; Qin, Y.; Hu, X.; Ren, L.; Zhang, C.; Wang, X.; Wang, W.; Zhang, Z.; Hao, J.; Guo, M.; et al. Melatonin protects in vitro matured porcine oocytes from toxicity of Aflatoxin B1. *J. Pineal Res.* **2019**, *66*, e12543. [CrossRef]
7. Xu, Q.; Shi, W.; Lv, P.; Meng, W.; Mao, G.; Gong, C.; Chen, Y.; Wei, Y.; He, X.; Zhao, J.; et al. Critical role of caveolin-1 in aflatoxin B1-induced hepatotoxicity via the regulation of oxidation and autophagy. *Cell Death Dis.* **2020**, *11*, 6. [CrossRef]
8. Wu, B.; Mughal, M.J.; Fang, J.; Peng, X. The Protective Role of Selenium Against AFB(1)-Induced Liver Apoptosis by Death Receptor Pathway in Broilers. *Biol. Trace Elem. Res.* **2019**, *191*, 453–463. [CrossRef]
9. Abdel-Hamid, A.A.; Firgany Ael, D. Vitamin E supplementation ameliorates aflatoxin B1-induced nephrotoxicity in rats. *Acta Histochem.* **2015**, *117*, 767–779. [CrossRef]
10. Reddy, L.; Odhav, B.; Bhoola, K. Aflatoxin B1-induced toxicity in HepG2 cells inhibited by carotenoids: Morphology, apoptosis and DNA damage. *Biol. Chem.* **2006**, *387*, 87–93. [CrossRef]
11. Choi, K.C.; Chung, W.T.; Kwon, J.K.; Yu, J.Y.; Jang, Y.S.; Park, S.M.; Lee, S.Y.; Lee, J.C. Inhibitory effects of quercetin on aflatoxin B1-induced hepatic damage in mice. *Food Chem. Toxicol.* **2010**, *48*, 2747–2753. [CrossRef]
12. Li, Y.; Shi, W.; Li, Y.; Zhou, Y.; Hu, X.; Song, C.; Ma, H.; Wang, C.; Li, Y. Neuroprotective effects of chlorogenic acid against apoptosis of PC12 cells induced by methylmercury. *Environ. Toxicol. Pharmacol.* **2008**, *26*, 13–21. [CrossRef]
13. Ali, N.; Rashid, S.; Nafees, S.; Hasan, S.K.; Shahid, A.; Majed, F.; Sultana, S. Protective effect of Chlorogenic acid against methotrexate induced oxidative stress, inflammation and apoptosis in rat liver: An experimental approach. *Chem. Biol. Interact.* **2017**, *272*, 80–91. [CrossRef]
14. Liu, Y.J.; Zhou, C.Y.; Qiu, C.H.; Lu, X.M.; Wang, Y.T. Chlorogenic acid induced apoptosis and inhibition of proliferation in human acute promyelocytic leukemia HL-60 cells. *Mol. Med. Rep.* **2013**, *8*, 1106–1110. [CrossRef]
15. Gong, W.; Li, J.; Zhu, G.; Wang, Y.; Zheng, G.; Kan, Q. Chlorogenic acid relieved oxidative stress injury in retinal ganglion cells through IncRNA-TUG1/Nrf2. *Cell Cycle* **2019**, *18*, 1549–1559. [CrossRef]
16. Buko, V.; Zavodnik, I.; Budryn, G.; Zaklos-Szyda, M.; Belonovskaya, E.; Kirko, S.; Zyzelewicz, D.; Zakrzeska, A.; Bakunovich, A.; Rusin, V.; et al. Chlorogenic Acid Protects against Advanced Alcoholic Steatohepatitis in Rats via Modulation of Redox Homeostasis, Inflammation, and Lipogenesis. *Nutrients* **2021**, *13*, 4155. [CrossRef]
17. Wang, X.; Fan, X.; Yuan, S.; Jiao, W.; Liu, B.; Cao, J.; Jiang, W. Chlorogenic acid protects against aluminium-induced cytotoxicity through chelation and antioxidant actions in primary hippocampal neuronal cells. *Food Funct.* **2017**, *8*, 2924–2934. [CrossRef]
18. Livak, K.J.; Schmittgen, T.D. Analysis of relative gene expression data using real-time quantitative PCR and the 2(-Delta Delta C(T)) Method. *Methods* **2001**, *25*, 402–408. [CrossRef]
19. Lennicke, C.; Cocheme, H.M. Redox metabolism: ROS as specific molecular regulators of cell signaling and function. *Mol. Cell* **2021**, *81*, 3691–3707. [CrossRef]
20. Zhang, Q.; Liu, J.; Duan, H.; Li, R.; Peng, W.; Wu, C. Activation of Nrf2/HO-1 signaling: An important molecular mechanism of herbal medicine in the treatment of atherosclerosis via the protection of vascular endothelial cells from oxidative stress. *J. Adv. Res.* **2021**, *34*, 43–63. [CrossRef]
21. Cohen, I.; Castedo, M.; Kroemer, G. Tantalizing Thanatos: Unexpected links in death pathways. *Trends Cell Biol.* **2002**, *12*, 293–295. [CrossRef]
22. Bertheloot, D.; Latz, E.; Franklin, B.S. Necroptosis, pyroptosis and apoptosis: An intricate game of cell death. *Cell Mol. Immunol.* **2021**, *18*, 1106–1121. [CrossRef]
23. Rushing, B.R.; Selim, M.I. Aflatoxin B1: A review on metabolism, toxicity, occurrence in food, occupational exposure, and detoxification methods. *Food Chem. Toxicol.* **2019**, *124*, 81–100. [CrossRef]
24. Rong, X.; Sun-Waterhouse, D.; Wang, D.; Jiang, Y.; Li, F.; Chen, Y.; Zhao, S.; Li, D. The Significance of Regulatory MicroRNAs: Their Roles in Toxicodynamics of Mycotoxins and in the Protection Offered by Dietary Therapeutics Against Mycotoxin-Induced Toxicity. *Compr. Rev. Food Sci. Food Saf.* **2019**, *18*, 48–66. [CrossRef]
25. Guo, Y.; Qin, X.; Tang, Y.; Ma, Q.; Zhang, J.; Zhao, L. CotA laccase, a novel aflatoxin oxidase from Bacillus licheniformis, transforms aflatoxin B(1) to aflatoxin Q(1) and epi-aflatoxin Q(1). *Food Chem.* **2020**, *325*, 126877. [CrossRef]
26. Guo, Y.; Zhao, L.; Ma, Q.; Ji, C. Novel strategies for degradation of aflatoxins in food and feed: A review. *Food Res. Int.* **2021**, *140*, 109878. [CrossRef]

27. Ma, J.; Liu, Y.; Guo, Y.; Ma, Q.; Ji, C.; Zhao, L. Transcriptional Profiling of Aflatoxin B1-Induced Oxidative Stress and Inflammatory Response in Macrophages. *Toxins* **2021**, *13*, 401. [CrossRef]
28. Goiran, T.; Duplan, E.; Rouland, L.; El Manaa, W.; Lauritzen, I.; Dunys, J.; You, H.; Checler, F.; Alves da Costa, C. Nuclear p53-mediated repression of autophagy involves PINK1 transcriptional down-regulation. *Cell Death Differ.* **2018**, *25*, 873–884. [CrossRef]
29. Mary, V.S.; Theumer, M.G.; Arias, S.L.; Rubinstein, H.R. Reactive oxygen species sources and biomolecular oxidative damage induced by aflatoxin B1 and fumonisin B1 in rat spleen mononuclear cells. *Toxicology* **2012**, *302*, 299–307. [CrossRef]
30. Upadhyay, R.; Mohan Rao, L.J. An outlook on chlorogenic acids-occurrence, chemistry, technology, and biological activities. *Crit. Rev. Food Sci. Nutr.* **2013**, *53*, 968–984. [CrossRef]
31. Xu, X.; Chang, J.; Wang, P.; Yin, Q.; Liu, C.; Li, M.; Song, A.; Zhu, Q.; Lu, F. Effect of chlorogenic acid on alleviating inflammation and apoptosis of IPEC-J2 cells induced by deoxyniyalenol. *Ecotoxicol. Environ. Saf.* **2020**, *205*, 111376. [CrossRef]
32. Schieber, M.; Chandel, N.S. ROS function in redox signaling and oxidative stress. *Curr. Biol.* **2014**, *24*, R453–R462. [CrossRef]
33. Li, S.; Muhammad, I.; Yu, H.; Sun, X.; Zhang, X. Detection of Aflatoxin adducts as potential markers and the role of curcumin in alleviating AFB1-induced liver damage in chickens. *Ecotoxicol. Environ. Saf.* **2019**, *176*, 137–145. [CrossRef]
34. Muhammad, I.; Wang, H.; Sun, X.; Wang, X.; Han, M.; Lu, Z.; Cheng, P.; Hussain, M.A.; Zhang, X. Dual Role of Dietary Curcumin Through Attenuating AFB(1)-Induced Oxidative Stress and Liver Injury via Modulating Liver Phase-I and Phase-II Enzymes Involved in AFB(1) Bioactivation and Detoxification. *Front. Pharmacol.* **2018**, *9*, 554. [CrossRef]
35. Cho, Y.H.; Bahuguna, A.; Kim, H.H.; Kim, D.I.; Kim, H.J.; Yu, J.M.; Jung, H.G.; Jang, J.Y.; Kwak, J.H.; Park, G.H.; et al. Potential effect of compounds isolated from Coffea arabica against UV-B induced skin damage by protecting fibroblast cells. *J. Photochem. Photobiol. B* **2017**, *174*, 323–332. [CrossRef]
36. Zhao, X.L.; Yu, L.; Zhang, S.D.; Ping, K.; Ni, H.Y.; Qin, X.Y.; Zhao, C.J.; Wang, W.; Efferth, T.; Fu, Y.J. Cryptochlorogenic acid attenuates LPS-induced inflammatory response and oxidative stress via upregulation of the Nrf2/HO-1 signaling pathway in RAW 264.7 macrophages. *Int. Immunopharmacol.* **2020**, *83*, 106436. [CrossRef]
37. Kobayashi, E.H.; Suzuki, T.; Funayama, R.; Nagashima, T.; Hayashi, M.; Sekine, H.; Tanaka, N.; Moriguchi, T.; Motohashi, H.; Nakayama, K.; et al. Nrf2 suppresses macrophage inflammatory response by blocking proinflammatory cytokine transcription. *Nat. Commun.* **2016**, *7*, 11624. [CrossRef]
38. Campbell, N.K.; Fitzgerald, H.K.; Dunne, A. Regulation of inflammation by the antioxidant haem oxygenase 1. *Nat. Rev. Immunol.* **2021**, *21*, 411–425. [CrossRef]
39. Yin, X.; He, X.; Wu, L.; Yan, D.; Yan, S. Chlorogenic Acid, the Main Antioxidant in Coffee, Reduces Radiation-Induced Apoptosis and DNA Damage via NF-E2-Related Factor 2 (Nrf2) Activation in Hepatocellular Carcinoma. *Oxid. Med. Cell Longev.* **2022**, *2022*, 4566949. [CrossRef]
40. Bao, L.; Li, J.; Zha, D.; Zhang, L.; Gao, P.; Yao, T.; Wu, X. Chlorogenic acid prevents diabetic nephropathy by inhibiting oxidative stress and inflammation through modulation of the Nrf2/HO-1 and NF-kB pathways. *Int. Immunopharmacol.* **2018**, *54*, 245–253. [CrossRef]
41. Shi, A.; Shi, H.; Wang, Y.; Liu, X.; Cheng, Y.; Li, H.; Zhao, H.; Wang, S.; Dong, L. Activation of Nrf2 pathway and inhibition of NLRP3 inflammasome activation contribute to the protective effect of chlorogenic acid on acute liver injury. *Int. Immunopharmacol.* **2018**, *54*, 125–130. [CrossRef]
42. McGarry, T.; Biniecka, M.; Veale, D.J.; Fearon, U. Hypoxia, oxidative stress and inflammation. *Free Radic. Biol. Med.* **2018**, *125*, 15–24. [CrossRef]
43. He, Y.; Hwang, S.; Ahmed, Y.A.; Feng, D.; Li, N.; Ribeiro, M.; Lafdil, F.; Kisseleva, T.; Szabo, G.; Gao, B. Immunopathobiology and therapeutic targets related to cytokines in liver diseases. *Cell Mol. Immunol.* **2021**, *18*, 18–37. [CrossRef]
44. Wu, K.; Jia, S.; Xue, D.; Rajput, S.A.; Liu, M.; Qi, D.; Wang, S. Dual effects of zearalenone on aflatoxin B1-induced liver and mammary gland toxicity in pregnant and lactating rats. *Ecotoxicol. Environ. Saf.* **2022**, *245*, 114115. [CrossRef]
45. Vukelic, I.; Detel, D.; Pucar, L.B.; Potocnjak, I.; Buljevic, S.; Domitrovic, R. Chlorogenic acid ameliorates experimental colitis in mice by suppressing signaling pathways involved in inflammatory response and apoptosis. *Food Chem. Toxicol.* **2018**, *121*, 140–150. [CrossRef]
46. Wang, D.; Tian, L.; Lv, H.; Pang, Z.; Li, D.; Yao, Z.; Wang, S. Chlorogenic acid prevents acute myocardial infarction in rats by reducing inflammatory damage and oxidative stress. *Biomed. Pharmacother.* **2020**, *132*, 110773. [CrossRef]
47. Zhou, X.; Zhang, B.; Zhao, X.; Lin, Y.; Wang, J.; Wang, X.; Hu, N.; Wang, S. Chlorogenic acid supplementation ameliorates hyperuricemia, relieves renal inflammation, and modulates intestinal homeostasis. *Food Funct.* **2021**, *12*, 5637–5649. [CrossRef]
48. Gray, C.M.; Remouchamps, C.; McCorkell, K.A.; Solt, L.A.; Dejardin, E.; Orange, J.S.; May, M.J. Noncanonical NF-kappaB signaling is limited by classical NF-kappaB activity. *Sci. Signal* **2014**, *7*, ra13. [CrossRef]
49. Sun, S.C. The noncanonical NF-kappaB pathway. *Immunol. Rev.* **2012**, *246*, 125–140. [CrossRef]
50. Dougan, M.; Dranoff, G.; Dougan, S.K. GM-CSF, IL-3, and IL-5 Family of Cytokines: Regulators of Inflammation. *Immunity* **2019**, *50*, 796–811. [CrossRef]
51. Topf, U.; Suppanz, I.; Samluk, L.; Wrobel, L.; Boser, A.; Sakowska, P.; Knapp, B.; Pietrzyk, M.K.; Chacinska, A.; Warscheid, B. Quantitative proteomics identifies redox switches for global translation modulation by mitochondrially produced reactive oxygen species. *Nat. Commun.* **2018**, *9*, 324. [CrossRef]

52. Davinelli, S.; De Stefani, D.; De Vivo, I.; Scapagnini, G. Polyphenols as Caloric Restriction Mimetics Regulating Mitochondrial Biogenesis and Mitophagy. *Trends Endocrinol. Metab.* **2020**, *31*, 536–550. [CrossRef]
53. Zhou, Y.; Zhou, L.; Ruan, Z.; Mi, S.; Jiang, M.; Li, X.; Wu, X.; Deng, Z.; Yin, Y. Chlorogenic acid ameliorates intestinal mitochondrial injury by increasing antioxidant effects and activity of respiratory complexes. *Biosci. Biotechnol. Biochem.* **2016**, *80*, 962–971. [CrossRef]
54. Lakhani, S.A.; Masud, A.; Kuida, K.; Porter, G.A., Jr.; Booth, C.J.; Mehal, W.Z.; Inayat, I.; Flavell, R.A. Caspases 3 and 7: Key mediators of mitochondrial events of apoptosis. *Science* **2006**, *311*, 847–851. [CrossRef]
55. Dey, D.K.; Kang, S.C. Aflatoxin B1 induces reactive oxygen species-dependent caspase-mediated apoptosis in normal human cells, inhibits Allium cepa root cell division, and triggers inflammatory response in zebrafish larvae. *Sci. Total Environ.* **2020**, *737*, 139704. [CrossRef]

antioxidants

MDPI

Article

Goji-Berry-Mediated Green Synthesis of Gold Nanoparticles and Their Promising Effect on Reducing Oxidative Stress and Inflammation in Experimental Hyperglycemia

Luminita David [1], Valentina Morosan [1], Bianca Moldovan [1,*], Gabriela Adriana Filip [2] and Ioana Baldea [2]

[1] Faculty of Chemistry and Chemical Engineering, "Babeş-Bolyai" University, 11 Arany Janos Street, 400028 Cluj-Napoca, Romania; luminita.david@ubbcluj.ro (L.D.); valentina.morosan@ubbcluj.ro (V.M.)

[2] Department of Physiology, "Iuliu Haţieganu" University of Medicine and Pharmacy, 1-3 Clinicilor Street, 400006 Cluj-Napoca, Romania; adrianafilip33@yahoo.com (G.A.F.); baldeaioana@gmail.com (I.B.)

* Correspondence: bianca.moldovan@ubbcluj.ro

Abstract: The present report focuses on a rapid and convenient method applicable in the green synthesis of gold nanoparticles (AuNPs) using goji berry (*Lycium barbarum*—LB) extracts rich in antioxidant compounds, as well as on the structural analysis and evaluation of the induced antioxidant protection and anti-inflammatory effects of the synthesized gold nanoparticles upon endothelial cells (HUVECs) exposed to hyperglycemia. The synthesized AuNPs were characterized using ultraviolet–visible (UV-Vis) spectroscopy and transmission electron microscopy (TEM), whereas the presence of bioactive compounds from the *L. barbarum* fruit extract on the surface of the nanoparticles was confirmed using Fourier transform infrared spectroscopy (FTIR). The antioxidant activity of the biosynthesized gold nanoparticles was evaluated on the HUVEC cell line. The results reveal that AuNPs with a predominantly spherical shape and an average size of 30 nm were obtained. The UV-Vis spectrum showed a characteristic absorption band at λ_{max} = 536 nm of AuNPs. FTIR analysis revealed the presence of phenolic acids, flavonoids and carotenoids acting as capping and stabilizing agents of AuNPs. Both the *L. barbarum* extract and AuNPs were well tolerated by HUVECs, increased the antioxidant defense and decreased the production of inflammatory cytokines induced via hyperglycemia-mediated oxidative damage.

Keywords: green synthesis; goji berries; gold nanoparticles; antioxidant activity

check for updates

Citation: David, L.; Morosan, V.; Moldovan, B.; Filip, G.A.; Baldea, I. Goji-Berry-Mediated Green Synthesis of Gold Nanoparticles and Their Promising Effect on Reducing Oxidative Stress and Inflammation in Experimental Hyperglycemia. *Antioxidants* **2023**, *12*, 1489. https://doi.org/10.3390/antiox12081489

Academic Editors: Baojun Xu and Bin Du

Received: 16 June 2023
Revised: 7 July 2023
Accepted: 14 July 2023
Published: 25 July 2023

1. Introduction

Nanotechnology is nowadays one of the most dynamic research fields of materials science. Among nanomaterials, metallic nanoparticles are widely used due to their specific properties, which confer them a large range of applications in various fields, such as optics, electronics, environmental remediation, catalysis, cosmetics and medicine [1,2]. Their large variety of utilizations has strongly increased interest in developing synthesis strategies to obtain high-purity, highly stable, monodisperse, innocuous and affordable metallic nanoparticles. The conventional chemical (chemical reduction and the sol–gel method) and physical (thermal decomposition and laser ablation) synthetic approaches generally involve the use of hazardous chemicals; require high energy and complex equipment; are labor-intensive; and deliver polydisperse, non-uniform, unstable and harmful nanoparticles. Therefore, searching for alternative green synthetic methods has become a primary concern. This has led to the substantial development of biological approaches for obtaining metallic nanoparticles methods that prevent the use of harmful chemicals and are reliable, cost-effective, easy to scale up and environmentally friendly. Furthermore, the biosynthesized nanoparticles usually possess enhanced stability and biocompatibility with no harmful effects. This makes them suitable for pharmaceutical and food applications [2–5]. The green strategies for the synthesis of metal nanoparticles exploit the tremendous potential of the biomass as a source of reducing and stabilizing agents for the metallic nanomaterials. To this end, various microorganisms and plants have been successfully used as nano-biofactories. The microbial-assisted synthesis of nanoparticles usually requires laborious purification in order to remove potential pathogenic or toxic compounds, especially when the nanoparticles are intended for biological applications. As a consequence, plant extracts rich in bioactive compounds such as amino acids, saccharides, terpenoids, polyphenols, flavonoids and alkaloids are by far the most investigated and applied agents in the green synthesis of metallic nanoparticles. Many plants have been noted for their ability to synthesize both silver and gold nanoparticles with biomedical applications, such as tea plant [6], milk thistle [7], cornelian cherry [8], European cranberry bush [9] and sage [1].

Goji berries (*L. barbarum* L.) are fruits well known for their high antioxidant potential. Goji berries are used in traditional Chinese medicine because of their numerous health benefits, including their ability to treat hypertension, infectious diseases, common illnesses and diabetes, as well as to regulate cholesterol and triglyceride levels; prevent tumors, aging effects and heart disease; and reduce anxiety, depression, fatigue and stress. All of these beneficial effects are mostly conferred by the presence of strong antioxidant compounds in these fruits, such as polyphenols, flavonoids, carotenoids and vitamins [10,11].

Because it is well documented that all these compounds possess a remarkable ability to reduce metal ions and to cap and stabilize the obtained zero valent metal nanoparticles, we decided to exploit these strong antioxidant fruits for the biosynthesis of biocompatible gold nanoparticles. To the best of our knowledge, this is the first report on the use of *L. barbarum* fruits' bioactive compounds for the green synthesis of AuNPs. Hence, the current study aimed to synthesize, characterize and investigate the antioxidant effect of gold nanoparticles in a biological environment against hyperglycemia-induced oxidative stress.

In diabetic patients, continuous or intermittent hyperglycemia generates free radical species through the protein kinase C (PKC) and NADPH oxidase pathways and involves mitochondrial respiratory chain enzymes, causing cytokine release and further inflammation, lack of angiogenesis control and cell death, which leads to complications such as retinopathy, neuropathy, nephropathy and cardiovascular diseases [12,13]. Moreover, hyperglycemia can induce the formation of advanced glycation end products (AGEs) that activate NF-kB and the transcription of proinflammatory genes [14]. The synthesis of cytokines, TGF-β, chemokines and vascular cell adhesion molecules (VCAMs) is responsible for the induction of endothelial cell inflammation and apoptosis [15]

There is a lot of research in this field with the aim of finding antioxidant agents that can interrupt these metabolic alterations in diabetes in order to delay the onset and evolution of ROS-induced complications.

The aim of this biological study was to find the possible beneficial effects of gold nanoparticles phytosynthesized using *L. barbarum* extract against high-glucose-environment-induced oxidative stress and inflammatory status. Therefore, here, we report the protective antioxidant and anti-inflammatory effects against hyperglycemia-induced oxidative stress by AuNPs and *L. barbarum* extract in vitro on endothelial cells with respect to oxidative-stress-induced lipid peroxidation antioxidant defense mechanisms and the synthesis of proinflammatory cytokines interleukin 1 α (IL1α) and interleukin 1β (IL1β).

2. Materials and Methods

2.1. Chemical Assays

2.1.1. Reagents

Tetrachloroauric acid trihydrate, sodium hydroxide, Folin–Ciocalteu reagent, 2,2′-azino-bis(3-ethylbenzothiazoline-6-sulfonic acid) (ABTS), 2-thiobarbituric acid and Trolox gallic acid were obtained from Merck (Darmstadt, Germany). Cytochrome c and *o*-phtalaldehyde were obtained from Sigma Aldrich, St. Louis, MO, USA.

2.1.2. Preparation of Goji Fruit Extract

L. barbarum L. (red goji) berries were obtained from a small ecological farm, "Goji Bio Oltenia", in Oltenia province (Romania). Fully ripened fruits of *L. barbarum*, Transilvania cultivar, were harvested in September 2022 in SV Romania, Bradesti (44°30′17″ N, 23°37′13″ E), Dolj County. The plant sample collection required no specific permissions, as it did not involve protected or endangered species. A voucher specimen (FCIC.2022.G1) was deposited in the Department of Chemistry, Babes-Bolyai University, Cluj-Napoca, Romania.

Two grams of powder-ground dried berries was subjected to successive extractions using 40 mL double-distilled water twice. After 2 h at 20 °C, the mixture was filtered through nr. 1 Whitman paper, and the obtained extract was stored at 2 °C until further analysis and use.

The phenolic content of the goji berry extract was evaluated using a slightly modified Folin–Ciocalteu method [16]. Thus, 1.2 mL of a 0.7 M sodium carbonate solution was added to a mixture of 250 μL of goji extract and 1.5 mL of Folin–Ciocalteu reagent, which was incubated for 5 min in the dark. The sample was allowed to react for 2 h at room temperature in the absence of light, and further, its absorbance was spectrophotometrically monitored at 765 nm. The total phenolic content of the extract was expressed in mg gallic acid equivalents/L using a calibration curve of the reference.

2.1.3. Preparation of Gold Nanoparticles Using the Green Synthetic Method

The goji berry extract obtained as described above was further employed in the green synthesis of gold nanoparticles. In a conical flask, 10 mL of 10-fold diluted fruit extract was treated with a 1 N NaOH solution until pH = 5.5 and then added to 10 mL of boiling $HAuCl_4$ 1 mM aqueous solution freshly prepared from tetrachloroauric acid trihydrate and distilled water. The reduction of Au^{3+} ions was visually confirmed by a color change from faint orange to dark purple-red. The goji berry phytocompounds were removed from the gold colloidal solution via centrifugation at 10.000 rpm for 60 min, followed by AuNPs' purification via pellet washing with deionized water and their resuspension via 10 min of sonication. The purified AuNPs were characterized using appropriate techniques and further used for biological assays.

2.1.4. Characterization of Green Synthesized AuNPs

UV–visible (UV-vis) spectroscopy was the first technique applied to monitor the synthesis of the gold nanoparticles. The UV-vis spectrum was recorded using a Perkin Elmer Lambda 25 spectrometer by scanning the absorbance in the range of 300–800 nm in a quartz cuvette of 1 cm using distilled water as a blank. The green synthesized AuNPs were morphologically characterized via transmission electron microscopy (TEM—using a Hitachi H-7650 automatic transmission electron microscope) at 120 kV on a carbon-coated copper grid. The mean diameter of the AuNPs was determined from the TEM images, using ImageJ 1.53t software, from at least 50 nanoparticles. The presence of the bioactive compounds from the goji extract on the surface of the AuNPs was confirmed by Fourier transform infrared spectroscopy (FTIR) using a Bruker Vector 22 FTIR spectrometer (Brucker, Rosenheim, Germany), in a KBr pellet, in the range of 4000–500 cm^{-1}. The hydrodynamic diameter and zeta potential of the biosynthesized AuNPs were assessed using a Zetasizer Nanoseries compact scattering spectrometer (Malvern Instruments Ltd.; Malvern, UK).

2.1.5. In Vitro Antioxidant Activity Assessment

The radical scavenging capacity of the goji extract and the green synthesized AuNPs was assessed using the $ABTS^{\cdot+}$ assay [17]. A volume of 3 mL of freshly prepared and properly diluted (absorbance at 734 nm 0.4–0.7) $ABTS^{\cdot+}$ solution was mixed with 50 μL of extract and stored for 15 min in the dark. The absorbance of the resulting sample was recorded spectrophotometrically at 734 nm, and the obtained values were converted to mM Trolox equivalents in order to express the antioxidant capacity of the samples.

The antioxidant ability of the fruit extract was also assessed using the FRAP assay. The FRAP reagent was prepared according to the method described by Benzie and Strain [18] using 1.25 mL of 0.3 M acetate buffer solution with pH = 3.6, 1.25 mL of 0.01 M TPTZ (2,4,6-tris (2-pyridyl)-s-triazine) solution and 1.25 mL of 0.02 M aqueous solution of iron chloride $FeCl_3 \cdot 6H_2O$. To 2850 µL of freshly prepared FRAP reagent heated to 37 °C, 150 µL of diluted extract (1/10, *v/v*) of goji berries was added, and the mixture was kept in the dark for 30 min. The absorbance of the resulting sample was measured at 593 nm, and the results were expressed in µM Trolox using a calibration curve of the standard.

2.2. *Biological Assays*

2.2.1. Cell Source

Human umbilical vein endothelial cells (HUVEC-ECCAC obtained from Sigma Aldrich, Steinheim am Albuch, Germany) and dermal fibroblasts (BJ CRL-2522™—purchased from ATCC, VA, USA) were cultured in RPMI and, respectively, Dulbecco's modified Eagle medium (DMEM) supplemented with 5% fetal calf serum, 50 µg/mL gentamicin and 5 ng/mL amphotericin, all procured from Biochrom AG (Berlin, Germany). Cultures were fed twice weekly and incubated in standard cell culture conditions. For experiments, the concentration of 2% FCS in medium was used.

2.2.2. Viability Assay

Cell viability was assessed using the CellTiter 96® AQueous Non-Radioactive Cell Proliferation Assay (Promega Corporation, Madison, WI, USA). Fibroblasts and endothelial cells at a density of 10^4/wells in 96-well plaques (TPP, Transadingen, Switzerland) were allowed to settle for 24 h and then exposed for 24 h to *L. barbarum* extract and, respectively, AuNPs in different concentrations, 0–70 µg/mL of polyphenols in *L. barbarum* extract or 0–70 µg/mL of AuNPs, suspended in medium.

High glucose exposure was induced by the supplementation of each cell culture medium with glucose up to 4.5 g/L. Cells in hyperglycemia conditions were treated with *L. barbarum* extract and AuNPs in similar concentrations [19].

Viability was measured colorimetrically using an ELISA plate reader (Tecan, Männedorf, Switzerland) set at 540 nm. All experiments were performed in triplicate. The control sample was represented by cells exposed solely to cell medium. The viability results were calculated as % of the control samples. The concentrations of AuNPs or *L. barbarum* extract which decreased the cell viability to less than 70% of the control were considered toxic. The concentration selected to test the antioxidant and anti-inflammatory effects was 6.5 µg/mL of AuNPs and *L. barbarum*, respectively.

2.2.3. Experimental Design

Assessments of inflammatory mediators and oxidative stress generation were performed on endothelial cells in order to assess the possible protective effect of the *L. barbarum* and AuNPs in a prooxidant and proinflammatory hyperglycemia environment. Cells were cultivated at a density of 10^5/cm^2 (TPP, Transadingen, Switzerland) for 24 h in medium containing 2% FCS, then exposed for 24 h to *L. barbarum* and AuNPs respectively, with and without (w/wt) hyperglycemia.

Six groups were made: (1) controls exposed to medium (C), (2) cells exposed to *L. barbarum*, (3) cells treated with AuNPs (NP), (4) high glucose exposure (4.5 g/L) (HG), concomitant exposure to (5) high glucose and *L. barbarum* extract (HG + LB) and (6) high glucose and AuNPs (HG + NP). After exposure, supernatant and cells were collected. All the experiments were carried out in triplicate.

2.2.4. Cells Lysis

Cell lysates were obtained as previously reported [20]. The total protein concentration was measured using a DC protein assay kit (Bio-Rad, Hercules, CA, USA) and serum γ globulin as standard.

2.2.5. Oxidative Stress Assessment

Malondialdehyde (MDA) was measured spectrophotometrically from cell lysates. Data were expressed as nM/mg protein [20,21].

The enzymatic activity assays of catalase (CAT) and superoxide dismutase (SOD) were determined spectrophotometrically from the cell lysates as previously described [21]. Data are reported as units/mg protein.

Reduced (GSH) and oxidized glutathione (GSSG) were measured fluorometrically from cell lysates, using *o*-phtalaldehyde, as reported. GSH and GSSH concentrations were determined using a standard curve. The results were expressed as nmoles/mg protein [22,23]. All reagents for the oxidative stress assessment were bought from Sigma.

2.2.6. Inflammation

The cytokine levels of IL1α and IL1β were determined from the cell supernatant via ELISA assays using the ElabScience ELISA kits (Houston, TX, USA), according to the producer's instructions. The results are expressed as pg/mg protein for IL1α and OD/mg protein for IL1β.

2.3. Statistical Analysis

Statistical differences between *L. barbarum*, AuNPs and control groups were assessed via one-way ANOVA and Bonferroni post-tests; data are expressed as mean ± standard deviation; significance = $p \leq 0.05$. Prism version 4.00 for Windows, GraphPad Software 8 (San Diego, CA, USA) was employed. For the calculation of the inhibitory concentration 50% (IC50), the soft "Quest Graph™ IC50 Calculator" was used [24].

3. Results and Discussion

3.1. Synthesis and Characterization of Gold Nanoparticles

Goji berries, also designated as "superfruits", are well known as a rich source of compounds with potent antioxidant capacities, which confers to these fruits important health-promoting properties, as they alleviate the negative effects of oxidative stress and prevent the damage of DNA, proteins or lipids caused by harmful free radicals [25]. The major bioactive compounds reported in goji fruits are phenolic derivatives, organic acids and carotenoids. The antioxidant capacity of the *L. barbarum* fruit extract, used in the green synthesis of gold nanoparticles, was evaluated using the ABTS radical scavenging assay, while its total phenolic content was assessed using the Folin–Ciocalteu method. The antioxidant capacity of the goji fruit extract was also evaluated using the FRAP method. The antioxidant activity of natural extracts, determined using the FRAP method, estimates the ability of antioxidants to reduce iron(III) to iron(II). The recorded high values of 714 ± 29.8 mg GAE/L and 39.3 ± 1.83 mM Trolox (ABTS) and 1.75 ± 0.07 mM Trolox (FRAP), respectively, indicated a high content of antioxidant metabolites, recommending the use of *L. barbarum* fruit extract both as a source of reducing agents for the Au^{3+} ions present in the $HAuCl_4$ aqueous solution and of stabilizing mediators in the synthesis of nanoparticles containing phytocompounds capable of enhancing their biological properties on their surface, especially their antioxidant and anti-inflammatory protective effects against hyperglycemia-induced oxidative stress. The successful preparation of AuNPs was primarily confirmed visually by a color change from faint orange to dark purple-red.

In order to investigate the shape and size of the obtained nanoparticles and confirm the presence of bioactive compounds from goji berries on the surface of metal nanoparticles, the most modern spectroscopic (UV-VIS and FTIR) and microscopic (transmission electron microscopy—TEM) techniques were used. The synthesis of gold nanoparticles has been validated using UV-vis spectroscopy. Metal nanoparticles possess unique optical properties due to the resonance of the surface plasmon (SPR). In the case of gold nanoparticles, the excitation of SPR in colloidal solution is responsible for producing absorption maxima situated at wavelength values in the range of 500–550 nm [26]. Figure 1 displays the UV-vis spectra of goji fruit extract and green synthesized AuNPs, depicting the presence of the characteristic gold SPR band at 536 nm.

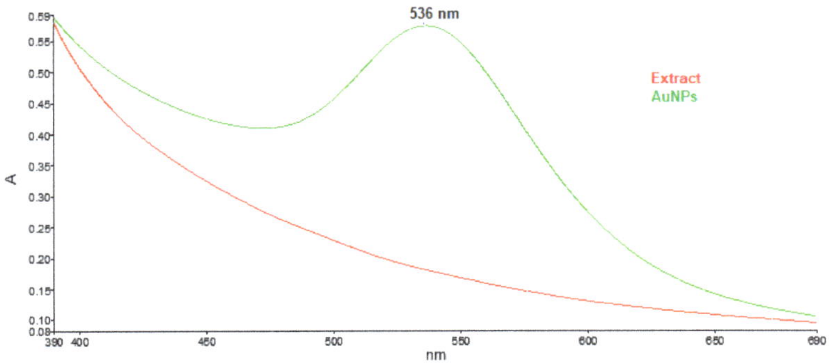

Figure 1. UV-vis spectra of goji fruit extract and gold nanoparticles.

The sizes and morphologies of gold nanoparticles synthesized with bioactive compounds from goji extract were analyzed using transmission electron microscopy (TEM) (Figure 2a). The TEM image confirms the presence of well-dispersed, mostly spherically shaped AuNPs, with an average diameter of 30 nm (Figure 2b).

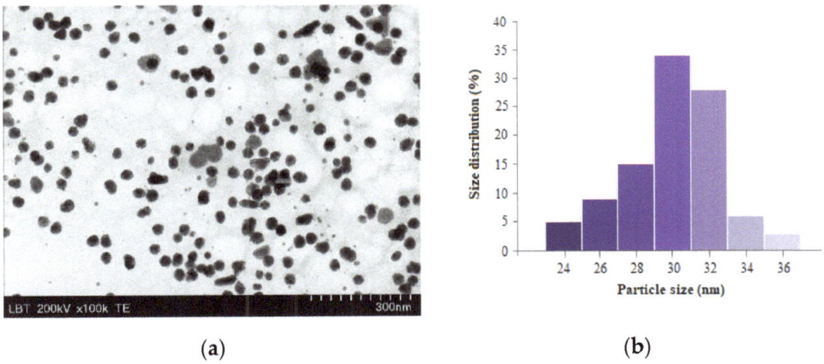

Figure 2. TEM image of AuNPs (**a**) and the corresponding size distribution (**b**).

FTIR analysis was applied in order to identify and confirm the presence of biomolecules from the fruit extract acting as stabilizing agents present at the surface of metal nanoparticles. The FTIR spectrum of the goji fruit extract (Figure 3a) shows a wide absorption band between 3090 and 3527 cm^{-1}, characteristic of the vibration of the OH groups in the phenolic compounds. The peak situated at 1631 cm^{-1} can be attributed to the vibration of the C=O bond (carbonyl) present in the flavonoid structure. The FTIR spectrum of AuNPs (Figure 3b) showed similar absorption bands: a wide peak at 3372 cm^{-1} corresponding to the characteristic vibration of the OH groups (phenolic or alcoholic) and a band at 2936 cm^{-1}, which can be attributed to the stretching vibration of C-H bond, while at 1651 cm^{-1}, a peak appeared, indicating the presence of the C=O bond in the biomolecules located on the surface of the metal nanoparticles. All these absorption bands observable in the FTIR spectra support the claim that phenolic phytocompounds such as phenolic acids and flavonoids from goji berries are present on the surface of phytosynthesized metallic nanoparticles, acting as coating ligands and being responsible for the effective stabilization of synthesized hybrid bionanomaterials.

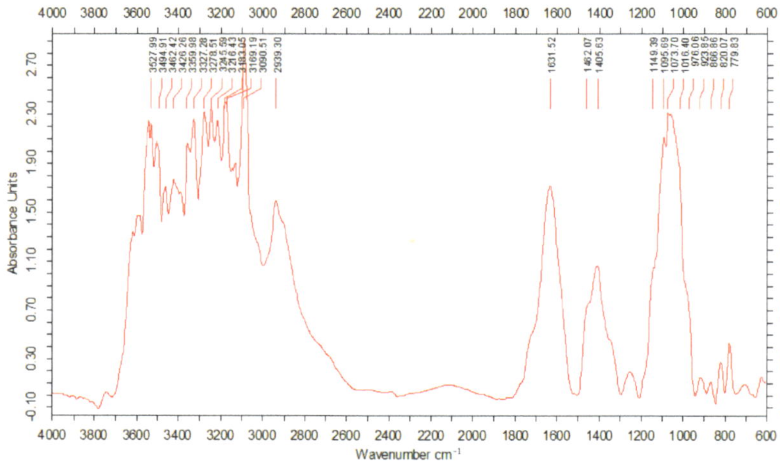

(**a**)

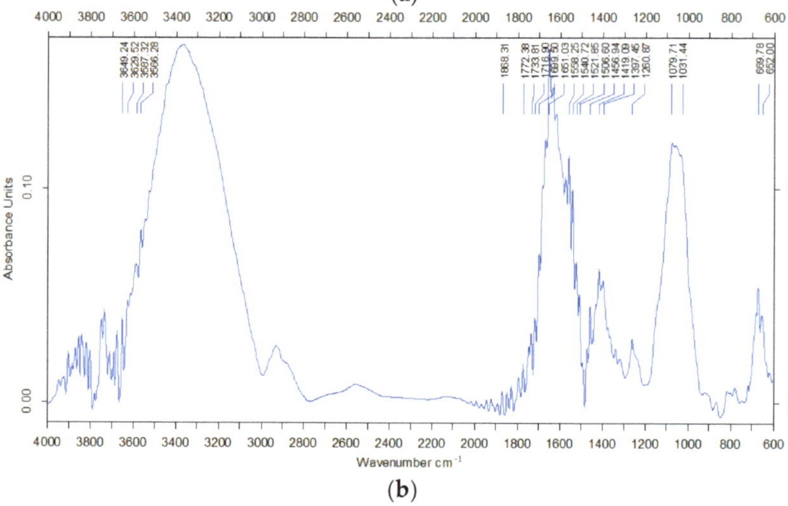

(**b**)

Figure 3. FTIR spectra of goji fruit extract (**a**) and gold nanoparticles (**b**).

The stability of gold nanoparticles synthesized with biomolecules from the goji berry extract was also investigated by determining the electrokinetic potential of the obtained colloidal systems (ζ-potential). Zeta potential value (surface charge) is a measure of the electrical charge on the surface of metallic nanoparticles, being an indirect measure of their physical stability. The value of the measured zeta potential of the synthesized AuNPs (Figure 4) was −11.8 mV, suggesting good stability, while their hydrodynamic diameter was 195.8 nm, thus indicating that anionic capping agents from the goji fruit extract are coordinated to the surface of silver nanoparticles' metallic shells.

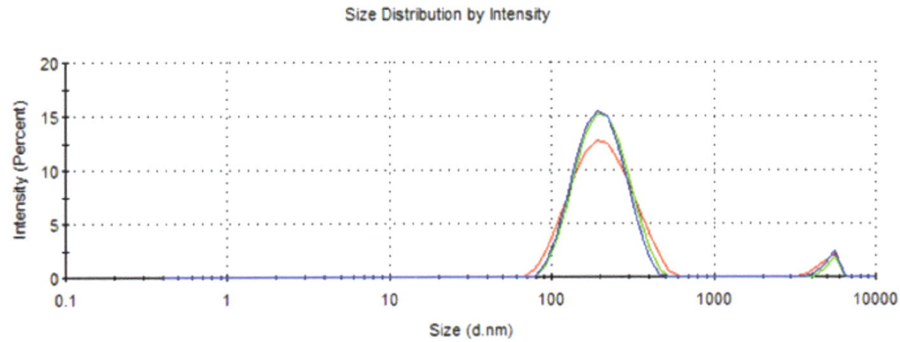

(a)

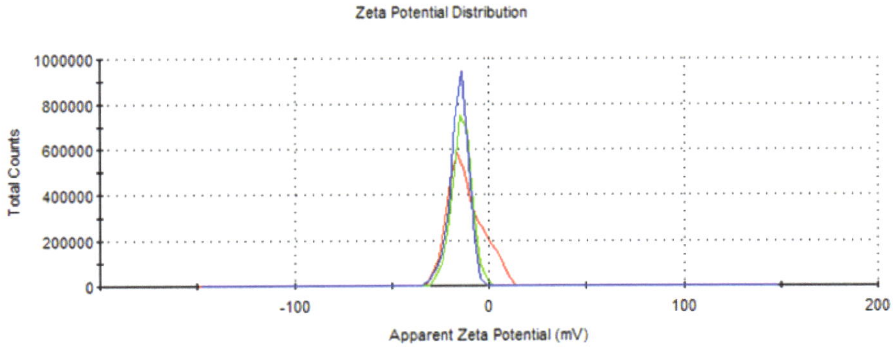

(b)

Figure 4. Hydrodynamic diameter (**a**) and zeta potential (**b**) of the obtained gold nanoparticles.

The free radical scavenging capacity of the goji-compound-mediated synthesized gold nanoparticles was assessed using the ABTS in vitro model. The quenching ability of the ABTS free radicals was visually observed through the discoloration of the green-blueish $ABTS^{\cdot+}$ solution, and the antioxidant capacity of AuNPs was spectrophotometrically measured and compared to the Trolox standard. The obtained value was 3.53 ± 0.14 mM Trolox, proving the ability of bioconjugated AuNPs to neutralize the free radicals.

3.2. Cell Viability

The viability of the endothelial cells and fibroblasts exposed to *L. barbarum* was not affected (Figure 5). When cells were exposed to hyperglycemia and *L. barbarum* extract, the cell viability was slightly increased for all concentrations in HUVECs and for the 25 µg/mL and 50 µg/mL in fibroblasts. The proliferation inhibitory concentration 50% (IC50) was calculated for HUVECs (IC50 = 41.53 µg/mL for *L. barbarum* and IC50 = 44.22 µg/mL for HG + LB) and for BJ fibroblasts (IC50 = 35.135 µg/mL for *L. barbarum* and IC50 = 76.26 µg/mL for HG + LB). The viability of HUVECs exposed to nanoparticles was maintained above the toxicity limit at concentrations up to 50 µg/mL. The viability was decreased by 70 µg/mL to almost 70% of the untreated control. When exposed to hyperglycemia, HUVECs showed strong proliferation at low concentrations of nanoparticles (6.25 and 12.5 µg/mL). At higher concentrations, there was similar behavior to normal glucose medium. The IC50 values for HUVECs treated with AuNPs were IC50 = 78.35 µg/mL for AuNPs and IC50 = 23.76 µg/mL for HG + NP.

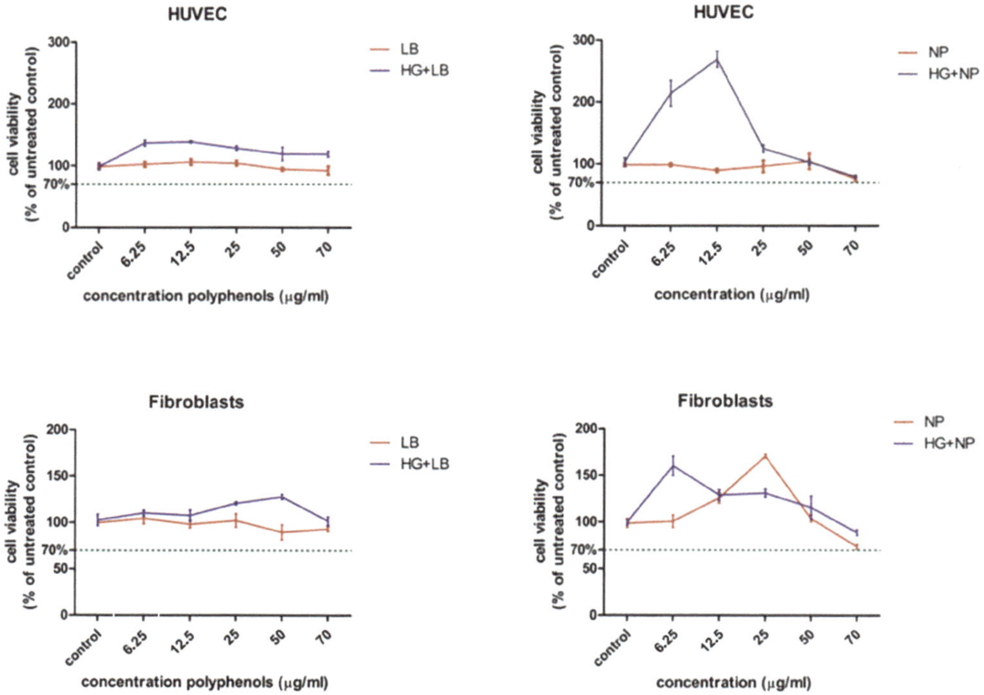

Figure 5. Comparative viability assessment of the cells (HUVECs—upper panels and fibroblasts—lower panels) exposed to *L. barbarum* extract and NP w/wt high glucose exposure. Data are presented as % of untreated control (media ± SD), n = 3.

In the fibroblasts, nanoparticles stimulated proliferation in a dose-dependent manner, with a maximum at 25 µg/mL. At 70 µg/mL, viability was slightly above the toxicity limit. A similar effect was obtained in fibroblasts exposed to both HG and AuNPs, but the proliferation stimulation was seen at a lower AuNP concentration (6.25 µg/mL). Higher concentrations induced lower proliferation. At the maximum concentration, the viability of the fibroblasts was similar to the controls. For fibroblasts, IC50 = 52.01 µg/mL for NP, and IC50 = 57.22 µg/mL for HG + NP. Following the viability test, the concentration of 6.5 µg/mL AuNPs and, respectively, *L. barbarum* extract was chosen for the oxidative stress and inflammatory cytokine assessment.

3.3. Oxidative Stress

Oxidative stress was assessed via MDA production, an indicator of the oxidative damage to the membrane lipids, the enzymatic activities of the antioxidant defense enzymes CAT and SOD and the content of GSH (reduced glutathione) and GSSG (oxidized glutathione). The GSH/GSSG ratio was calculated and represented a marker of oxidative stress (Figure 6). Malondialdehyde production was decreased by the exposure to *L. barbarum* extract and not to AuNPs. However, AuNP exposure did not increase the MDA content. In high-glucose conditions, all groups showed significantly increased MDA levels compared to the controls. *L. barbarum* did not significantly decrease the MDA production. One-way ANOVA showed the significant interaction of hyperglycemia and the increase in MDA production ($p < 0.0001$).

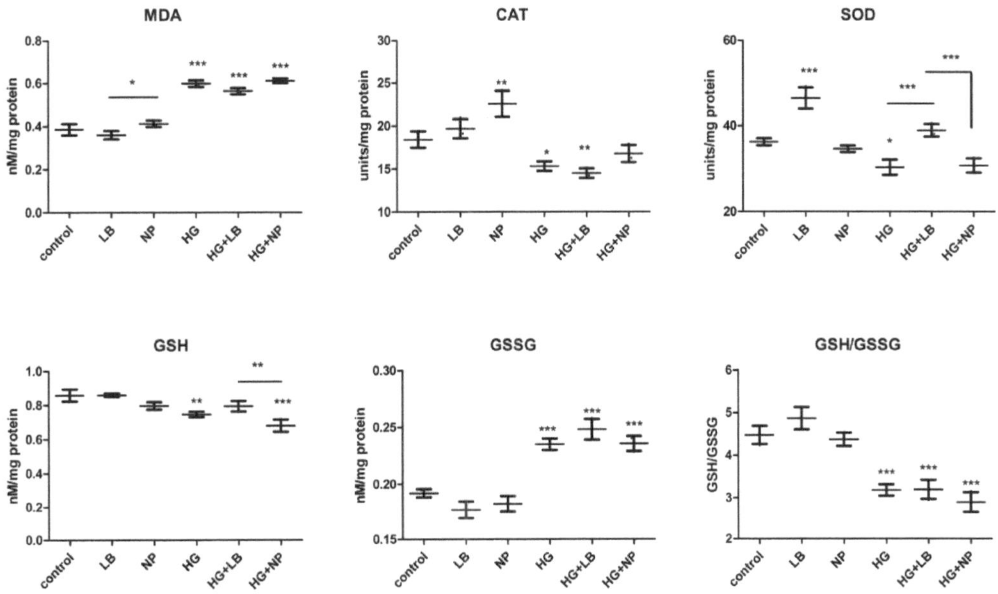

Figure 6. Oxidative stress assessment. MDA = malondyaldehyde, CAT = catalase, SOD = superoxide-dismutase, GSH = reduced glutathione, GSSG = oxidized glutathione. GSH/GSSG ratio was calculated. Bars represent mean ± SD (n = 3); * = $p < 0.05$, ** = $p < 0.001$, *** = $p < 0.0001$; control versus treated group, respectively, between each two selected groups. $p < 0.05$ was considered significant; only significant differences are presented.

The antioxidant enzymes activity and catalase and superoxide dismutase activities were both significantly decreased by hyperglycemia. CAT activities were significantly increased by AuNP exposure compared to controls. HG + AuNP partially increased the CAT activity, almost to the control. SOD activity was strongly increased by *L. barbarum* extract w/wt high glucose. AuNP-treated cells showed similar SOD activity to the control in high-glucose conditions. One-way ANOVA showed a significant decreasing effect of hyperglycemia on the CAT and SOD enzymatic activities ($p < 0.0001$).

GSH content was lower in all treated groups compared to the control. The decrease was significant for HG and HG + AuNP. There was no significant difference between the HG and HG + AuNP groups. Oxidized glutathione, GSSG content, was slightly decreased by both *L. barbarum* and AuNP exposure. Hyperglycemia significantly increased the GSSG content in all groups. The ratio between reduced and oxidized glutathione levels is considered as an indicator of oxidative stress. In the *L. barbarum* group, the ratio was increased, while in the AuNP group, it was not altered. High glucose strongly diminished the antioxidant levels in all groups ($p < 0.0001$, One-way ANOVA)

Overall, *L. barbarum* extract decreased oxidative stress induced by exposure to hyperglycemia by reducing lipid peroxidation and increased the antioxidant pool with an increased GSH/GSSG ratio and antioxidant superoxide dismutase activity. Since hyperglycemia induces the production of oxygen superoxide anions, previously reported as being involved in microvasculature damage in diabetes [12,27], the increase in the SOD enzymatic activity via *L. barbarum* extract would be highly beneficial for the oxidative statuses of the cells. Moreover, GSH content was decreased via hyperglycemia as previously reported [27], but the higher level of GSH induced by the *L. barbarum* extract led to higher scavenging ROS activity, which can normalize the oxidative balance and may contribute to a sustainable metabolic change effective against ROS induced by hyperglycemia.

The AuNPs induced higher catalase activity but did not increase the antioxidant reserve and had no impact on the lipid peroxidation induced by hyperglycemia. The level of catalase is important since it neutralizes the H_2O_2 produced by the activity of SOD. Enhanced H_2O_2 levels were reported in hyperglycemia, and they promote cell cycle arrest by decreasing Cdc25C [28]. This can lead to the release of proinflammatory cytokines via NF-kB pathway activation, autophagy or necroptosis [12].

There are reports that show the importance of antioxidants such as vitamin E, which inhibits phosphokinase C activation following hyperglycemia and retinal neoangiogenesis in diabetic rats [12,27]. Also, diabetic mouse embryopathy was prevented by treatment with vitamin C, E, SOD, N-acetyl-cysteine or glutathione ethyl ester [27]. However, in patients with diabetes mellitus who underwent the administration of antioxidants, such as vitamin E 400 UI/day, for 4 years in a clinical study, Heart Outcomes Prevention Evaluation (HOPE) provided no preventive effects against cardiovascular complications, while shorter studies showed better outcomes with vitamin E supplementation on retinal neoangiogenesis or nephropathy [27]. Therefore, other types of antioxidants that increase antioxidant enzymatic activity and antioxidant reserve, such as GSH, can have a higher impact on the oxidative balance of cells, leading to a better outcome.

3.4. Inflammatory Cytokines

IL1α production was significantly decreased by *L. barbarum* w/wt HG compared to the controls. The NP groups showed no proinflammatory response w/wt HG. High glucose induced a slight, not significant increase of IL1α level (Figure 7).

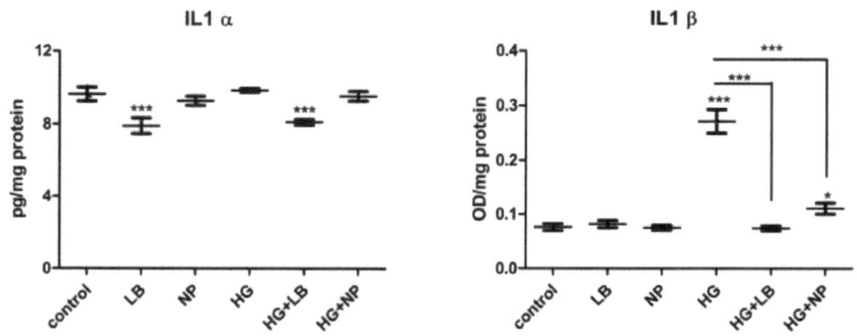

Figure 7. Inflammatory cytokines. $p < 0.05$ was considered significant. Bars represent mean ± SD (n = 3); * = $p < 0.05$, *** = $p < 0.0001$; control versus treated group, respectively, between each two selected groups.

IL1β production was slightly increased by *L. barbarum*. In the high glucose group, there was a strong increase of the IL1β content, which was not seen in the HG + LB, moreover, the difference between these two groups is strongly significant. HG + NP group showed increased level of IL1β, compared to control, but strongly decreased it compared to the HG alone. These data show an anti-inflammatory effect of both *L. barbarum* extract and NP in high glucose conditions.

The cytokine family of IL1 contains 11 members with multiple roles in inflammation and malignancies. Both IL1α and IL1β use the same receptor IL1R and have proinflammatory functions. IL1α is an unusual cytokine, leading to a sterile inflammatory status in human diseases like autoimmune diseases and cancer. IL1α is expressed constitutively in epithelial and endothelial stromal cells. It can be stimulated in other cells by Toll-like receptor agonists [29], inflammatory cytokines [30–32] and oxidative stress [33–35], which is reviewed in [36]. Hyperglycemia exposure triggers the release of proinflammatory cytokines, leading to an inflammatory status, responsible for some of the complications in diabetes, such as retinopathy, neuropathy and cardiovascular diseases. Therefore, combined antioxidant and anti-inflammatory activity would be beneficial in hyperglycemic conditions to delay the onset of complications and their further aggravation. Overall, the *L. barbarum* extract, and to a lesser extent, the AuNPs synthesized using *L. barbarum* extract were able, at non-toxic, low concentrations, to decrease the proinflammatory effect of hyperglycemia, leading to the release of IL1β.

IL1α can bind AP1 and NF-kB, leading to their activation during inflammation. IL1β is mainly expressed in immune cells and also endothelial cells; its level is increased by IL1α signaling through AP1 and NF-kB. IL1β was found to be increased in patients with type 1 diabetes mellitus having poor disease therapy and also in the early stage. Moreover, increased IL1β production was responsible for the inflammatory statuses of patients with diabetes mellitus type 1 [37].

Increased IL1β levels were also reported in obese and diabetic patients (type II) and were involved in β-cell dysfunction, leading to deregulated insulin secretion and the promotion of the disease and acting as a link between immunity and metabolism. IL1β signaling induces the acute phase response, vasodilatation, fever, low blood pressure and the further activation of general inflammatory mechanisms. In hyperglycemic patients, IL1β increased levels are produced as a result of inflammasome and particularly NLRP3 signaling. The NLRP3 inflammasome is made from several proteins: a sensor (NLRP3); an adaptor (ASC)—apoptosis-associated speck-like protein—which comprises a domain for caspase recruitment; and procaspase 1, the effector, which activates IL1β and IL18 synthesis [38]. The decrease in IL1β induced by hyperglycemia as a result in cells treated with *L. barbarum* or AuNPs is an important anti-inflammatory effect leading to possible beneficial effects in both the prevention and therapy of hyperglycemia.

Both compounds tested, *L. barbarum* extract and AuNPs, were well tolerated by the HUVECs and dermal fibroblasts. Overall, in the hyperglycemia-stimulated endothelial cell model employed in the current study, *L. barbarum* extract showed important protective antioxidant effects against hyperglycemia-induced oxidative alterations. *L. barbarum* decreased lipoperoxidation, increased SOD activity and increased antioxidant balance GSH/GSSH. Moreover, proinflammatory cytokine IL1α and β levels were decreased. AuNPs synthesized using the *L. barbarum* extract partially recuperated the antioxidant activity of the SOD and especially CAT enzymes, reduced by hyperglycemia exposure and decreased IL1β, strongly augmented by hyperglycemia to almost control levels. Therefore, both compounds show promising antioxidant and anti-inflammatory effects in endothelial cells exposed to hyperglycemia.

4. Conclusions

The present study reports an eco-friendly, one-pot synthetic method of gold nanoparticles by effectively exploiting the antioxidant properties of *L. barbarum* fruits, which successfully acted as reducing and capping agents of the obtained nanoparticles. The obtained bioconjugated AuNPs were mostly regular in shape, spherical, with an average diameter of 30 nm.

In the endothelial cells, AuNPs showed protective effects against the oxidative stress and proinflammatory status induced by hyperglycemia, responsible for endothelial injuries, leading to several diabetes complications. AuNPs partially recuperated the antioxidant defense, particularly catalase activity, and inhibited the IL1β proinflammatory cytokine release induced by hyperglycemia. Therefore, the AuNPs retained some of the antioxidant and anti-inflammatory properties of the *L. barbarum* extract and showed promising results against hyperglycemia's deleterious effects.

Author Contributions: Conceptualization, L.D. and G.A.F.; methodology, B.M. and I.B.; validation, V.M. and I.B.; formal analysis, I.B.; investigation, V.M., B.M. and I.B.; resources, G.A.F. and I.B.; data curation, B.M. and I.B.; writing—original draft preparation, L.D., B.M., I.B. and G.A.F.; writing—review and editing, B.M.; supervision, L.D. and G.A.F.; project administration, B.M. and I.B.; funding acquisition, B.M. All authors have read and agreed to the published version of the manuscript.

Funding: This research was funded by CNCS-UEFISCDI, Romania, project number PN-III_P1.1.1-TE-2021-0511.

Institutional Review Board Statement: Not applicable.

Informed Consent Statement: Not applicable.

Data Availability Statement: The data presented in this study are contained in the article.

Conflicts of Interest: The authors declare no conflict of interest. The funders had no role in the design of the study; in the collection, analyses, or interpretation of data; in the writing of the manuscript; or in the decision to publish the results.

References

1. Takci, D.K.; Ozdenefe, M.S.; Genc, S. Green synthesis of silver nanoparticles with an antibacterial activity using *Salvia officinalis* aqueous extract. *J. Cryst. Growth* **2023**, *614*, 127–239. [CrossRef]
2. Schroffel, A.; Kratosova, G.; Safarik, I.; Safarikova, M.; Raska, I.; Shol, L.M. Applications of biosynthesized metallic nanoparticles —A review. *Acta Biomater.* **2014**, *10*, 4023–4042. [CrossRef] [PubMed]
3. Sandhya, M.V.S.; Rashkumar, K.; Burgula, S. Efficient eco-friendly approach towards bimetallic nanoparticles synthesis and characterization using *Exiguobacterium aestuarii* by statistical optimization. *Green Chem. Lett. Rev.* **2019**, *12*, 420–434. [CrossRef]
4. Das, C.G.A.; Kumar, V.D.; Dharani, G.; Dhas, T.S.; Karthik, V.; Kumar, C.M.B.; Embrandiri, A. Macroalgae-associated halotolerant marine bacteria *Exiguobacterium aestuarii* A.D.C.G. SYST 3 synthesized gold nanoparticles and its anticancer activity in breast cancer cell line (MCF–7). *J. Mol. Liq.* **2023**, *383*, 122061.
5. Filip, G.A.; Florea, A.; Olteanu, D.; Clichici, S.; David, L.; Moldovan, B.; Cenariu, M.; Potara, M.; Scrobota, I.; Baldea, I. Biosynthesis of silver nanoparticles using *Sambucus nigra* L. fruit extract for targeting cell death in oral dysplastic cells. *Mater. Sci. Eng. C* **2021**, *123*, 111974. [CrossRef]
6. Vaseeharan, B.; Ramasamy, P.; Chen, J.C. Antibacterial activity of silver nanoparticles (AgNPs) synthesized by tea leaf extracts against pathogenic *Vibrio harveyi* and its protective efficacy on juvenile *Fenneropenaeus indicus*. *Lett. Appl. Microbiol.* **2010**, *50*, 352–356. [CrossRef]
7. Clichici, S.; David, L.; Moldovan, B.; Baldea, I.; Olteanu, D.; Filip, M.; Nagy, A.; Luca, V.; Crivii, C.; Mircea, P.; et al. Hepatoprotective effects of silymarin coated gold nanoparticles in experimental tumors. *Mater. Sci. Eng. C* **2020**, *115*, 111117. [CrossRef]
8. Filip, A.G.; Potara, M.; Florea, A.; Baldea, I.; Olteanu, D.; Bolfa, P.; Clichici, S.; David, L.; Moldovan, B.; Olenic, L.; et al. Comparative evaluation by scanning confocal Raman spectroscopy and transmission electron microscopy of therapeutic effects of noble metal nanoparticles in experimental acute inflammation. *RSC Adv.* **2015**, *5*, 67435–67448. [CrossRef]
9. Bidian, C.; Filip, G.A.; David, L.; Florea, A.; Moldovan, B.; Popa Robu, D.; Olteanu, D.; Radu, T.; Clichici, S.; Mitrea, D.-R.; et al. The impact of silver nanoparticles phytosynthesized with *Viburnum opulus* L. extract on the ultrastrastructure and cell death in the testis of offspring rats. *Food Chem. Toxicol.* **2021**, *150*, 112053. [CrossRef]
10. Potterat, O. Goji (*Lycium barbarum* and *L. chinense*): Phytochemistry, pharmacology and safety in the perspective of traditional uses and recent popularity. *Planta Medica* **2010**, *76*, 7–19. [CrossRef]
11. Cui, B.; Liu, S.; Lin, X.; Wang, J.; Li, S.; Wang, Q.; Li, S. Effects of *Lycium barbarum* aqueous and ethanol extracts on high-fed-diet induced oxidative stress in rat liver tissue. *Molecules* **2011**, *16*, 9116–9128. [CrossRef] [PubMed]
12. Volpe, C.M.O.; Villar-Delfino, P.H.; Ferreira dos Anjos, P.M.; Nogueira-Machado, J.A. Cellular death, reactive oxygen species (ROS) and diabetic complications. *Cell Death Dis.* **2018**, *9*, 119. [CrossRef] [PubMed]
13. Fiorentino, T.V.; Prioletta, A.; Zuo, P.; Folli, F. Hyperglycemia-induced oxidative stress and its role in diabetes mellitus related cardiovascular diseases. *Curr. Pharm. Des.* **2013**, *19*, 5695–5703. [CrossRef] [PubMed]
14. Lv, X.; Lv, G.H.; Dai, G.Y.; Sun, H.M.; Xu, H.Q. Food-advanced glycation end products aggravate the diabetic vascular complications via modulating the AGEs/RAGE pathway. *Chin. J. Nat. Med.* **2016**, *14*, 844–855. [CrossRef]
15. Suryavanshi, S.V.; Kulkarni, Y.A. NF-κβ: A potential target in the management of vascular complications of diabetes. *Front. Pharmacol.* **2017**, *8*, 798. [CrossRef]
16. Perde-Schrepler, M.; David, L.; Olenic, L.; Potara, M.; Fischer-Fodor, E.; Virag, P.; Imre-Lucaci, F.; Brie, I.; Florea, A. Gold nanoparticles with a polyphenol-rich extract from Cornelian cherry (*Cornus mas*) fruits: Effects on human skin cells. *Synth. J. Nanomater.* **2016**, *2016*, 6986370. [CrossRef]
17. Re, R.; Pellegrini, N.; Proteggente, A.; Pannala, A.; Yang, M.; Rice-Evans, C. Antioxidant activity applying an improved ABTS radical cation decolorization assay. *Free. Radic. Biol. Med.* **1999**, *26*, 1231–1237. [CrossRef]
18. Benzie, I.F.; Strain, J.J. The ferric reducing ability of plasma (FRAP) as a measure of "antioxidant power": The FRAP assay. *Anal. Biochem.* **1996**, *239*, 70–76. [CrossRef]
19. Login, C.C.; Bâldea, I.; Tiperciuc, B.; Benedec, D.; Vodnar, D.C.; Decea, N.; Suciu, S. A novel thiazolyl Schiff base: Antibacterial and antifungal effects and in vitro oxidative stress modulation on human endothelial cells. *Oxid. Med. Cell. Longev.* **2019**, *2019*, 1607903. [CrossRef]

20. Baldea, I.; Olteanu, D.E.; Bolfa, P.; Ion, R.M.; Decea, N.; Cenariu, M.; Baniu, M.; Sesarman, A.V.; Filip, A.G. Efficiency of photodynamic therapy on WM35 melanoma with new synthetic porphyrins: Role of chemical structure, intracellular targeting and antioxidant defense. *J. Photochem. Photobiol. B Biol.* **2015**, *151*, 142–152. [CrossRef]
21. Baldea, I.; Costin, G.-E.; Shellman, Y.; Kechris, K.; Olteanu, E.D.; Filip, A.; Cosgarea, M.R.; Norris, D.A.; Birlea, S.A. Biphasic pro-melanogenic and pro-apoptotic effects of all-trans- retinoic acid (ATRA) on human melanocytes: Time-course study. *J. Dermatol. Sci.* **2013**, *72*, 168–176. [CrossRef]
22. Filip, A.; Daicoviciu, D.; Clichici, S.; Bolfa, P.; Catoi, C.; Baldea, I.; Bolojan, L.; Olteanu, D.; Muresan, A.; Postescu, I.D. The effects of grape seeds polyphenols on SKH-1 mice skin irradiated with multiple doses of UV-B. *J. Photochem. Photobiol. B* **2011**, *105*, 133–142. [CrossRef]
23. Olteanu, D.; Nagy, A.; Dudea, M.; Filip, A.; Muresan, A.; Catoi, C.; Mircea, P.A.; Clichici, S. Hepatic and systemic effects of rosuvastatin on an experimental model of bile duct ligation in rats. *J. Physiol. Pharmacol.* **2012**, *63*, 483–496.
24. Quest Graph™ IC50 Calculator. AAT Bioquest, Inc. 4 May 2023. Available online: https://www.aatbio.com/tools/ic50-calculator (accessed on 4 May 2023).
25. Donno, D.; Peccaru, G.L.; Mellano, M.G.; Cerutti, A.K.; Bounous, G. Goji berry fruit (*Lycium* spp.): Antioxidant compound fingerprint and bioactivity evaluation. *J. Funct. Food* **2015**, *18*, 1070–1085. [CrossRef]
26. Bidian, C.; Filip, G.A.; David, L.; Moldovan, B.; Baldea, I.; Olteanu, D.; Filip, M.; Bolfa, P.; Potara, M.; Toader, A.M.; et al. *Viburnum opulus* fruit extract-capped gold nanoparticles attenuated oxidative stress and acute inflammation in carrageenan-induced paw edema model. *Green Chem. Lett. Rev.* **2022**, *15*, 319–335. [CrossRef]
27. King, G.L.; Loeken, M.R. Hyperglycemia-induced oxidative stress in diabetic complications. *Histochem. Cell Biol.* **2004**, *122*, 333–338. [CrossRef]
28. Savitsky, P.A.; Finkel, T. Redox regulation of Cdc25C. *J. Biol. Chem.* **2002**, *277*, 20535–20540. [CrossRef] [PubMed]
29. Garlanda, C.; Dinarello, C.A.; Mantovani, A. The interleukin-1 family: Back to the future. *Immunity* **2013**, *39*, 1003–1018. [CrossRef] [PubMed]
30. Palomo, J.; Dietrich, D.; Martin, P.; Palmer, G.; Gabay, C. The interleukin (IL)-1 cytokine family–Balance between agonists and antagonists in inflammatory diseases. *Cytokine* **2015**, *76*, 25–37. [CrossRef] [PubMed]
31. Weber, A.; Wasiliew, P.; Kracht, M. Interleukin-1 (IL-1) pathway. *Sci. Signal.* **2010**, *3*, cm1. [CrossRef]
32. Kimura, H.; Inukai, Y.; Takii, T.; Furutani, Y.; Shibata, Y.; Hayashi, H.; Sakurada, S.; Okamoto, T.; Inoue, J.; Oomoto, Y.; et al. Molecular Analysis of Constitutive Il-1α Gene Expression in Human Melanoma Cells: Authocrine Stimulation Through Nf-Kb Activation By Endogenous IL-1α. *Cytokine* **1998**, *10*, 872–879. [CrossRef] [PubMed]
33. Rider, P.; Kaplanov, I.; Romzova, M.; Bernardis, L.; Braiman, A.; Voronov, E.; Apte, R.N. The transcription of the alarmin cytokine interleukin-1 alpha is controlled by hypoxia inducible factors 1 and 2 alpha in hypoxic cells. *Front. Immunol.* **2012**, *3*, 290. [CrossRef]
34. McCarthy, D.A.; Ranganathan, A.; Subbaram, S.; Flaherty, N.L.; Patel, N.; Trebak, M.; Hempel, N.; Melendez, J.A. Redox-control of the alarmin, interleukin-1α. *Redox Biol.* **2013**, *1*, 218–225. [CrossRef] [PubMed]
35. McCarthy, D.A.; Clark, R.R.; Bartling, T.R.; Trebak, M.; Melendez, J.A. Redox control of the senescence regulator interleukin-1α and the secretory phenotype. *J. Biol. Chem.* **2013**, *288*, 32149–32159. [CrossRef]
36. Malik, A.; Kanneganti, T.D. Function and regulation of IL-1α in inflammatory diseases and cancer. *Immunol. Rev.* **2018**, *281*, 124–137. [CrossRef] [PubMed]
37. Cano-Cano, F.; Gómez-Jaramillo, L.; Ramos-García, P.; Arroba, A.I.; Aguilar-Diosdado, M. IL-1β Implications in Type 1 Diabetes Mellitus Progression: Systematic Review and Meta-Analysis. *J. Clin. Med.* **2022**, *11*, 1303. [CrossRef]
38. Alfadul, H.; Sabico, S.; Al-Daghri, N.M. The role of interleukin-1b in type 2 diabetes mellitus: A systematic review and meta-analysis. *Front. Endocrinol.* **2022**, *13*, 901616. [CrossRef]

antioxidants

MDPI

Systematic Review

Effects of Dietary Nitrate Supplementation on Performance during Single and Repeated Bouts of Short-Duration High-Intensity Exercise: A Systematic Review and Meta-Analysis of Randomised Controlled Trials

Nehal S. Alsharif [1,2], Tom Clifford [1], Abrar Alhebshi [1,3], Samantha N. Rowland [1] and Stephen J. Bailey [1,*]

[1] School of Sport, Exercise and Health Sciences, Loughborough University, Loughborough LE11 3TU, UK; n.alsharif@lboro.ac.uk (N.S.A.); t.clifford@lboro.ac.uk (T.C.); a.alhebshi@lboro.ac.uk (A.A.); s.rowland@lboro.ac.uk (S.N.R.)

[2] Department of Clinical Nutrition, Faculty of Applied Medical Sciences, King Abdulaziz University, Jeddah 21589, Saudi Arabia

[3] Department of Clinical Nutrition, Faculty of Applied Medical Sciences, Umm Al-Qura University, Makkah 24382, Saudi Arabia

* Correspondence: s.bailey2@lboro.ac.uk

Abstract: Inorganic nitrate (NO_3^-) has emerged as a potential ergogenic aid over the last couple of decades. While recent systematic reviews and meta-analyses have suggested some small positive effects of NO_3^- supplementation on performance across a range of exercise tasks, the effect of NO_3^- supplementation on performance during single and repeated bouts of short-duration, high-intensity exercise is unclear. This review was conducted following PRISMA guidelines. MEDLINE and SPORTDiscus were searched from inception to January 2023. A paired analysis model for cross-over trials was incorporated to perform a random effects meta-analysis for each performance outcome and to generate standardized mean differences (SMD) between the NO_3^- and placebo supplementation conditions. The systematic review and meta-analysis included 27 and 23 studies, respectively. Time to reach peak power (SMD: 0.75, $p = 0.02$), mean power output (SMD: 0.20, $p = 0.02$), and total distance covered in the Yo-Yo intermittent recovery level 1 test (SMD: 0.17, $p < 0.0001$) were all improved after NO_3^- supplementation. Dietary NO_3^- supplementation had small positive effects on some performance outcomes during single and repeated bouts of high-intensity exercise. Therefore, athletes competing in sports requiring single or repeated bouts of high-intensity exercise may benefit from NO_3^- supplementation.

Keywords: nitric oxide; beetroot; exercise performance

check for updates

Citation: Alsharif, N.S.; Clifford, T.; Alhebshi, A.; Rowland, S.N.; Bailey, S.J. Effects of Dietary Nitrate Supplementation on Performance during Single and Repeated Bouts of Short-Duration High-Intensity Exercise: A Systematic Review and Meta-Analysis of Randomised Controlled Trials. *Antioxidants* **2023**, *12*, 1194. https://doi.org/10.3390/antiox12061194

Academic Editor: Reto Asmis

Received: 13 May 2023
Revised: 25 May 2023
Accepted: 27 May 2023
Published: 31 May 2023

1. Introduction

Inorganic nitrate (NO_3^-) has been conventionally considered an environmental carcinogen and inert end-product of endogenous nitric oxide (NO) oxidation [1]. More recent research challenges these assertions and has revealed various potential health benefits afforded by increased dietary NO_3^- intake [2]. Over the last couple of decades, dietary NO_3^- supplementation has emerged as a potential nutritional strategy to improve exercise performance in healthy and moderately trained individuals [3,4]. The ergogenic effects of NO_3^- supplementation have been attributed to its stepwise reduction to nitrite (NO_2^-) and the subsequent reduction of NO_2^- to NO [2,5]. Although initially recognised for its vasodilatory properties [6], it is now appreciated that NO can positively modulate a plethora of physiological responses in skeletal muscle [7–9], the conflation of which is likely to underpin improved exercise performance following dietary NO_3^- supplementation [5].

Initial studies assessing the potential efficacy of NO_3^- supplementation to enhance physiological and performance responses during exercise revealed improvements in exercise economy and exercise tolerance [10–12]. These improvements in endurance exercise performance parameters after NO_3^- supplementation were initially linked to a lower adenosine triphosphate (ATP) cost of muscle force production (improved contractile efficiency), an associated blunting in the perturbation to high-energy phosphate substrates and metabolites [13], and to a lower mitochondrial adenosine diphosphate/oxygen ratio (P/O ratio; a lower O_2 cost of ATP resynthesis), reflecting improved mitochondrial respiratory efficiency [14]. However, the mechanisms by which NO_3^- supplementation can improve exercise economy and endurance exercise performance are still to be resolved in human skeletal muscle [15,16].

Following on from the initial human studies, experiments conducted using murine models indicated potential fibre-type-specific effects of NO_3^- supplementation on physiological responses [17]. Indeed, NO_3^- supplementation was initially reported to increase calcium (Ca^{2+}) handling proteins and evoked force production in type II skeletal muscle, but not slow-twitch (type I) skeletal muscle, in mice [18]. Subsequently, NO_3^- supplementation increased hindlimb blood flow in exercising rats, with this additional blood flow shunted towards more fast-twitch (type II) muscle fibres [19]. The potential for enhanced efficacy of NO_3^- supplementation to improve physiological and performance responses in murine type II muscle is consistent with data from human studies demonstrating enhanced pulmonary O_2 uptake ($\dot{V}O_2$) and muscle deoxyhaemoglobin + deoxymyoglobin kinetics in exercise settings that evoke greater type II muscle fibre recruitment compared to exercise settings that evoke mostly type I muscle fibre recruitment [20]. Moreover, cross-sectional data have revealed that NO_3^- supplementation is less likely to improve exercise economy and endurance performance as aerobic fitness increases [21], an effect that has been attributed, at least in part, to a lower % and proportional recruitment of type II muscle fibres in endurance-trained participants with a more aerobic phenotype [22]. On this basis, NO_3^- supplementation may have greater potential as an ergogenic aid in exercise settings which evoke greater type II muscle fibre recruitment.

It is well documented that type II skeletal muscle fibres are recruited in an intensity-dependent manner, with greater recruitment of type II muscle fibres at higher exercise intensities [23–25]. In addition, the reduction of NO_2^- to NO is enhanced in conditions of acidosis and hypoxia [26–28]. The partial pressures of O_2 (PO_2) and pH are lower in contracting type II than type I muscles [29,30] and progressively decline with increasing exercise intensity [31]. Therefore, high-intensity exercise, which is supramaximal with regards to the power output required to elicit $\dot{V}O_{2max}$, and evokes significant recruitment of type II muscle fibres and declines in muscle pH and PO_2, appears to have greater potential to elicit an ergogenic effect from NO_3^- supplementation compared to continuous submaximal endurance exercise. There is also evidence to suggest that NO_3^- supplementation is more effective at improving physiological and functional responses at higher, compared to lower, movement velocities [32,33]. In addition, NO_3^- supplementation has been reported to increase the peak contractile velocity of, and power output generated by, contracting skeletal muscle [33,34], and to lower the time taken to achieve peak power output [35,36]. Collectively, these improvements in skeletal muscle contractile function after NO_3^- supplementation would be expected to translate into enhanced single and repeated sprint performances. However, whilst there is some evidence to support an ergogenic effect of NO_3^- supplementation on single and repeated bouts of short-duration large muscle mass exercise in humans (e.g., [37,38]), the existing evidence basis is equivocal (e.g., [39–41]). In part, these interstudy discrepancies may be attributable to disparate NO_3^- supplementation and high-intensity exercise protocols, which complicates interpretation of the ergogenic potential of NO_3^- supplementation for high-intensity exercise.

Although the effects of NO_3^- supplementation on performance in a variety of exercise performance tests have been systematically reviewed and have undergone meta-analyses before [42–48], these have not yet considered the effects of NO_3^- supplementation on single and repeated bouts of short-duration large muscle mass exercise in humans. This is important to address to help improve understanding of the exercise settings in which NO_3^-

supplementation is ergogenic and to inform recommendations for NO_3^- supplementation to improve exercise performance. Therefore, the purpose of this study was to conduct a systematic review and meta-analysis of the effects of NO_3^- supplementation on single and repeated bouts of short-duration large muscle mass exercise in healthy humans. A secondary purpose was to conduct sub-analyses to evaluate the influence of the NO_3^- supplementation dose and duration, participant sex, exercise type (single vs. repeated sprints), exercise duration, and plasma NO_3^- and NO_2^- concentrations ($[NO_3^-]$ and $[NO_2^-]$, respectively) to further refine understanding of the experimental conditions in which NO_3^- supplementation is more likely to enhance single and repeated bouts of short-duration large muscle mass exercise.

2. Materials and Methods

This systematic review and meta-analysis was reported according to Preferred Reporting items for Systematic Reviews and Meta-Analysis (PRISMA) guidelines [49]. The study protocol was registered with the Center for Open Science organisation (registration number: 10.17605/OSF.IO/JSGKM).

2.1. Inclusion and Exclusion Criteria

Three researchers (N.S.A., S.J.B., and T.C.) agreed on the inclusion and exclusion criteria. These were based on a Population, Intervention, Comparator, Outcome, Study design (PICOS) methodology (see Online Supplementary Material). Briefly, studies were included if they met the following criteria: (1) participants were healthy adults ≥16 years old; (2) they administered oral inorganic NO_3^- supplements such as beetroot juice or sodium/potassium NO_3^- salts and provided information about the dose, frequency, and duration of supplementation; (3) they included exercise that recruited a large muscle mass such as running, cycling, and kayaking; (4) the exercise test included ≥1 high-intensity effort ($\geq\dot{V}O_{2peak}$), with each effort ≤60 s; (5) they measured performance as completion time, total distance covered, maximal or mean power output, total work performed, or maximal number of repetitions. Studies were excluded if participants were <16 years old or had a chronic medical condition; NO_3^- was administered with another dietary supplement; there was insufficient information about the dose, frequency, and duration of supplementation; exercise was submaximal ($\leq\dot{V}O_{2max}$) or if any single effort was ≥60 s; and if exercise was performed in hypoxic or hot conditions.

2.2. Search Strategy

We searched Medline and SPORTdiscus databases for English language papers from inception to January 2023. Our search strategy was based on our PICOS methodology and the full search terms for both databases are presented in the Online Supplementary Material. The reference lists of eligible full text articles were also searched to identify any other potential studies for inclusion.

2.3. Study Selection

The search results were downloaded into Rayyan software, a web tool for screening abstracts [50]. After removing duplicates, two researchers (N.S.A. and S.N.R.) independently screened titles and abstracts for inclusion. Full texts of studies deemed eligible were retrieved and compared against the predefined PICOS criteria. Where there was disagreement on whether a study should be included or excluded from the systematic review and meta-analysis, this was discussed with, and resolved by, a third researcher (S.J.B.). The study selection process is summarised in Figure 1.

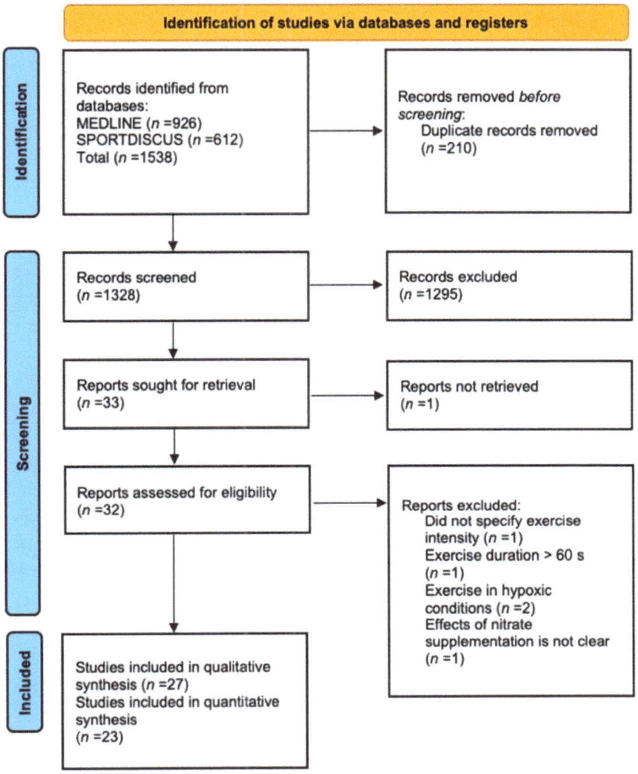

Figure 1. Preferred reporting items for systematic review and meta-analysis PRISMA flow diagram for study selection process. Nitrate; NO_3^-.

2.4. Data Extraction

Data were extracted into a Microsoft Excel Spreadsheet by one researcher (N.S.A.) and substantiated by a second researcher (S.N.R.). The spreadsheet was designed and trialled by three authors (N.S.A., T.C., and S.J.B.) and refined prior to extraction. The following data and information were extracted: study design, sample size, participant characteristics (age, training status, $\dot{V}O_{2peak/max}$), supplementation protocol (type, dose, frequency, duration, timing of last dose relative to exercise onset, total exposure, placebo, and washout period between trials), exercise protocol (mode, intensity, duration, recovery between bouts, and number of repetitions), and mean $\pm$ SD of relevant outcomes, including the mean of all peak power outputs (PP), PP during the first sprint (PP_{First}), PP during the last sprint (PP_{Last}), time to reach PP (PP_{Time}), mean power output from all repetitions (MP), MP during the first sprint (MP_{First}), MP during the last sprint (MP_{Last}), minimum power (P_{Min}), total work performed in repeated cycling efforts (TWD), and total distance covered in the Yo-Yo IR1 running test (TDC). When standard error of the mean (SEM) was reported, SD was calculated as SD = SEM $\times$ $\sqrt{n}$, where n represents the sample size. Authors of studies included in the meta-analysis were contacted to retrieve individual participants' data for the calculation of pooled SD and correlation coefficient. For 15 studies, data for individual participants were provided [35–38,41,51–60]. The correlation coefficient (Corr) was imputed for the studies with available individual participant data using the following formula:

$$\mathrm{Corr} = \mathrm{SD_E}^2 + \mathrm{SD_C}^2 - \mathrm{SD^2}_{\mathrm{diff}}/2 \times \mathrm{SD_E} \times \mathrm{SD_C},$$

where:

Corr = correlation, SD_E = standard deviation for the NO_3^- trial, SD_C = standard deviation for the placebo trial, SD_{diff} = the difference between the standard deviation for the NO_3^- trial and standard deviation for the placebo trial.

Subsequently, the standard error of the SMD (SE(SMD)) was calculated using the formula:

$$\mathrm{SE(SMD)} = \sqrt{1/n + \mathrm{SMD}^2/2n} \times \sqrt{2(1 - \mathrm{Corr})},$$

where:

SE(SMD) = the standard error for the standardised mean difference, n = sample size, and Corr = correlation coefficient.

For the remaining studies (n = 10) [34,39,61–67], Corr was estimated as the average Corr from the studies in which individual data were available.

2.5. Quality Assessment

Risk of bias of included studies was assessed using the Revised Cochrane Collaboration risk of bias tool (ROB2) for crossover trials [68], which assesses studies based on five specific domains: (1) randomisation process; (2) deviations from the intended outcome; (3) missing outcome data; (4) measurement of the outcome; and (5) selection of the reported results. This was performed on the Cochrane excel tool available at https://www.riskofbias.info (accessed on 31 January 2022), which allows an entry for each domain in a risk of bias table rated as "low risk", "some concerns", or "high risk". Two researchers (N.S.A., and A.A.) independently evaluated the risk of bias for each study and any discrepancies were resolved through discussion. As previously recommended [69], funnel plot asymmetry was visually inspected to assess publication bias for meta-analyses that included ≥10 studies.

2.6. Statistical Analysis

Quantitative synthesis was only performed if ≥2 studies measured the same outcome. The meta-analysis was conducted using RevMan 5.4v [70]. A separate meta-analysis was performed for each of the following continuous outcomes: PP, PP_{First}, PP_{Last}, MP, MP_{First}, MP_{Last}, PP_{Time}, TWD, and TDC. Data are presented as forest plots with 95% confidence intervals. Due to significant between-study heterogeneity, effect sizes were calculated with an inverse variance random-effects model using the DerSimonian–Laird method [71]. Effect sizes were interpreted according to Cohen's guidelines where an SMD of 0.2, 0.5, and 0.8, respectively, reflect small, medium, and large effects [72]. Heterogeneity was assessed using the Chi^2 and I^2 statistics. A value of $p \leq 0.10$ on the Chi^2 test was considered significant. The I^2 was interpreted as follows: <25%, low risk; 25–75%, moderate risk; and >75% high risk [69]. Additionally, forest plots were visually inspected to check for observable differences in study results. A sensitivity analysis was conducted by using a correlation coefficient of 0.5 for all studies [73], removing studies that had a high risk of bias for at least one domain, and those with elite endurance athletes, as previous studies have reported that dietary NO_3^- supplementation is less effective in this population [60,63]. For sub-group analysis, the influence of the NO_3^- supplementation dose (<8 mmol vs. ≥8 mmol) and duration (single day vs. multiple days supplementation), exercise type (single vs. repeated sprints), and exercise duration (≤15 s vs. >15 s–≤30 s) were assessed. Due to the low number of studies that measured plasma [NO_3^-] and [NO_2^-] and included female participants, a sub-group analysis on the influence of plasma [NO_3^-] and [NO_2^-] and biological sex could not be performed. Studies recruiting well-trained endurance athletes were omitted from sub-group analyses on the basis that this population group does not exhibit an ergogenic effect after NO_3^- supplementation [60,63]. Statistical significance was accepted at $p < 0.05$.

3. Results

A total of 1538 articles were retrieved from the two databases; after duplicates were removed, 1328 articles remained. No studies were identified through searching the reference lists of included studies. Following initial screening of titles and abstracts, thirty-two full-text articles were retrieved, of which five were excluded for failing to meet the inclusion criteria. Twenty-seven studies were identified as eligible for the systematic review and twenty-five for the meta-analysis. Results of the search strategy are presented in Figure 1.

3.1. Study Characteristics

Table 1 provides a summary of the studies included in the systematic review and meta-analysis. All studies employed a randomised, double (n = 23) [34–41,51,53–57,59–62,64,66,67,74,75] or single (n = 4) [52,58,63,65] blind, placebo controlled, crossover design. Studies were published between 2013 and 2022. The sample size varied between studies (range: 7–52 participants). Participants' ages ranged from 17 to 31 years. Participant training status was described as healthy or recreationally active (n = 4) [39,52,57,65], competing at a recreational or amateur standard (n = 18) [35–38,40,41,51,54–56,58,59,61,62,66,67,74,75], highly competitive (n = 5) [34,36,40,53,64], or elite (n = 3) [36,60,63]. Participants were involved in different types of sports, including team sports (n = 13) [34,37,38,41,51,52,55,56,62,65,66,74,75], cycling (n = 3) [34,36,60], resistance training (n = 4) [35,54,57,67], tennis (n = 2) [34,40], mixed martial arts (n = 1) [64], kayaking (n = 1) [53], speed skating (n = 1) [36], CrossFit (n = 1) [59], and sprinting (n = 1) [61]. The dose, duration, and type of NO_3^- supplementation varied between studies. NO_3^- supplementation was administered as beetroot juice (n = 24) [34–41,51,53–58,60–66,74,75], potassium NO_3^- (n = 1) [59], pomegranate extract (n = 1) [67], or as a high NO_3^- diet (n = 1) [52]. The dose of NO_3^- supplementation ranged from 4.8 to 16.4 mmol/day (mean; 8.5 mmol/day). Fifteen studies administered NO_3^- supplementation as a single dose 2.5–3 h before exercise [34,35,39,40,53,54,56,57,61,62,64,66,67,74,75] and twelve studies as repeated doses over 2–7 days [36–38,41,51,52,55,58–60,63,65]. In these latter studies, the last dose was administered 40–180 min before (n = 11) [36–38,41,51,52,55,58,60,63,65] or ≥24 h before exercise (n = 1) [59]. Total NO_3^- exposure in all studies ranged between 4.8 and 77.4 mmol. Most of the included studies recruited exclusively male participants (n = 22) [35,37–41,51–65,75], four studies recruited male and female participants [34,36,66,67], and one study recruited only female participants [74]. Of the 410 participants included in the review, 354 participants (86%) were reported as male, with 56 participants (14%) reported as female. The most frequent modality of exercise was cycling (n = 19) [34–36,39,41,51,52,54,56–61,63–67], followed by running (n = 7) [37,38,40,55,62,74,75] and kayaking (n = 1) [53]. Studies used different exercise protocols to assess performance: repeated all-out sprints with a fixed number of repetitions (n = 13) [34,36,39,41,51–53,56,60,64,67,74,75], high-intensity intervals (n = 3) [63,65,66], the 30 s Wingate test (n = 7) [34–36,54,57,59,61], and the Yo-Yo intermittent recovery level 1 test (Yo-Yo IR1) (n = 4) [37,38,55,62]. Different assessment methods were used to evaluate exercise performance, with each study measuring 1–4 performance variables. Performance variables included PP (n = 11) [34,35,41,54,56–60,64,66], PP during a single sprint (n = 7) [36,39,52,53,60,64,67], time to reach PP (n = 4) [35,36,54,57], MP (n = 13) [35,41,54,56–61,63–66], MP during a single sprint (n = 11) [36,39,41,54,56–58,60,61,64,67], TWD (n = 6) [34,51,56,61,65,66], minimum power (n = 3) [35,54,57], optimal pedalling cadence (n = 1) [34], number of completed repetitions (n = 3) [63,65,66], TDC (n = 4) [37,38,55,62], sprint time (n = 3) [40,74,75], best sprint time (n = 2) [74,75], slowest sprint time (n = 1) [75], and fatigue index (n = 6) [34,53,54,57,58,64]. Of the twenty-seven studies included, only eight studies measured plasma [NO_3^-] and [NO_2^-] [36–38,51–53,55,65], two studies measured only plasma [NO_3^-] [74,75], and one study only measured plasma [NO_2^-] [41].

Table 1. Summary of studies included in the systematic review and meta-analysis that examined the effects of nitrate supplementation on exercise performance during single and repeated bouts of short duration high-intensity exercise.

Study	Participants			Supplementation Protocol							
	No. (♂, ♀)	Health/Training Status	Age (Years)	Type/Volume	NO_3^- Dose (mmol)	Duration	Time before Trial	Placebo	Exercise Protocol	Performance Variables	Results
Aucouturier et al. (2015) [65]	17 ♂	Healthy, active in team sports	23 ± 3	BR juice/500 mL	10.9	3 D	3 h	Apple-black currant juice	15 s cycling at 170% of MAP to exhaustion, interspersed with 30 s	MP, TWD, reps, exercise duration	ND in MP, improved TWD, reps and exercise duration
Bender et al. (2018) [39]	16 ♂	Healthy, recreationally active	17 ± 1	BR shot/2 × 70 mL	12.9	Single D	3 h	NR-depleted BR shot	4 × 20 s all-out WAnT, interspersed with 240 s	PP, MP	ND in PP and MP
Bernardi et al. (2018) [64]	10 ♂	Well-trained mixed martial arts athletes	25 ± 5	BR juice/400 mL	9.3	Single D	2 h	Black current juice	20 × 6 s all-out cycling interspersed with 24 s	PP, MP, FI	ND in PP, MP, and FI
Buck et al. (2015) [74]	13 ♀	Team sport players	26 ± 2	BR shot/1 × 70 mL	6	Single D	3 h	NR-depleted BR shot	6 × 20 m all-out effort running, interspersed with 25 s recovery	ST, best ST	ND in ST and best ST
Christensen et al. (2013) [60]	10 ♂	Elite cyclists	29 ± 4	BR juice/500 mL	8	4 D	3 h	Apple-black currant juice	6 × 20 s cycling at 0.75 N/kg, interspersed with 100 s	PP, MP	ND in PP and MP
Corry et al. (2015) [58]	10 ♂	Recreationally active	20 ± 1	BR shot/2 × 70 mL	8	2 D	40 min	Black current juice	30 s all-out WanT	PP, MP, FI	Improved MP, ND in PP and FI
Cuenca et al. (2018) [54]	15 ♂	Resistance trained	22 ± 2	BR shot/1 × 70 mL	6	Single D	3 h	NR-depleted BR juice	30 s all-out WAnT	PP, MP, PP_{Time}, P_{Min}, FI	Improved PP, MP and PP_{Time}, ND in FI
Domínguez et al. (2017) [57]	15 ♂	Healthy trained	22 ± 2	BR shot/1 × 70 mL	5.6	Single D	3 h	NR-depleted BR juice	30 s all-out WAnT	PP, MP, PP_{Time}, P_{Min}, FI	Improved PP and MP, ND in PP_{Time}, P_{Min}, FI
Dumar et al. (2021) [61]	10 ♂	National level sprinters	20.3 ± 2	BR shot/1 × 70 mL	6.4	Single D	2 h	Black current juice	3 × 15 s all-out WAnT	MP and TWD	Improved MP and TWD
Esen et al. (2022) [62]	12 ♂	Recreational active	27 ± 10	BR shot/1 × 140 mL	12.8	Single D	3 h	BR shot/1 × 70 mL	Yo-Yo IR1 test	TDC	Longer TDC
Jodra et al. (2020) [35]	15 ♂	Resistance trained	23 ± 2	BR shot/1 × 70 mL	6.4	Single D	2.5-3 h	NR-depleted BR juice	30 s all-out WAnT	PP, MP, PP_{Time}, P_{Min}	Improved PP and PP_{Time}, ND in MP and P_{Min}
Jonvik et al. (2018) [36]	29 ♂ 23 ♀	Recreational cyclists (*n* = 20), national talent speed skaters (*n* = 23), Olympic-level track cyclists (*n* = 10)	♂= 22 ± 5 ♀= 26 ± 8	BR shot/2 × 70 mL	12.9	6 D	3 h	NR-depleted BR juice	3 × 30 s all-out WAnT interspersed with 240 s recovery	PP, MP, PPTime	ND in PP and MP, improved PPTime
Kramer et al. (2016) [59]	12 ♂	CrossFit athletes	23 ± 5	KNR/2 capsules	8	6 D	≥24 h	KCL capsules	30 s all-out WAnT	PP, MP	Improved PP, ND in MP
López-Samanes et al. (2020) [40]	13 ♂	Highly competitive tennis players	25 ± 5	BR shot/1 × 70 mL	6.4	Single D	3 h	NR-depleted BR juice	10 m Sprint	ST	ND in ST
Martin et al. (2014) [66]	9 ♂ 7 ♀	Moderately trained team sport athletes	♂= 22 ± 2 ♀= 21 ± 1	BR shot/1 × 70 mL	4.8	Single D	2 h	NR-depleted BR shot	8 s high intensity cycling to exhaustion interspersed with 30 s	PP, MP, TWD, no of reps	ND in PP, MP, TWD, no of reps

Table 1. *Cont.*

Study	Participants			Supplementation Protocol							
	No. (♂, ♀)	Health/Training Status	Age (Years)	Type/Volume	NO_3^- Dose (mmol)	Duration	Time before Trial	Placebo	Exercise Protocol	Performance Variables	Results
Muggeridge et al. (2013) [53]	8 ♂	Trained kayakers	31 ± 15	BR shot/1 × 70 mL	5	Single D	3 h	Tomato juice	5 × 10 s maximum effort kayaking, interspersed with 50 s recovery	PP, FI	ND in PP and FI
Nyakayiru et al. (2017) [55]	32 ♂	Soccer players	23 ± 1	BR shot/2 × 70 mL	12.9	6 D	3 h	NR-depleted BR shot	Yo-Yo IR1 test	TDC	Longer TDC
Pawlak-Chaouch et al. (2019) [63]	11 ♂	Elite endurance athletes	22 ± 4	BR juice/500 mL	5.5	3 D	3 h	Apple-black currant juice	15 s cycling at 170% of MAP to exhaustion interspersed with 30 s	MP, TWD and no of reps	ND in MP, TWD and no of reps
Porcelli et al. (2016) [52]	7 ♂	Healthy recreationally active	25 ± 2	High NR diet	8.2	6 D	3 h	Control diet ~2.9 mmol NR/day	5 × 6 s all-out cycling, interspersed with 24 s recovery	PP	Improved PP
Reynolds et al. (2020) [75]	16 ♂	Team sport athletes	21 ± 2	BR shot/1 × 70 mL	6	Single D	3 h	NR-depleted BR shot	10 × 40 m all-out running interspersed with 30 s recovery	ST, fastest ST, slowest ST	ND in ST, fastest ST and slowest ST
Rimer et al. (2017) [34]	11 ♂ 2 ♀	Competitively trained athletes	26 ± 8	BR shot/2 × 70 mL	11.2	Single D	2.5 h	NR-depleted BR shot	4 × 3–4 s all-out cycling interspersed with 120 s. Followed by 30 s WAnT after 300 s rest.	PP, TW, optimal pedalling rate, FI	Improved PP and optimal pedalling rate during 4 × 3–4 s test. ND in PP, TW, and FI during 30 s Wingate test
Roelofs et al. (2017) [67]	10 ♂ 11 ♀	Recreationally resistance-trained	22 ± 2	Pomegranate extract/capsule	6.8	Single D	-	Maltodextrin capsule	10 × 6 s all-out, interspersed with 30 s	PP, MP	Improved PP and MP
Smith et al. (2019) [56]	12 ♂	Recreationally trained, team sport athletes	22 ± 4	BR shot/1 × 70 mL	6.2	Single D	2.5 h	NR-depleted BR shot	2 halves of 20 × 6 s all out cycling interspersed with 114 s recovery	PP, MP, TWD	ND in PP, MP, TW
Thompson et al. (2016) [37]	32 ♂	Team-sport players	24 ± 4	BR shot/1 × 70 mL	6.4	5 D	2.5 h	NR-depleted BR shot	Yo-Yo IR1	TDC, 20 m sprint time, 5, 10, 5–10, 10–20 m split time	Longer TDC, improved 5, 10, 5–10 m split time, ND in 10-20 m split time
Thompson et al. (2015) [51]	16 ♂	Recreational team-sport players	24 ± 5	BR shot/2 × 70 mL	12.8	7 D	2.5 h	NR-depleted BR shot	2 halves of 20 × 6 s all out cycling interspersed with 114 s recovery	TWD	Improved TWD
Wylie et al. (2016) [41]	10 ♂	Recreational team-sport players	21 ± 1	BR shot/2 × 70 mL	8.2	3 D	2.5 h	NR-depleted BR shot	D3: 24 × 6 s all out cycling interspersed with 24 s D4: 7 × 30 s all-out cycling interspersed with 240 s D5: 6 × 60 s interspersed with 60 s	PP, MP	Improved PP and MP during 24 × 6 s. ND in PP and MP during 7 × 30 s and 6 × 60 s
Wylie et al. (2013) [38]	14 ♂	Recreational team-sport players	22 ± 2	Day 1, BR shot/ 4 × 70 mL Day 2, BR shot/ 3 × 70 mL	D1:16.4 D2:12. 3	2 D	1.5 h	NR-depleted BR shot	Yo-Yo IR1	TDC	Longer TDC

NR, nitrate; BR, beetroot; PL, placebo; PP, peak power; PP_{Time}, time to peak power; MP, mean power output; P_{Min}, minimum power; TWD, total work done; TDC, total distance covered; ST; sprint time; TT; time trial; reps, number of repetitions; FI, fatigue index; MAP, maximal aerobic power, HI, high intensity; Yo-Yo IR1; Yo-Yo intermittent recovery level 1 test; WAnT. Wingate anaerobic test; resistance, kg; kilograms; D, day; h, hour; s, second; min, minutes; ND, no difference; KNR, potassium nitrate; KCL, potassium chloride; -, no information provided; no., number of participants; ♂, male biological sex; ♀, female biological sex.

3.2. *Quality Assessment*

Five studies had a low risk of bias in the overall bias domain [34,39,54,59,67], fifteen studies had some concerns [35,37,38,40,41,51,53,55,56,61,62,64,66,74,75], and seven studies had a high risk of bias [36,52,57,58,60,63,65]. Seven studies had a low risk of bias in the randomisation process [34,36,39,54,59,63,67] and the remaining twenty studies had some concerns [35,37,38,40,41,51–53,55–58,60–62,64–66,74,75]. All studies had a low risk of bias for bias arising from period and carryover effects [34–41,51–67,74,75]. Twenty-five studies had low risk of bias [34–41,51–57,59–64,66,67,74,75], one study had a low risk of bias [58], and one study had some concerns [65] in the deviation from the intended intervention domain. For missing outcome data, twenty-five studies had a low risk of bias [34,35,37–41,51–59,61–67,74,75] and two studies had a high risk [36,60]. In the measurement of the outcome domain, six studies had a high risk of bias [52,57,58,60,63,65] and the remaining twenty-one had a low risk of bias [34–41,51,53–56,59,61,62,64,66,67,74,75]. One study had a low risk of bias [57] and twenty-six studies had some concerns [34–41,51–56,58–67,74,75] in the selection of reported results domain. A summary of risk of bias for crossover trials is presented in Figure 2 and a risk of bias assessment for individual studies is presented in Figure S1 in the online Supplementary Materials. Funnel plots suggest little evidence of publication bias, as presented in the online Supplementary Materials (Figures S2–S5).

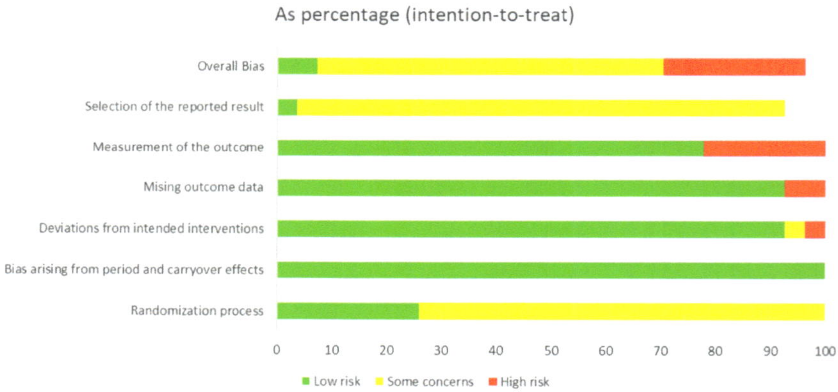

Figure 2. Summary risk of bias graph for crossover trials evaluating the effects of nitrate supplementation on different performance outcomes during single and repeated bouts of short-duration high-intensity exercise.

3.3. *Meta-Analysis*

3.3.1. Time to Reach Peak Power

NO_3^- supplementation lowered PP_{Time} compared to placebo (SMD: 0.75, 95% CI: −1.38 to 0.11, $p = 0.02$) (Figure 3). There was a high risk of statistical heterogeneity between studies ($Chi^2 = 23.29$; $I^2 = 87\%$, $p < 0.0001$). Removing a study with a high risk of bias [57] did not remove statistical heterogeneity but slightly changed the pooled SMD (SMD: 0.88, 95% CI: −1.90 to 0.13, $p = 0.09$).

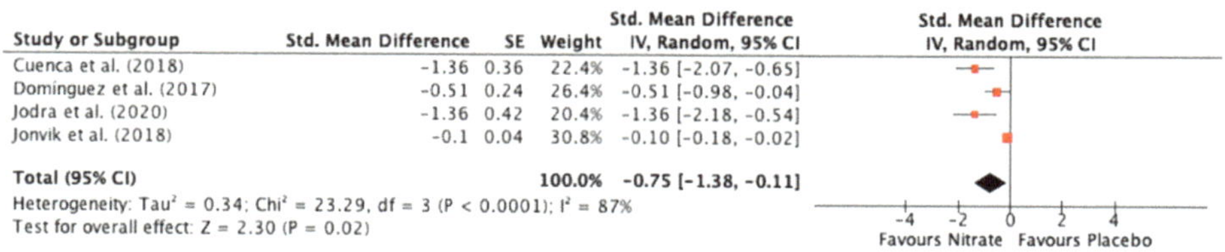

Figure 3. Forest plot for time to reach peak power in the nitrate and placebo trials [35,36,54,57].

3.3.2. Peak Power

There was no difference between dietary NO_3^- and placebo supplementation in PP (SMD: 0.01, 95% CI: −0.06 to 0.08, $p = 0.75$) (Figure S6a), PP_{First} (SMD: 0.05, 95% CI: −0.05 to 0.15, $p = 0.36$) (Figure S6b), and PP_{Last} (SMD: 0.10, 95% CI: −0.06 to 0.27, $p = 0.23$) (Figure S6c).

3.3.3. Mean Power

Both MP (SMD: 0.20, 95% CI: 0.03 to 0.36, $p = 0.02$) (Figure 4a) and MP_{First} (SMD: 0.11, 95% CI: 0.02 to 0.21, $p = 0.02$) (Figure 4b) were greater after dietary NO_3^- compared to placebo supplementation, with no significant difference between dietary NO_3^- and placebo supplementation in MP_{Last} (SMD: 0.06, 95% CI: −0.05 to 0.18, $p = 0.29$) (Figure 4c). There was a high risk of statistical heterogeneity between studies ($Chi^2 = 57.13$; $I^2 = 79\%$, $p < 0.00001$) measuring MP. Sensitivity analyses revealed that excluding studies in elite athletes [60,63] slightly increased the pooled SMD (SMD: 0.24, 95% CI: 0.06 to 0.42, $p = 0.009$) and reduced the statistical heterogeneity ($Chi^2 = 32.89$; $I^2 = 70\%$, $p < 0.0003$), while excluding studies with a high risk of bias [57,58,60] slightly reduced statistical heterogeneity ($Chi^2 \leq 31.44$; $I^2 = 71\%$, $p < 0.0002$) and the pooled SMD (SMD: 0.18, 95% CI: −0.01 to 0.36, $p = 0.07$). When the influence of NO_3^- dose was isolated, MP was greater after NO_3^- compared to placebo supplementation with high NO_3^- doses ≥ 8 mmol (SMD: 0.27, 95% CI: 0.01 to 0.54, $p = 0.04$), but there were no differences between NO_3^- and placebo supplementation when a NO_3^- dose < 8 mmol was administered (SMD: 0.19, 95% CI: −0.02 to 0.40, $p = 0.08$) (Figure S7a). There was no difference in MP between NO_3^- and placebo supplementation when a single-day supplementation protocol was adopted (SMD: 0.12, 95% CI: −0.03 to 0.26, $p = 0.11$), but the increase in MP after NO_3^- compared to placebo supplementation approached statistical significance when multiple-day supplementation was adopted (SMD: 0.27, 95% CI: 0.01 to 0.54, $p = 0.05$) (Figure S7b). When the influence of exercise type and duration was evaluated, MP was improved after NO_3^- compared to placebo supplementation during a single sprint (SMD: 0.31, 95% CI: 0.10 to 0.51, $p = 0.004$), but not during repeated sprints (SMD: 0.14, 95% CI: −0.04 to 0.32, $p = 0.13$) (Figure S7c) and when sprint time was >15 s–≤30 (SMD: 0.31, 95% CI: 0.12 to 0.50, $p = 0.001$), but not when sprint time ≤ 15 s (SMD: 0.14, 95%, CI: −0.05 to 0.34, $p = 0.15$) (Figure S7d). There were no differences in any of these comparisons for $MP_{First.}$

3.3.4. Total Work Done

NO_3^- supplementation did not alter TWD compared to placebo (SMD: 0.06, 95% CI: −0.13 to 0.26, $p = 0.52$) (Figure S8). There was a high risk of statistical heterogeneity between studies ($Chi^2 = 34.40$; $I^2 = 85\%$, $p < 0.00001$). Sensitivity analyses did not remove statistical heterogeneity or change the pooled SMD. Sub-group analysis on supplementation dose revealed a significant sub-group difference ($p = 0.03$) between high NO_3^- doses ≥ 8 mmol (SMD: 0.23, 95% CI: −0.03 to 0.49, $p = 0.08$) and low NO_3^- doses < 8 mmol (SMD: −0.14, 95% CI: −0.37 to 0.09, $p = 0.22$) (Figure 5a). The sub-group analysis on supplementation duration revealed a significant difference ($p = 0.004$) between multiple-day supplementation (SMD: 0.34, 95% CI: 0.09 to 0.60, $p = 0.008$) and single-day supplementation (SMD: −0.10, 95% CI: −0.28 to 0.07, $p = 0.24$) (Figure 5b).

3.3.5. Total Distance Covered

NO_3^- supplementation increased TDC compared to placebo (SMD: 0.17, 95% CI: 0.09 to 0.24, $p < 0.0001$) (Figure 6). There was a low risk of statistical heterogeneity between studies ($Chi^2 = 4.01$; $I^2 = 25\%$, $p = 0.26$). Sub-group and sensitivity analyses could not be performed due to an insufficient number of studies measuring TDC ($n = 4$).

(a)

Study or Subgroup	Std. Mean Difference	SE	Weight	Std. Mean Difference IV, Random, 95% CI
Aucouturier et al. (2015)	0.09	0.2	7.2%	0.09 [−0.30, 0.48]
Bernardi et al. (2018)	0.44	0.23	6.3%	0.44 [−0.01, 0.89]
Christensen et al. (2013)	0	0.04	11.9%	0.00 [−0.08, 0.08]
Corry et al. (2015)	0.53	0.07	11.2%	0.53 [0.39, 0.67]
Cuenca et al. (2018)	0.26	0.12	9.7%	0.26 [0.02, 0.50]
Domínguez et al. (2017)	0.35	0.23	6.3%	0.35 [−0.10, 0.80]
Dumar et al. (2021)	0.67	0.24	6.1%	0.67 [0.20, 1.14]
Jodra et al. (2020)	0.14	0.25	5.8%	0.14 [−0.35, 0.63]
Kramer et al. (2016)	0.02	0.2	7.2%	0.02 [−0.37, 0.41]
Martin et al. (2014)	0.03	0.19	7.5%	0.03 [−0.34, 0.40]
Pawlak-Chaouch et al. (2019)	0.01	0.23	6.3%	0.01 [−0.44, 0.46]
Smith et al. (2019)	−0.07	0.1	10.4%	−0.07 [−0.27, 0.13]
Wylie et al. (2016)	0.21	0.34	4.0%	0.21 [−0.46, 0.88]
Total (95% CI)			**100.0%**	**0.20 [0.03, 0.36]**

Heterogeneity: $Tau^2 = 0.06$; $Chi^2 = 57.13$, df = 12 ($P < 0.00001$); $I^2 = 79\%$
Test for overall effect: $Z = 2.35$ ($P = 0.02$)

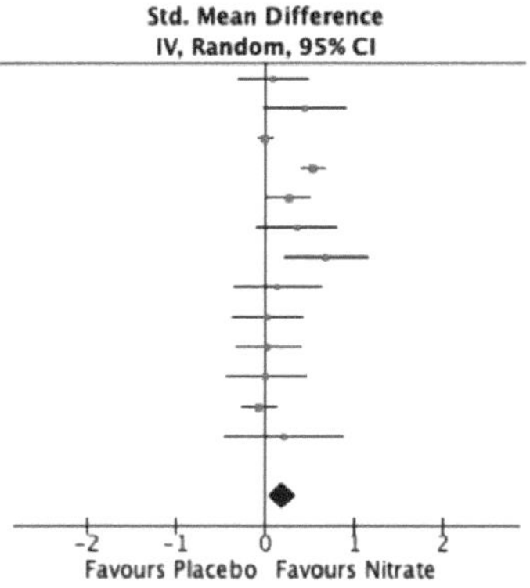

(b)

Study or Subgroup	Std. Mean Difference	SE	Weight	Std. Mean Difference IV, Random, 95% CI
Bender et al. (2018)	0.13	0.15	7.8%	0.13 [−0.16, 0.42]
Bernardi et al. (2018)	0.44	0.32	2.3%	0.44 [−0.19, 1.07]
Christensen et al. (2013)	0.01	0.13	9.5%	0.01 [−0.24, 0.26]
Corry et al. (2015)	0	0.24	3.7%	0.00 [−0.47, 0.47]
Cuenca et al. (2018)	0.28	0.13	9.5%	0.28 [0.03, 0.53]
Domínguez et al. (2017)	0.45	0.18	6.0%	0.45 [0.10, 0.80]
Dumar et al. (2021)	0.57	0.34	2.0%	0.57 [−0.10, 1.24]
Jonvik et al. (2018)	0.01	0.02	24.9%	0.01 [−0.03, 0.05]
Roelofs et al. (2017)	0.02	0.12	10.5%	0.02 [−0.22, 0.26]
Smith et al. (2019)	−0.01	0.07	17.3%	−0.01 [−0.15, 0.13]
Wylie et al. (2016)	0.38	0.17	6.5%	0.38 [0.05, 0.71]
Total (95% CI)			**100.0%**	**0.11 [0.02, 0.21]**

Heterogeneity: $Tau^2 = 0.01$; $Chi^2 = 19.34$, df = 10 ($P = 0.04$); $I^2 = 48\%$
Test for overall effect: $Z = 2.26$ ($P = 0.02$)

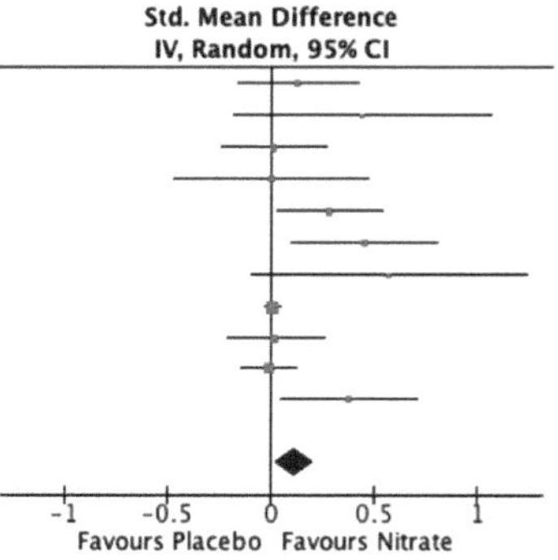

(c)

Study or Subgroup	Std. Mean Difference	SE	Weight	Std. Mean Difference IV, Random, 95% CI
Bender et al. (2018)	−0.19	0.18	8.0%	−0.19 [−0.54, 0.16]
Bernardi et al. (2018)	−0.18	0.2	6.8%	−0.18 [−0.57, 0.21]
Christensen et al. (2013)	0.11	0.09	16.5%	0.11 [−0.07, 0.29]
Corry et al. (2015)	0.28	0.22	5.9%	0.28 [−0.15, 0.71]
Cuenca et al. (2018)	0.25	0.26	4.5%	0.25 [−0.26, 0.76]
Domínguez et al. (2017)	0.2	0.26	4.5%	0.20 [−0.31, 0.71]
Dumar et al. (2021)	0.66	0.22	5.9%	0.66 [0.23, 1.09]
Jonvik et al. (2018)	−0.01	0.04	23.1%	−0.01 [−0.09, 0.07]
Roelofs et al. (2017)	0.08	0.15	10.1%	0.08 [−0.21, 0.37]
Smith et al. (2019)	−0.23	0.17	8.6%	−0.23 [−0.56, 0.10]
Wylie et al. (2016)	0.2	0.22	5.9%	0.20 [−0.23, 0.63]
Total (95% CI)			**100.0%**	**0.06 [−0.05, 0.18]**

Heterogeneity: $Tau^2 = 0.01$; $Chi^2 = 18.03$, df = 10 ($P = 0.05$); $I^2 = 45\%$
Test for overall effect: $Z = 1.06$ ($P = 0.29$)

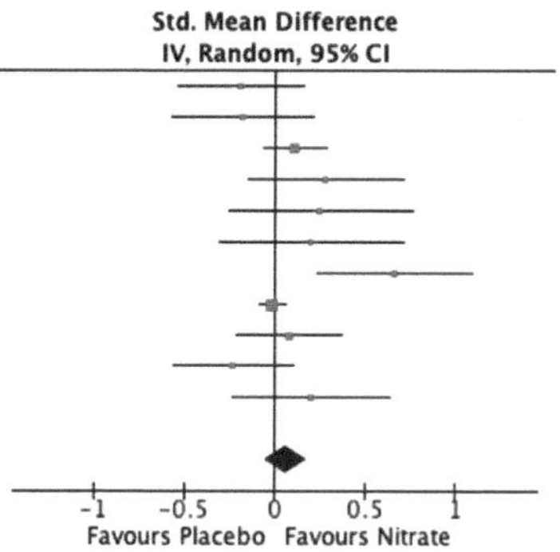

Figure 4. Forest plot for mean power from all sprints (**a**), mean power during the first sprint (**b**), and mean power during the last sprint (**c**) in the nitrate and placebo trials [35,36,39,41,54,56–61,63–67].

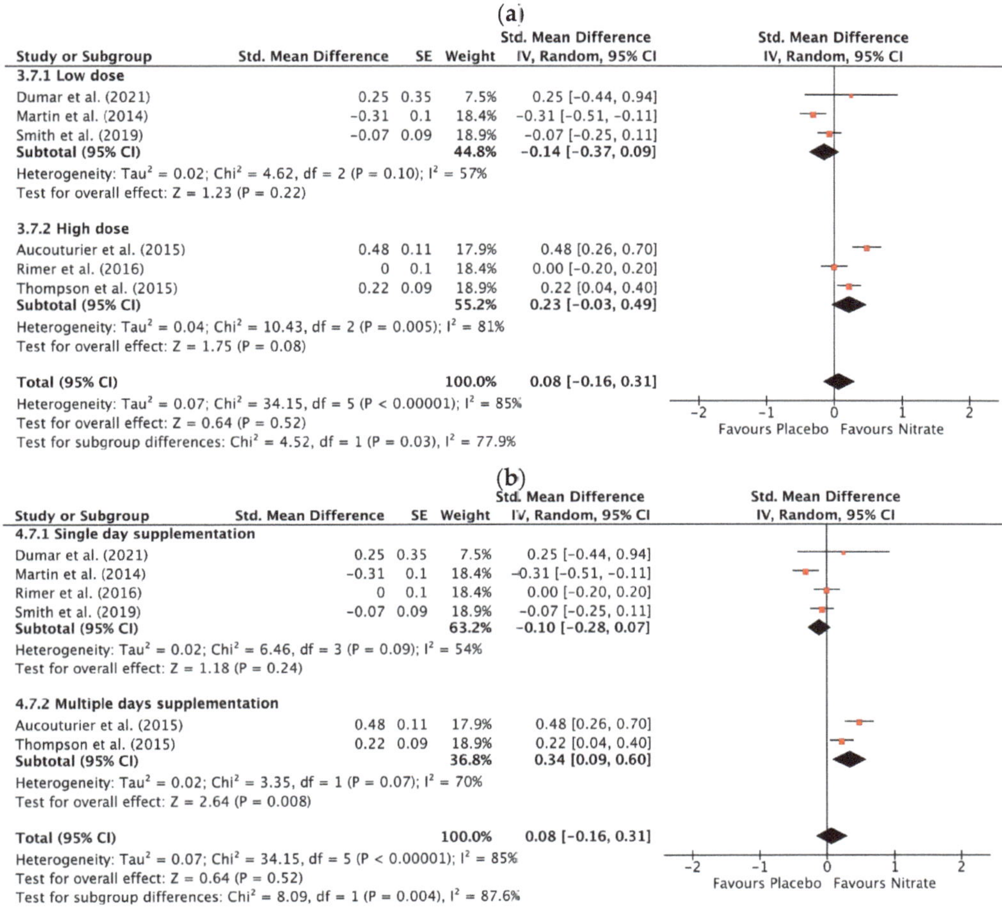

Figure 5. Forest plot of total work performed in the nitrate and placebo trials (**a**), low nitrate dose (<8 mmol/day) compared to high nitrate dose (≥8 mmol/day) (**b**), multiday nitrate supplementation compared to single day nitrate supplementation [34,51,56,61,65,66].

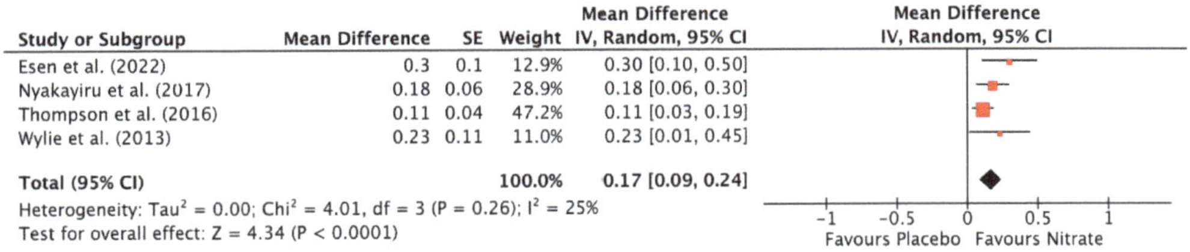

Figure 6. Forest plot for total distance covered in the nitrate and placebo trials [37,38,55,62].

4. Discussion

The principal observations of this systematic review and meta-analyses are that, compared to a placebo condition, NO_3^- supplementation lowered PP_{Time} without impacting PP, increased MP and MP_{First}, and increased TDC in the Yo-Yo IR1 test. The improvement in MP after NO_3^- supplementation was more likely to occur when NO_3^- was administered for multiple days at a dose ≥ 8 mmol as opposed to an acute serving of <8 mmol during a single bout rather than repeated bouts of high-intensity exercise, and when the high-intensity exercise duration was >15 s–≤30 s versus ≤15 s. The sub-group analysis also revealed that NO_3^- supplementation was more likely to improve TWD in a high-intensity repeated bout protocol when NO_3^- was administered at a dose ≥ 8 mmol and was supplemented for multiple days as opposed to an acute serving or a dose < 8 mmol. These observations improve our understanding of the effects of NO_3^- supplementation on single and repeated bouts of short-duration, high-intensity, large muscle mass exercise, and reveal two apparently distinct and supplementation-strategy-dependent effects of dietary NO_3^- on high-intensity exercise performance. Firstly, NO_3^- supplementation appears to improve PP_{Time} and MP_{First}, with the improvements in these variables not necessarily requiring multiple-day supplementation with ≥8 mmol NO_3^-, as such effects appear to be achievable after acute supplementation with ~6 mmol NO_3^-. Secondly, TWD in a repeated sprint protocol was more likely to be improved when NO_3^- was administered at a dose ≥ 8 mmol, and was supplemented for multiple days, consistent with the NO_3^- supplementation regime administered in the studies assessing TDC in the Yo-Yo IR1 test, all of which reported improved performance. Therefore, it appears that a single bout of high-intensity exercise can be enhanced by acute NO_3^- supplementation, with high-intensity intermittent exercise performance more likely to improve after multiple day supplementation, with a NO_3^- dose ≥8 mmol. These findings may have implications for future study design and for improving performance in athletes participating in sports that require high-intensity bouts of exercise.

Although there are some examples of enhanced PP after NO_3^- supplementation [35,52,54,57–59,67], the current meta-analysis indicates that most previous studies did not report improved PP, PP_{First}, or PP_{Last} after NO_3^- supplementation [36,39,41,53,56,60,64,66]. However, whilst PP was not altered, PP_{Time} was lowered after NO_3^- supplementation with all four studies assessing this variable observing a lower PP_{Time} after NO_3^- supplementation [35,36,54,57], with three of these studies administering an acute NO_3^- dose of ~6 mmol [35,54,57]. This observation is compatible with an increase in muscle contractile velocity, which would be expected to contribute to lower PP_{Time} after acute NO_3^- supplementation [33,34]. With regard to MP variables, MP and MP_{First}, but not MP_{Last}, were improved after NO_3^- supplementation. When the improvement in MP after NO_3^- supplementation was explored further, MP was improved after NO_3^- supplementation when doses ≥ 8 mmol were administered [41,64,65], when multiple day supplementation protocols were adopted [41,64,65], and when a single sprint >15 s–≤30 s was performed [35,54,58,59]. The improvements in PP_{Time} and MP_{First} were exhibited after acute supplementation with ~6 mmol NO_3^- [35,39,54,57]. All four studies assessing the effect of NO_3^- supplementation on TDC in the Yo-Yo IR1 test revealed a greater TDC after NO_3^- supplementation [37,38,55,62]. While TWD during high-intensity intermittent exercise was not improved after NO_3^- supplementation, the sub-group analysis revealed that TWD was increased when the NO_3^- dose was ≥8 mmol compared to <8 mmol [51,65], and with multiple-day supplementation compared to acute supplementation [51,65]. Importantly, the four studies reporting improved TDC in the Yo-Yo IR1 test all adopted a multiple-day supplementation protocol with a NO_3^- dose of >8 mmol [37,38,55,62]. Therefore, it appears that a multiple-day supplemental protocol with a NO_3^- dose of >8 mmol is important to elicit an ergogenic effect on repeated bouts of high-intensity exercise after NO_3^- supplementation but that performance in single sprints (lower PP_{Time} and higher MP) can be enhanced after acute ingestion of ~6 mmol NO_3^-.

The ergogenic effect of NO_3^- supplementation has been attributed to its stepwise reduction to NO_2^- and the subsequent reduction of NO_2^- to NO [2,5]. It is now recognised that ~25% of ingested NO_3^- is extracted from the circulation by the salivary glands [76] via the NO_3^-/H^+ cotransporter, sialin [77]. NO_3^- is subsequently concentrated within salivary glands [78] with excreted salivary NO_3^- undergoing reduction to NO_2^- by certain species of the oral micobiome [79–81]. NO_2^--rich saliva is then swallowed and subsequently reduced to NO and various reactive nitrogen intermediates, including S-nitrosothiols

(RSNO) within the stomach [2,78], but it is also clear that circulating plasma [NO_2^-] and [RSNO] are increased post NO_3^- supplementation [78,82,83]. Circulating plasma NO_2^- can undergo a one-electron reduction to NO in a reaction catalysed by numerous NO_2^- reductases [84,85]. Although the relationship between exercise performance and plasma [NO_3^-] is unclear, exercise responses are positively associated with the increases in plasma [NO_2^-] [82,86], muscle [NO_3^-], and muscle NO_3^- utilisation [87] after NO_3^- supplementation.

It is increasingly appreciated that skeletal muscle can serve as an important store of NO_3^- and NO_2^- for subsequent NO synthesis, as evidenced by higher [NO_3^-] and [NO_2^-] in skeletal muscle than blood [88,89]. The NO_3^- transporter, sialin, has been identified in skeletal muscle [89,90] which, together with chloride channel 1 [90], facilitate the concentration of NO_3^- within skeletal muscle. Therefore, a portion of the increased circulating blood NO_3^- after NO_3^- supplementation, which is not extracted by the kidney for clearance in the urine or absorbed by the salivary glands for subsequent oral reduction to NO_2^-, can be accrued in skeletal muscle. Indeed, skeletal muscle [NO_3^-] and [NO_2^-] are increased following NO_3^- supplementation with duration-dependent increases at least up to 7 days of supplementation [88]. In addition to its role as a NO_2^- reductase [91], xanthine oxidoreductase (XOR) can function as a NO_3^- reductase to increase NO_2^- synthesis [92] and is present in skeletal muscle [89,90]. It has been reported that the increase in skeletal muscle [NO_2^-] after NO_3^- administration is enhanced by exercise and, as muscle pH is lowered, with both NO_3^- reduction to NO_2^- and NO_2^- reduction to NO abolished after XOR inhibition [93]. It is, therefore, possible that increased XOR activity during exercise, particularly high-intensity exercise [94], could contribute to enhanced muscle NO_3^- and NO_2^- reduction in such settings. Indeed, the increase in skeletal muscle [NO_3^-] after NO_3^- supplementation is lowered following the completion of exhaustive cycling exercise [89] and maximal knee extensor contractions [87], suggesting that this elevated muscle NO_3^- pool is utilised as a substrate for sequential reduction to NO_2^- and then NO. There is also a positive arterial-venous difference in plasma [NO_3^-] and [NO_2^-] across contracting skeletal muscles after NO_3^- supplementation [95]. Since NO_2^- reduction to NO is augmented in hypoxia and acidosis [26–28], and given that such conditions develop within the muscle microvasculature during exercise in an intensity-dependent manner [31], elevating circulating plasma [NO_2^-] is likely to increase NO synthesis in the muscle microvasculature during high-intensity exercise. Based on the existing evidence, NO_3^- and NO_2^- can be increased systemically and within skeletal muscle following dietary NO_3^- supplementation with the potential to enhance NO synthesis, particularly during the hypoxic and acidic conditions that develop during high-intensity exercise, which might underpin the improvements in high-intensity exercise performance variables reported in this manuscript.

The improvements in PP_{Time} and MP_{First} during an all-out sprint after NO_3^- supplementation are likely mediated by mechanisms intrinsic to the myocytes. The initial stages of a short-duration all-out sprint, during which PP_{Time} will be determined, will involve maximal recruitment of, and proportion contribution to force production from, type II skeletal muscle fibres [96,97]. Previous research has indicated that 7 days NO_3^- supplementation can increase calcium (Ca^{2+}) handling proteins and evoke force production in type II skeletal muscle, but not slow-twitch (type I) skeletal muscle, in mice [18]. However, three [35,54,57] of the four [35,36,54,57] studies reporting improved PP_{Time}, and six [39,54,56,57,64,67] of the eleven [36,39,41,54,56–58,60,61,64,67] studies reporting improved MP_{First} after NO_3^- supplementation administered NO_3^- acutely, and it has been reported that increased evoked muscle force production can occur independently of changes in Ca^{2+} handling proteins in human skeletal muscle [98]. Therefore, the improvements in PP_{Time} and MP_{First} after NO_3^- supplementation are likely to be underpinned by NO-cyclic guanosine monophosphate (cGMP)-mediated signalling and/or post-translational modification of protein thiols [99].

In contrast to the NO_3^- supplementation regime required to improve PP_{Time} and MP_{First}, TWD during high-intensity intermittent exercise was improved after NO_3^- supplementation when the NO_3^- dose was ≥8 mmol, but not <8 mmol, and only with multiple-day supplementation. There was also a greater TDC in the Yo-Yo IR1 after NO_3^- supplementation with all studies reporting this ergogenic effect employing multiple-day NO_3^- supplementation at a daily dose ≥8 mmol. Greater ergogenic effects during high-intensity intermittent exercise after multiple-day, higher dose NO_3^- supplementation might be linked to the greater time course to increase muscle [NO_2^-] after NO_3^- supplementation as unlike muscle [NO_3^-], muscle [NO_2^-] is not increased after acute NO_3^- ingestion [87,89] but can be increased after 7 days of NO_3^- ingestion [88]. Indeed, when mouse single myocytes were acutely exposed to increased NO_2^-, contractile function and Ca^{2+} handling were not altered in the earlier stages of a fatigue-inducing contraction protocol, whereas time to task failure was extended as a result of better maintenance of myocyte contractility, Ca^{2+} sensitivity, and Ca^{2+} pumping towards the latter stages of the protocol [100] In human skeletal muscle, greater potential for improved muscle contractile responses during a fatigue-inducing 60 maximum voluntary contraction protocol has been reported during the initial contractions after acute NO_3^- ingestion [87] and following completion of the fatiguing protocol after multiple-day NO_3^- supplementation [101]. Skeletal muscle [NO_3^-] and [NO_2^-] increase in a duration-dependent manner following NO_3^- supplementation [88], and muscle [NO_3^-] declines during sustained high-intensity exercise [87,89] and is correlated with improved muscle force production [87]. Therefore, multiple-day NO_3^- supplementation with a NO_3^- dose exceeding 8 mmol may be more effective at improving MP during a single 15–30 s bout of high-intensity exercise or at improving TWD or TDC during high-intensity intermittent exercise by eliciting greater increases in muscle [NO_3^-] and [NO_2^-] to support greater NO_3^- reduction and NO generating potential during these high-intensity exercise settings. As such, NO_3^- may impact skeletal muscle contractile function in a supplementation-strategy-dependent manner that may be mediated by different muscle exposures to NO_3^- and NO_2^-.

Although the findings of the current study may have implications for improving NO_3^- supplementation strategies to bolster performance in different types of high-intensity exercise, there are several limitations of, and experimental considerations from, the studies included in this systematic review and meta-analysis. Firstly, the SMD was typically small across all variables that did exhibit an ergogenic effect after NO_3^- supplementation, which underscores the importance of assessing the translational potential of these findings to improve in-competition performance in sports where performance outcomes are dictated by the capability to perform high-intensity exercise. Moreover, the meta-analysis conducted on PP_{Time} and MP exhibited high heterogeneity, indicating a substantial variation in the results of the included studies. Since a limited number of studies assessed plasma [NO_3^-] and [NO_2^-] and included female participants, not all planned sub-analyses could be completed. There was also limited assessment of the physiological mechanisms for any improvement in high-intensity exercise performance in the studies included in the current systematic review and meta-analyses. Therefore, further research is required to resolve the putative mechanisms for improved performance during single and repeated bouts of short duration high-intensity exercise and the extent to which such mechanisms are influenced by acute and multiple-day NO_3^- ingestion and mediated by plasma and muscle [NO_3^-] and [NO_2^-] and different population groups

5. Conclusions

The current study conducted a systematic review and completed several meta analyses to evaluate the effect of dietary NO_3^- supplementation of different aspects of high-intensity exercise performance, with sub-analyses conducted to provide wider contextual insight. It was observed that NO_3^- supplementation lowered PP_{Time}, increased MP and MP_{First}, and increased TDC in the Yo-Yo IR1 test, supporting the ergogenic potential of dietary NO_3^- supplementation for some aspects of high-intensity exercise performance. Sub-group analyses revealed that MP was more likely to be improved during a single >15 s–≤30 s versus ≤15 s bout rather than repeated bouts of high-intensity exercise, and that MP, TWD, and TDC were more likely to be improved after multiple-day supplementation with a daily NO_3^- dose ≥8 mmol compared to acute ingestion of <8 mmol NO_3^-. These findings improve our understanding of the ergogenic potential of dietary NO_3^- supplementation for high-intensity exercise and can help inform NO_3^- supplementation strategies to improve high-intensity exercise performance.

Supplementary Materials: The following supporting information can be downloaded at: https://www.mdpi.com/article/10.3390/antiox12061194/s1: Search strategy; Table S1: Population, Intervention, Comparator, Outcome, Study design (PICOS) framework for study eligibility; Figure S1: Risk of bias summary for individual for crossover trials; Figure S2: Funnel plot evaluating publication bias of trials assessing mean peak power output (n = 12); Figure S3: Funnel plot evaluating publication bias of trials assessing mean of the mean power output (n = 12); Figure S4: Funnel plot evaluating publication bias of trials assessing mean power output during the first sprint (n = 10); Figure S5: Funnel plot evaluating publication bias of trials assessing mean power output during the last sprint (n = 10); Figure S6: Forrest plot for mean peak power output (a), peak power during the first sprint (b), and peak power during the last sprint (c) in the nitrate and placebo trials; Figure S7: Forrest plot for mean power output sub-group analyse; low nitrate dose <8 mmol compared to high nitrate dose ≥8 mmol (a), single day nitrate supplementation compared multiple days nitrate supplementation (b), single sprint compared to repeated sprints (c), exercise duration ≤ 15 s compared to exercise duration >15 s–≤30 s (d); Figure S8: Forrest plot for total work done in the nitrate and placebo trials.

Author Contributions: Conceptualization, S.J.B., N.S.A. and T.C.; methodology, S.J.B., N.S.A. and T.C; formal analysis, N.S.A., S.N.R., A.A. and T.C.; writing—original draft preparation, S.J.B. and N.S.A.; writing-review and editing, S.J.B., N.S.A., T.C., S.N.R. and A.A. All authors have read and agreed to the published version of the manuscript.

Funding: This research was supported by the NIHR Leicester Biomedical Research Centre.

Institutional Review Board Statement: Not Applicable.

Informed Consent Statement: Not Applicable.

Data Availability Statement: Data can be provided at reasonable request from corresponding authors.

Conflicts of Interest: The authors declare no conflict of interest.

References

1. Lundberg, J.O.; Weitzberg, E.; Gladwin, M.T. The Nitrate-Nitrite-Nitric Oxide Pathway in Physiology and Therapeutics. *Nat. Rev. Drug Discov.* **2008**, *7*, 156–167. [CrossRef] [PubMed]
2. Kapil, V.; Khambata, R.S.; Jones, D.A.; Rathod, K.; Primus, C.; Massimo, G.; Fukuto, J.M.; Ahluwalia, A. The Noncanonical Pathway for in Vivo Nitric Oxide Generation: The Nitrate-Nitrite-Nitric Oxide Pathway. *Pharmacol. Rev.* **2020**, *72*, 692–766. [CrossRef] [PubMed]
3. Maughan, R.J.; Burke, L.M.; Dvorak, J.; Larson-Meyer, D.E.; Peeling, P.; Phillips, S.M.; Rawson, E.S.; Walsh, N.P.; Garthe, I.; Geyer, H.; et al. IOC Consensus Statement: Dietary Supplements and the High-Performance Athlete. *Int. J. Sport Nutr. Exerc. Metab.* **2018**, *52*, 439–455. [CrossRef]
4. Peeling, P.; Binnie, M.J.; Goods, P.S.R.; Sim, M.; Burke, L.M. Evidence-Based Supplements for the Enhancement of Athletic Performance. *Int. J. Sport Nutr. Exerc. Metab.* **2018**, *28*, 178–187. [CrossRef]
5. Jones, A.M.; Vanhatalo, A.; Seals, D.R.; Rossman, M.J.; Piknova, B.; Jonvik, K.L. Dietary Nitrate and Nitric Oxide Metabolism: Mouth, Circulation, Skeletal Muscle, and Exercise Performance. *Med. Sci. Sports Exerc.* **2021**, *53*, 280–294. [CrossRef]
6. Ignarro, L.J.; Buga, G.M.; Wood, K.S.; Byrns, R.E. Artery and Vein Is Nitric Oxide. *Sci. York* **1987**, *84*, 9265–9269.
7. Moon, Y.; Balke, J.E.; Madorma, D.; Siegel, M.P.; Knowels, G.; Brouckaert, P.; Buys, E.S.; Marcinek, D.J.; Percival, J.M. Nitric Oxide Regulates Skeletal Muscle Fatigue, Fiber Type, Microtubule Organization, and Mitochondrial ATP Synthesis Efficiency Through CGMP-Dependent Mechanisms. *Antioxid. Redox Signal.* **2017**, *26*, 966–985. [CrossRef]
8. Stamler, J.S.; Meissner, G. Physiology of Nitric Oxide in Skeletal Muscle. *Physiol. Rev.* **2001**, *81*, 209–237. [CrossRef]
9. Suhr, F.; Gehlert, S.; Grau, M.; Bloch, W. Skeletal Muscle Function during Exercise-Fine-Tuning of Diverse Subsystems by Nitric Oxide. *Int. J. Mol. Sci.* **2013**, *14*, 7109–7139. [CrossRef]
10. Bailey, S.J.; Winyard, P.; Vanhatalo, A.; Blackwell, J.R.; DiMenna, F.J.; Wilkerson, D.P.; Tarr, J.; Benjamin, N.; Jones, A.M. Dietary Nitrate Supplementation Reduces the O2 Cost of Low-Intensity Exercise and Enhances Tolerance to High-Intensity Exercise in Humans. *J. Appl. Physiol.* **2009**, *107*, 1144–1155. [CrossRef]
11. Larsen, F.J.; Weitzberg, E.; Lundberg, J.O.; Ekblom, B. Effects of Dietary Nitrate on Oxygen Cost during Exercise. *Acta Physiol.* **2007**, *191*, 59–66. [CrossRef]

12. Bailey, S.; Vanhatalo, A.; Winyard, P.; Jones, A. The Nitrate-Nitrite-Nitric Oxide Pathway: Its Role in Human Exercise Physiology. *Eur. J. Sport Sci.* **2012**, *12*, 309–320. [CrossRef]
13. Bailey, S.J.; Fulford, J.; Vanhatalo, A.; Winyard, P.G.; Blackwell, J.R.; DiMenna, F.J.; Wilkerson, D.P.; Benjamin, N.; Jones, A.M. Dietary Nitrate Supplementation Enhances Muscle Contractile Efficiency during Knee-Extensor Exercise in Humans. *J. Appl. Physiol.* **2010**, *109*, 135–148. [CrossRef]
14. Larsen, F.J.; Schiffer, T.A.; Borniquel, S.; Sahlin, K.; Ekblom, B.; Lundberg, J.O.; Weitzberg, E. Dietary Inorganic Nitrate Improves Mitochondrial Efficiency in Humans. *Cell Metab.* **2011**, *13*, 149–159. [CrossRef]
15. Betteridge, S.; Bescós, R.; Martorell, M.; Pons, A.; Garnham, A.P.; Stathis, C.C.; McConell, G.K. No Effect of Acute Beetroot Juice Ingestion on Oxygen Consumption, Glucose Kinetics, or Skeletal Muscle Metabolism during Submaximal Exercise in Males. *J. Appl. Physiol.* **2016**, *120*, 391–398. [CrossRef]
16. Whitfield, J.; Ludzki, A.; Heigenhauser, G.J.F.; Senden, J.M.G.; Verdijk, L.B.; van Loon, L.J.C.; Spriet, L.L.; Holloway, G.P. Beetroot Juice Supplementation Reduces Whole Body Oxygen Consumption but Does Not Improve Indices of Mitochondrial Efficiency in Human Skeletal Muscle. *J. Physiol.* **2016**, *594*, 421–435. [CrossRef]
17. Jones, A.M.; Ferguson, S.K.; Bailey, S.J.; Vanhatalo, A.; Poole, D.C. Fiber Type-Specific Effects of Dietary Nitrate. *Exerc. Sport Sci. Rev.* **2016**, *44*, 53–60. [CrossRef]
18. Hernández, A.; Schiffer, T.A.; Ivarsson, N.; Cheng, A.J.; Bruton, J.D.; Lundberg, J.O.; Weitzberg, E.; Westerblad, H. Dietary Nitrate Increases Tetanic [Ca 2+] i and Contractile Force in Mouse Fast-Twitch Muscle. *J. Physiol.* **2012**, *590*, 3575–3583. [CrossRef]
19. Ferguson, S.K.; Hirai, D.M.; Copp, S.W.; Holdsworth, C.T.; Allen, J.D.; Jones, A.M.; Musch, T.I.; Poole, D.C. Impact of Dietary Nitrate Supplementation via Beetroot Juice on Exercising Muscle Vascular Control in Rats. *J. Physiol.* **2013**, *591*, 547–557. [CrossRef]
20. Breese, B.C.; Mcnarry, M.A.; Marwood, S.; Blackwell, J.R.; Bailey, S.J.; Jones, A.M. Beetroot Juice Supplementation Speeds O2 Uptake Kinetics and Improves Exercise Tolerance during Severe-Intensity Exercise Initiated from an Elevated Metabolic Rate. *Am. J. Physiol. Regul. Integr. Comp. Physiol.* **2013**, *305*, R1441–R1450. [CrossRef]
21. Porcelli, S.; Ramaglia, M.; Bellistri, G.; Pavei, G.; Pugliese, L.; Montorsi, M.; Rasica, L.; Marzorati, M. Aerobic Fitness Affects the Exercise Performance Responses to Nitrate Supplementation. *Med. Sci. Sports Exerc.* **2015**, *47*, 1643–1651. [CrossRef] [PubMed]
22. Jones, A.M. Dietary Nitrate Supplementation and Exercise Performance. *Sport. Med.* **2014**, *44*, 35–45. [CrossRef] [PubMed]
23. Krustrup, P.; Söderlund, K.; Mohr, M.; Bangsbo, J. The Slow Component of Oxygen Uptake during Intense, Sub-Maximal Exercise in Man Is Associated with Additional Fibre Recruitment. *Pflugers Arch. Eur. J. Physiol.* **2004**, *447*, 855–866. [CrossRef] [PubMed]
24. Krustrup, P.; Söderlund, K.; Relu, M.U.; Ferguson, R.A.; Bangsbo, J. Heterogeneous Recruitment of Quadriceps Muscle Portions and Fibre Types during Moderate Intensity Knee-Extensor Exercise: Effect of Thigh Occlusion. *Scand. J. Med. Sci. Sport.* **2009**, *19*, 576–584. [CrossRef]
25. Vøllestad, N.K.; Blom, P.C.S. Effect of Varying Exercise Intensity on Glycogen Depletion in Human Muscle Fibres. *Acta Physiol. Scand.* **1985**, *125*, 395–405. [CrossRef]
26. Castello, P.R.; David, P.S.; McClure, T.; Crook, Z.; Poyton, R.O. Mitochondrial Cytochrome Oxidase Produces Nitric Oxide under Hypoxic Conditions: Implications for Oxygen Sensing and Hypoxic Signaling in Eukaryotes. *Cell Metab.* **2006**, *3*, 277–287. [CrossRef]
27. Li, H.; Cui, H.; Kundu, T.K.; Alzawahra, W.; Zweier, J.L. Nitric Oxide Production from Nitrite Occurs Primarily in Tissues Not in the Blood: Critical Role of Xanthine Oxidase and Aldehyde Oxidase. *J. Biol. Chem.* **2008**, *283*, 17855–17863. [CrossRef]
28. Modin, A.; Björne, H.; Herulf, M.; Alving, K.; Weitzberg, E.; Lundberg, J.O.N. Nitrite-Derived Nitric Oxide: A Possible Mediator of "acidic-Metabolic" Vasodilation. *Acta Physiol. Scand.* **2001**, *171*, 9–16. [CrossRef]
29. Harkema, S.J.; Adams, G.R.; Meyer, R.A. Acidosis Has No Effect on the ATP Cost of Contraction in Cat Fast- and Slow-Twitch Skeletal Muscles. *Am. J. Physiol. Cell Physiol.* **1997**, *272*, 485–490. [CrossRef]
30. McDonough, P.; Behnke, B.J.; Padilla, D.J.; Musch, T.I.; Poole, D.C. Control of Microvascular Oxygen Pressures in Rat Muscles Comprised of Different Fibre Types. *J. Physiol.* **2005**, *563*, 903–913. [CrossRef]
31. Richardson, R.S.; Noyszewski, E.A.; Kendrick, K.F.; Leigh, J.S.; Wagner, P.D. Myoglobin O2 Desaturation during Exercise: Evidence of Limited O2 Transport. *J. Clin. Investig.* **1995**, *96*, 1916–1926. [CrossRef]
32. Bailey, S.J.; Varnham, R.L.; DiMenna, F.J.; Breese, B.C.; Wylie, L.J.; Jones, A.M. Inorganic Nitrate Supplementation Improves Muscle Oxygenation, O2 Uptake Kinetics, and Exercise Tolerance at High but Not Low Pedal Rates. *J. Appl. Physiol.* **2015**, *118*, 1396–1405. [CrossRef]
33. Coggan, A.R.; Leibowitz, J.L.; Kadkhodayan, A.; Thomas, D.P.; Ramamurthy, S.; Spearie, C.A.; Waller, S.; Farmer, M.; Peterson, L.R. Effect of Acute Dietary Nitrate Intake on Maximal Knee Extensor Speed and Power in Healthy Men and Women. *Nitric. Oxide Biol. Chem.* **2015**, *48*, 16–21. [CrossRef]
34. Rimer, E.G.; Peterson, L.R.; Coggan, A.R.; Martin, J.C. Acute Dietary Nitrate Supplementation Increases Maximal Cycling Power in Athletes. *Physiol. Behav.* **2017**, *176*, 139–148. [CrossRef]
35. Jodra, P.; Domínguez, R.; Sánchez-Oliver, A.J.; Veiga-Herreros, P.; Bailey, S.J. Effect of Beetroot Juice Supplementation on Mood, Perceived Exertion, and Performance during a 30-Second Wingate Test. *Int. J. Sports Physiol. Perform.* **2020**, *15*, 243–248. [CrossRef]
36. Jonvik, K.L.; Nyakayiru, J.; Van Dijk, J.W.; Maase, K.; Ballak, S.B.; Senden, J.M.G.; Van Loon, L.J.C.; Verdijk, L.B. Repeated-Sprint Performance and Plasma Responses Following Beetroot Juice Supplementation Do Not Differ between Recreational, Competitive and Elite Sprint Athletes. *Eur. J. Sport Sci.* **2018**, *18*, 524–533. [CrossRef]

37. Thompson, C.; Vanhatalo, A.; Jell, H.; Fulford, J.; Carter, J.; Nyman, L.; Bailey, S.J.; Jones, A.M. Dietary Nitrate Supplementation Improves Sprint and High-Intensity Intermittent Running Performance. *Nitric. Oxide Biol. Chem.* **2016**, *61*, 55–61. [CrossRef]
38. Wylie, L.J.; Mohr, M.; Krustrup, P.; Jackman, S.R.; Ermιdis, G.; Kelly, J.; Black, M.I.; Bailey, S.J.; Vanhatalo, A.; Jones, A.M. Dietary Nitrate Supplementation Improves Team Sport-Specific Intense Intermittent Exercise Performance. *Eur. J. Appl. Physiol.* **2013**, *113*, 1673–1684. [CrossRef]
39. Bender, D.; Townsend, J.R.; Vantrease, W.C.; Marshall, A.C.; Henry, R.N.; Heffington, S.H.; Johnson, K.D. Acute Beetroot Juice Administration Improves Peak Isometric Force Production in Adolescent Males. *Appl. Physiol. Nutr. Metab.* **2018**, *43*, 816–821. [CrossRef]
40. López-Samanes, Á.; Pérez-López, A.; Moreno-Pérez, V.; Nakamura, F.Y.; Acebes-Sánchez, J.; Quintana-Milla, I.; Sánchez-Oliver, A.J.; Moreno-Pérez, D.; Fernández-Elías, V.E.; Domínguez, R. Effects of Beetroot Juice Ingestion on Physical Performance in Highly Competitive Tennis Players. *Nutrients* **2020**, *12*, 584. [CrossRef]
41. Wylie, L.J.; Bailey, S.J.; Kelly, J.; Blackwell, J.R.; Vanhatalo, A.; Jones, A.M. Influence of Beetroot Juice Supplementation on Intermittent Exercise Performance. *Eur. J. Appl. Physiol.* **2016**, *116*, 415–425. [CrossRef] [PubMed]
42. Alvares, T.S.; de Oliveira, G.V.; Volino-Souza, M.; Conte-Junior, C.A.; Murias, J.M. Effect of Dietary Nitrate Ingestion on Muscular Performance: A Systematic Review and Meta-Analysis of Randomized Controlled Trials. *Crit. Rev. Food Sci. Nutr.* **2021**, *62*, 5284–5306. [CrossRef]
43. Coggan, A.R.; Baranauskas, M.N.; Hinrichs, R.J.; Liu, Z.; Carter, S.J. Effect of Dietary Nitrate on Human Muscle Power: A Systematic Review and Individual Participant Data Meta-Analysis. *J. Int. Soc. Sports Nutr.* **2021**, *18*, 66. [CrossRef] [PubMed]
44. Van De Walle, G.P.; VUKOVICH, M.D.; VanDeWalle, G.P.; VUKOVICH, M.D. The Effect of Nitrate Supplementation on Exercise Tolerance and Performance: A Systematic Review and Meta-Analysis. *J. Strength Cond. Res.* **2018**, *32*, 1796–1808. [CrossRef] [PubMed]
45. Gao, C.; Gupta, S.; Adli, T.; Hou, W.; Coolsaet, R.; Hayes, A.; Kim, K.; Pandey, A.; Gordon, J.; Chahil, G.; et al. The Effects of Dietary Nitrate Supplementation on Endurance Exercise Performance and Cardiorespiratory Measures in Healthy Adults: A Systematic Review and Meta-Analysis. *J. Int. Soc. Sports Nutr.* **2021**, *18*, 55. [CrossRef]
46. McMahon, N.F.; Leveritt, M.D.; Pavey, T.G. The Effect of Dietary Nitrate Supplementation on Endurance Exercise Performance in Healthy Adults: A Systematic Review and Meta-Analysis. *Sport. Med.* **2017**, *47*, 735–756. [CrossRef] [PubMed]
47. Campos, H.O.; Drummond, L.R.; Rodrigues, Q.T.; Machado, F.S.M.; Pires, W.; Wanner, S.P.; Coimbra, C.C. Nitrate Supplementation Improves Physical Performance Specifically in Non-Athletes during Prolonged Open-Ended Tests: A Systematic Review and Meta-Analysis. *Br. J. Nutr.* **2018**, *119*, 636–657. [CrossRef]
48. Senefeld, J.W.; Wiggins, C.C.; Regimbal, R.J.; Dominelli, P.B.; Baker, S.E.; Joyner, M.J.; Wiggins, C.C.; Regimbal, R.J.; Dominelli, P.B.; Baker, S.E.; et al. Ergogenic Effect of Nitrate Supplementation: A Systematic Review and Meta-Analysis. *Med. Sci. Sport. Exerc.* **2020**, *52*, 2250–2261. [CrossRef]
49. Moher, D.; Liberati, A.; Tetzlaff, J.; Altman, D.G. Preferred Reporting Items for Systematic Reviews and Meta-Analyses: The PRISMA Statement. *BMJ* **2009**, *339*, 332–336. [CrossRef]
50. Ouzzani, M.; Hammady, H.; Fedorowicz, Z.; Elmagarmid, A. Rayyan—a Web and Mobile App for Systematic Reviews. *Syst. Rev.* **2016**, *5*, 210. [CrossRef]
51. Thompson, C.; Wylie, L.J.; Fulford, J.; Kelly, J.; Black, M.I.; McDonagh, S.T.J.; Jeukendrup, A.E.; Vanhatalo, A.; Jones, A.M. Dietary Nitrate Improves Sprint Performance and Cognitive Function during Prolonged Intermittent Exercise. *Eur. J. Appl. Physiol.* **2015**, *115*, 1825–1834. [CrossRef]
52. Porcelli, S.; Pugliese, L.; Rejc, E.; Pavei, G.; Bonato, M.; Montorsi, M.; La Torre, A.; Rasica, L.; Marzorati, M. Effects of a Short-Term High-Nitrate Diet on Exercise Performance. *Nutrients* **2016**, *8*, 534. [CrossRef]
53. Muggeridge, D.J.; Howe, C.C.F.; Spendiff, O.; Pedlar, C.; James, P.E.; Easton, C. The Effects of a Single Dose of Concentrated Beetroot Juice on Performance in Trained Flatwater Kayakers. *Int. J. Sport Nutr. Exerc. Metab.* **2013**, *23*, 498–506. [CrossRef]
54. Cuenca, E.; Jodra, P.; Pérez-López, A.; González-Rodríguez, L.G.; da Silva, S.F.; Veiga-Herreros, P.; Domínguez, R.; Fernandes da Silva, S.; Veiga-Herreros, P.; Domínguez, R. Effects of Beetroot Juice Supplementation on Performance and Fatigue in a 30-s All-Out Sprint Exercise: A Randomized, Double-Blind Cross-Over Study. *Nutrients* **2018**, *10*, 1222. [CrossRef]
55. Nyakayiru, J.; Jonvik, K.L.; Trommelen, J.; Pinckaers, P.J.M.; Senden, J.M.; van Loon, L.J.C.; Verdijk, L.B. Beetroot Juice Supplementation Improves High-Intensity Intermittent Type Exercise Performance in Trained Soccer Players. *Nutrients* **2017**, *9*, 314. [CrossRef]
56. Smith, K.; Muggeridge, D.J.; Easton, C.; Ross, M.D. An Acute Dose of Inorganic Dietary Nitrate Does Not Improve High-Intensity, Intermittent Exercise Performance in Temperate or Hot and Humid Conditions. *Eur. J. Appl. Physiol.* **2019**, *119*, 723–733. [CrossRef]
57. Domínguez, R.; Garnacho-Castaño, M.V.; Cuenca, E.; García-Fernández, P.; Muñoz-González, A.; de Jesús, F.; Lozano-Estevan, M.D.C.; da Silva, S.F.; Veiga-Herreros, P.; Maté-Muñoz, J.L.; et al. Effects of Beetroot Juice Supplementation on a 30-s High-Intensity Inertial Cycle Ergometer Test. *Nutrients* **2017**, *9*, 1360. [CrossRef]
58. Corry, L.R.; Gee, T.I. Dietary Nitrate Enhances Power Output During the Early Phases of Maximal Intensity Sprint Cycling. *Int. J. Coach. Sci.* **2015**, *9*, 87–97.
59. Kramer, S.J.; Baur, D.A.; Spicer, M.T.; Vukovich, M.D.; Ormsbee, M.J. The Effect of Six Days of Dietary Nitrate Supplementation on Performance in Trained CrossFit Athletes. *J. Int. Soc. Sport. Nutr.* **2016**, *13*, 39. [CrossRef]

60. Christensen, P.M.; Nyberg, M.; Bangsbo, J. Influence of Nitrate Supplementation on VO2 Kinetics and Endurance of Elite Cyclists. *Scand. J. Med. Sci. Sport.* **2013**, *23*, 21–31. [CrossRef] [PubMed]
61. Dumar, A.M.; Huntington, A.F.; Rogers, R.R.; Kopec, T.J.; Williams, T.D.; Ballmann, C.G. Acute Beetroot Juice Supplementation Attenuates Morning-Associated Decrements in Supramaximal Exercise Performance in Trained Sprinters. *Int. J. Environ. Res. Public Health* **2021**, *18*, 412. [CrossRef] [PubMed]
62. Esen, O.; Domínguez, R.; Karayigit, R. Acute Beetroot Juice Supplementation Enhances Intermittent Running Performance but Does Not Reduce Oxygen Cost of Exercise among Recreational Adults. *Nutrients* **2022**, *14*, 2839. [CrossRef]
63. Pawlak-Chaouch, M.; Boissière, J.; Munyaneza, D.; Gamelin, F.X.; Cuvelier, G.; Berthoin, S.; Aucouturier, J. Beetroot Juice Does Not Enhance Supramaximal Intermittent Exercise Performance in Elite Endurance Athletes. *J. Am. Coll. Nutr.* **2019**, *38*, 729–738. [CrossRef]
64. Bernardi, B.B.; Schoenfeld, B.J.; Alves, R.C.; Urbinati, K.S.; McAnulty, S.R.; Junior, T.P.S. Acute Supplementation with Beetroot Juice Does Not Enhance Exercise Performance among Well-Trained Athletes: A Randomized Crossover Study. *J. Exerc. Physiol. Online* **2018**, *21*, 1–12.
65. Aucouturier, J.; Boissière, J.; Pawlak-Chaouch, M.; Cuvelier, G.; Gamelin, F.-X.X. Effect of Dietary Nitrate Supplementation on Tolerance to Supramaximal Intensity Intermittent Exercise. *Nitric Oxide Biol. Chem.* **2015**, *49*, 16–25. [CrossRef]
66. Martin, K.; Smee, D.; Thompson, K.G.; Rattray, B. No Improvement of Repeated-Sprint Performance with Dietary Nitrate. *Int. J. Sport. Physiol. Perform.* **2014**, *9*, 845–850. [CrossRef]
67. Roelofs, E.J.; Smith-Ryan, A.E.; Trexler, E.T.; Hirsch, K.R.; Mock, M.G. Effects of Pomegranate Extract on Blood Flow and Vessel Diameter after High-Intensity Exercise in Young, Healthy Adults. *Eur. J. Sport Sci.* **2017**, *17*, 317–325. [CrossRef] [PubMed]
68. Sterne, J.A.C.; Savović, J.; Page, M.J.; Elbers, R.G.; Blencowe, N.S.; Boutron, I.; Cates, C.J.; Cheng, H.Y.; Corbett, M.S.; Eldridge, S.M.; et al. RoB 2: A Revised Tool for Assessing Risk of Bias in Randomised Trials. *BMJ* **2019**, *366*, 14898. [CrossRef]
69. Higgins, J.P.T.; Thompson, S.G.; Deeks, J.J.; Altman, D.G. Measuring Inconsistency in Meta-Analyses. *Br. Med. J.* **2003**, *327*, 557–560. [CrossRef]
70. The Cochrane Collaboration Review Manager Web (RevMan Web) 5.4 2020. Available online: revman.cochrane.org (accessed on 31 January 2022).
71. DerSimonian, R.; Laird, N. Meta-Analysis in Clinical Trials. *Control. Clin. Trials* **1986**, *7*, 177–188. [CrossRef]
72. Cohen, J. *Statistical Power Analysis for the Behavioral Sciences*; Erlbaum: Hillsdale, NJ, USA, 1988; pp. 11–13.
73. Elbourne, D.R.; Altman, D.G.; Higgins, J.P.T.; Curtin, F.; Worthington, H.V.; Vail, A. Meta-Analyses Involving Cross-over Trials: Methodological Issues. *Int. J. Epidemiol.* **2002**, *31*, 140–149. [CrossRef]
74. Buck, C.L.; Henry, T.; Guelfi, K.; Dawson, B.; McNaughton, L.R.; Wallman, K. Effects of Sodium Phosphate and Beetroot Juice Supplementation on Repeated-Sprint Ability in Females. *Eur. J. Appl. Physiol.* **2015**, *115*, 2205–2213. [CrossRef] [PubMed]
75. Reynolds, C.M.E.E.; Evans, M.; Halpenny, C.; Hughes, C.; Jordan, S.; Quinn, A.; Hone, M.; Egan, B. Acute Ingestion of Beetroot Juice Does Not Improve Short-Duration Repeated Sprint Running Performance in Male Team Sport Athletes. *J. Sport. Sci.* **2020**, *38*, 2063–2070. [CrossRef]
76. Spiegelhalder, B.; Eisenbrand, G.; Preussmann, R. Influence of Dietary Nitrate on Nitrite Content of Human Saliva: Possible Relevance to in Vivo Formation of N-Nitroso Compounds. *Food Cosmet. Toxicol.* **1976**, *14*, 545–548. [CrossRef]
77. Qin, L.; Liu, X.; Sun, Q.; Fan, Z.; Xia, D.; Ding, G.; Ong, H.L.; Adams, D.; Gahl, W.A.; Zheng, C.; et al. Sialin (SLC17A5) Functions as a Nitrate Transporter in the Plasma Membrane. *Proc. Natl. Acad. Sci. USA* **2012**, *109*, 13434–13439. [CrossRef] [PubMed]
78. Lundberg, J.O.; Govoni, M. Inorganic Nitrate Is a Possible Source for Systemic Generation of Nitric Oxide. *Free Radic. Biol. Med.* **2004**, *37*, 395–400. [CrossRef]
79. Burleigh, M.C.; Liddle, L.; Monaghan, C.; Muggeridge, D.J.; Sculthorpe, N.; Butcher, J.P.; Henriquez, F.L.; Allen, J.D.; Easton, C. Salivary Nitrite Production Is Elevated in Individuals with a Higher Abundance of Oral Nitrate-Reducing Bacteria. *Free Radic. Biol. Med.* **2018**, *120*, 80–88. [CrossRef]
80. Hyde, E.R.; Andrade, F.; Vaksman, Z.; Parthasarathy, K.; Jiang, H.; Parthasarathy, D.K.; Torregrossa, A.C.; Tribble, G.; Kaplan, H.B.; Petrosino, J.F.; et al. Metagenomic Analysis of Nitrate-Reducing Bacteria in the Oral Cavity: Implications for Nitric Oxide Homeostasis. *PLoS ONE* **2014**, *9*, e88645. [CrossRef]
81. Vanhatalo, A.; Blackwell, J.R.; Heureux, J.E.L.; Williams, D.W.; Smith, A.; Van Der Giezen, M.; Winyard, P.G.; Kelly, J.; Jones, A.M. Free Radical Biology and Medicine Nitrate-Responsive Oral Microbiome Modulates Nitric Oxide Homeostasis and Blood Pressure in Humans. *Free Radic. Biol. Med.* **2018**, *124*, 21–30. [CrossRef]
82. Wylie, L.J.; Kelly, J.; Bailey, S.J.; Blackwell, J.R.; Skiba, P.F.; Winyard, P.G.; Jeukendrup, A.E.; Vanhatalo, A.; Jones, A.M. Beetroot Juice and Exercise: Pharmacodynamic and Dose-Response Relationships. *J. Appl. Physiol.* **2013**, *115*, 325–336. [CrossRef]
83. Abu-Alghayth, M.; Vanhatalo, A.; Wylie, L.J.; McDonagh, S.T.; Thompson, C.; Kadach, S.; Kerr, P.; Smallwood, M.J.; Jones, A.M.; Winyard, P.G. S-Nitrosothiols, and Other Products of Nitrate Metabolism, Are Increased in Multiple Human Blood Compartments Following Ingestion of Beetroot Juice. *Redox Biol.* **2021**, *43*, 101974. [CrossRef] [PubMed]
84. Bender, D.; Schwarz, G. Nitrite-Dependent Nitric Oxide Synthesis by Molybdenum Enzymes. *FEBS Lett.* **2018**, *592*, 2126–2139. [CrossRef] [PubMed]
85. van Faassen, E.E.; Bahrami, S.; Feelisch, M.; Hogg, N.; Kelm, M.; Kim-Shapiro, D.B.; Kozlov, A.V.; Li, H.; Lundberg, J.O.; Mason, R.; et al. Nitrite as Regulator of Hypoxic Signaling in Mammalian Physiology. *Med. Res. Rev.* **2009**, *29*, 683–741. [CrossRef]

86. Coggan, A.R.; Broadstreet, S.R.; Mahmood, K.; Mikhalkova, D.; Madigan, M.; Bole, I.; Park, S.; Leibowitz, J.L.; Kadkhodayan, A.; Thomas, D.P.; et al. Dietary Nitrate Increases VO 2 Peak and Performance but Does Not Alter Ventilation or Efficiency in Patients With Heart Failure With Reduced Ejection Fraction. *J. Card. Fail.* **2018**, *24*, 65–73. [CrossRef]
87. Kadach, S.; Won, J.; Zdravko, P.; Matthew, S.; Vanhatalo, A.; Burnley, M.; Walter, P.J.; Cai, H.; Schechter, A.N.; Piknova, B.; et al. 15 N-Labeled Dietary Nitrate Supplementation Increases Human Skeletal Muscle Nitrate Concentration and Improves Muscle Torque Production. *Acta Physiol.* **2023**, *237*, e13924. [CrossRef]
88. Gilliard, C.N.; Lam, J.K.; Cassel, K.S.; Park, J.W.; Schechter, A.N.; Piknova, B.; Branch, M. Effect of Dietary Nitrate Levels on Nitrate Fluxes in Rat Skeletal Muscle and Liver. *Nitric. Oxide* **2018**, *75*, 1–7. [CrossRef]
89. Wylie, L.J.; Park, J.W.; Vanhatalo, A.; Kadach, S.; Black, M.I.; Stoyanov, Z.; Schechter, A.N.; Jones, A.M.; Piknova, B. Human Skeletal Muscle Nitrate Store: Influence of Dietary Nitrate Supplementation and Exercise. *J. Physiol.* **2019**, *597*, 5565–5576. [CrossRef]
90. Srihirun, S.; Park, J.W.; Teng, R.; Sawaengdee, W.; Piknova, B.; Schechter, A.N.; Branch, M.; Diseases, K. Nitrate Uptake and Metabolism in Human Skeletal Muscle Cell Cultures. *Nitric. Oxide Biol. Chem.* **2020**, *94*, 1–8. [CrossRef]
91. Zhang, Z.; Naughton, D.; Winyard, P.G.; Benjamin, N.; Blake, D.R.; Symons, M. Generation of Nitric Oxide by a Nitrite Reductase Activity of Xanthine Oxidase: A Potential Pathway for Nitric Oxide Formation in the Absence of Nitric Oxide Synthase Activity. *Biochem. Biophys. Res. Commun.* **1998**, *249*, 767–772. [CrossRef]
92. Jansson, E.A.; Huang, L.; Malkey, R.; Govoni, M.; Nihlén, C.; Olsson, A.; Stensdotter, M.; Petersson, J.; Holm, L.; Weitzberg, E.; et al. A Mammalian Functional Nitrate Reductase That Regulates Nitrite and Nitric Oxide Homeostasis. *Nat. Chem. Biol.* **2008**, *4*, 411–417. [CrossRef]
93. Piknova, B.; Park, J.W.; Kwan Jeff Lam, K.; Schechter, A. Nitrate as a Source of Nitrite and Nitric Oxide during Exercise Hyperemia in Rat Skeletal Muscle. *Nitric. Oxide* **2016**, *55–56*, 54–61. [CrossRef]
94. Hellsten, Y.; Frandsen, U.; Ørthenblad, N.; Sjødin, B.; Richter, E.A. Xanthine Oxidase in Human Skeletal Muscle Following Eccentric Exercise: A Role in Inflammation. *J. Physiol.* **1997**, *498*, 239–248. [CrossRef]
95. Nyberg, M.; Christensen, P.M.; Blackwell, J.R.; Hostrup, M.; Jones, A.M.; Bangsbo, J. Nitrate-Rich Beetroot Juice Ingestion Reduces Skeletal Muscle O2 Uptake and Blood Flow during Exercise in Sedentary Men. *J. Physiol.* **2021**, *599*, 5203–5214. [CrossRef]
96. Smith, J.C.; Hill, D.W. Contribution of Energy Systems during a Wingate Power Test. *Br. J. Sport. Med.* **1991**, *25*, 196–199. [CrossRef]
97. Bogdanis, G.C.; Nevill, M.E.; Boobis, L.H.; Lakomy, H.K.A. Contribution of Phosphocreatine and Aerobic Metabolism to Energy Supply during Repeated Sprint Exercise. *J. Appl. Physiol.* **1996**, *80*, 876–884. [CrossRef]
98. Whitfield, J.; Gamu, D.; Heigenhauser, G.J.F.; Van Loon, L.J.C.; Spriet, L.L.; Tupling, A.R.; Holloway, G.P. Beetroot Juice Increases Human Muscle Force without Changing Ca2+-Handling Proteins. *Med. Sci. Sport. Exerc.* **2017**, *49*, 2016–2024. [CrossRef]
99. Coggan, A.R.; Peterson, L.R. Dietary Nitrate Enhances the Contractile Properties of Human Skeletal Muscle. *Exerc. Sport Sci. Rev.* **2018**, *46*, 254–261. [CrossRef]
100. Bailey, S.J.; Gandra, P.G.; Jones, A.M.; Hogan, M.C.; Nogueira, L. Incubation with Sodium Nitrite Attenuates Fatigue Development in Intact Single Mouse Fibres at Physiological PO2. *J. Physiol.* **2019**, *597*, 5429–5443. [CrossRef]
101. Tillin, N.A.; Moudy, S.; Nourse, K.M.; Tyler, C.J. Nitrate Supplement Benefits Contractile Forces in Fatigued but Not Unfatigued Muscle. *Med. Sci. Sport. Exerc.* **2018**, *50*, 2122–2131. [CrossRef]

MDPI AG
Grosspeteranlage 5
4052 Basel
Switzerland
Tel.: +41 61 683 77 34

Antioxidants Editorial Office
E-mail: antioxidants@mdpi.com
www.mdpi.com/journal/antioxidants

www.ingramcontent.com/pod-product-compliance
Lightning Source LLC
Chambersburg PA
CBHW041400210926
51850CB00013B/3135